Jerry Bauer

About the Author

JAMES C. DAVIS taught history at the University of Pennsylvania for thirty-four years. He is the author of four other books dealing with Venice, the early history of European nations, and the lives of peasants and blue-collar workers.

OTHER BOOKS BY JAMES C. DAVIS

Rise from Want: A Peasant Family in the Machine Age

A Venetian Family and Its Fortune, 1500–1900: The Donà and the Conservation of Their Wealth

Pursuit of Power: Venetian Ambassadors' Reports on Turkey, France, and Spain in the Age of Philip II, 1560–1600
(editor, translator)

The Decline of the Venetian Nobility as a Ruling Class

THE HUMAN STORY

Our History, from the Stone Age to Today

JAMES C. DAVIS

HARPER
PERENNIAL

HARPER PERENNIAL

A hardcover edition of this book was published in 2004 by HarperCollins Publishers.

HarperCollins books may be purchased for educational, business, or sales promotional use. For information, please e-mail the Special Markets Department at SPsales@harpercollins.com.

FIRST HARPER PERENNIAL EDITION PUBLISHED 2005.

Designed by Cassandra J. Pappas

The Library of Congress has catalogued the hardcover edition as follows:

Davis, James C. (James Cushman)
The human story: our history, from the Stone Age to today / James C. Davis.—1st ed.
p. cm.
Includes bibliographical references and index.
ISBN 0-06-051619-4 (alk. paper)
1. World history. I. Title.
D23.D38 2004
909—dc22

2003067570

ISBN-10: 0-06-051620-8 (pbk.)
ISBN-13: 978-0-06-051620-8 (pbk.)

16 ❖/RRD 20 19 18 17 16 15 14 13 12

To our children and theirs

Contents

Maps and Illustrations

Maps

Illustrations

To the Reader

This book tells how ancient wandering peoples settled down, and how they founded cities, conquered neighbors, formed religions, found out who they were and where among the stars they lived, did some good and many wrongs, thrived, and journeyed into space.

I never told a soul that I was writing a book about the human past without his asking, "What's your slant, your point of view?" If I have one it is this: In spite of all we hear and say, the world has been improving for a good long time.

My hardest task was leaving out. Writing the human story is like packing a suitcase; you can't find space for everything. I regret that this book seldom mentions the deeds of women. The human past was often like a play produced in Shakespeare's time; men took all the roles. Inescapably the book has much to say in the final chapters about the United States, while it never mentions many of the nearly 200 nations in the world. Since so many nations have often behaved badly, not being mentioned here probably reflects well on them.

I welcome your comments and suggestions.

For their generosity and help, let me warmly thank these friends and fellow students of the past. Most of them are former colleagues at the University of Pennsylvania. Strange as it may seem, some of them wouldn't know me if they saw me. But all of them were kind enough to lend a hand.

My thanks to Wendy Ashmore, Tom Austin, James Baker, Richard Balkin, Richard Beeman, Tom Boyd, Lee Cassanelli, David Chaplin-Loebell, Thomas Childers, Frank Conaway, Hilary Conroy, five Davises (Daniel, David, Elda, Susan, and William, the last of whom greatly improved the whole manuscript), Richard Dunn, Ann Farnsworth-Alvear, Jeffrey Fear, Robert Forster, Louis Girifalco, Avery Goldstein, Ward Goodenough, Samuel Humes, Jeremy Jackson, Margaret Jacob, Christopher Jones, Robert Kraft, Bruce Kuklick, John and Miriam Lally, Lynn Lees, Walter Levy, and Paul Liebman.

Also: Mia Macintosh, Victor Mair, Alan Mann, Joyce Martin, Walter McDougall, Cynthia Merman, Allyn Miner, Sue Naquin, Benjamin Nathans, Martin Ostwald, Robert Palmer, Ivo Panjek, Edward Peters, Sumathi Ramaswamy, Robert Regan, Frankie Rubinstein, Jerry Ruderman (who improved the entire manuscript), Madeline Sauvion, Selig and Jacqueline Savits, Barbara von Schlegell, Gino Segré, Benjamin Shen, David Silverman, Nathan Sivin, Ake Sjoberg, Bernard Steinberg, Nancy Steinhardt, Yvonne Surh, Emidio Sussi, Henry Teune, Jeffrey Tigay, Robert Turner, Étienne Van de Walle, Hugh Van Dusen, Susan Watkins, Martin and Dotty Wolfe, Charles Wright, Vikash Yadav, Sally Zigmond, and an expert copyeditor who chooses to remain anonymous.

THE HUMAN STORY

CHAPTER 1

We fill the earth.

OUR TALE BEGINS when humans much like us evolved and filled the earth.

Before that happened other humans had already come and gone. The most important of our forebears was *Homo erectus,* or Upright Men, so named because they stood on their two feet. They evolved in Africa about two million years ago and wandered into Asia. They sometimes lived in caves and sometimes in the open, and they chipped their simple tools from stone and learned the use of fire. *Erectus* had heavy brows and flatter skulls than we do, and if one were to enter a bus today the other riders probably would stealthily slip out.

Before *erectus* vanished perhaps 300,000 years ago, they begat the species we belong to. We of course are *Homo sapiens,* or Wise Men. Immodestly we gave ourselves that name because we have larger brains, encased in higher skulls, than *erectus.* In spite of having larger brains, the early *sapiens* humans may not have had the gift of language.

• • •

THEY CHANGE their minds every time they find an ancient skull, but anthropologists are fairly sure that our own subspecies evolved from *sapiens* about 160,000 years ago. We probably evolved in Africa, below the Sahara Desert. To indicate that we are a subspecies of *sapiens,* we call ourselves *Homo sapiens sapiens,* or Wise Wise Men. We are now the only variety of humans on earth.

We evolved in different ways. Some of those in Africa developed tall, thin bodies that exposed a lot of skin and that air could therefore cool more easily. Dark pigment in their skin protected them from the tropical sun's ultraviolet rays, and their tight-curled hair protected their heads from the heat. But humans who lived in Europe and Asia, coping with the long, dark winters, had other needs. To keep their bones from weakening, they needed sunlight to stimulate vitamin D production. Dark skin would have blocked out too much sun, so they developed pink or sallow skin with little pigment.

Prehistorians have learned a lot about the life of our *sapiens sapiens* ancestors, especially those who lived in southwest Europe about thirty thousand years ago. For example, individuals took as much pleasure in looking different from each other as modern humans do. In a cave in the Pyrenees Mountains between France and Spain, an artist scratched on the walls more than a hundred sketches of what appear to be real people. Some of them wore their hair long, and others short; some had it in braids, others in buns. Some men had beards and mustaches, while others were clean-shaven.

At some point, but the time is much debated, humans learned to speak to one another. They may have done this because they were developing a richer culture that depended on communication. They must have often hunted and collected food in groups, and they probably worked together when they fashioned fishing boats and sheltered entrances to caves.

They had clever hands. They could light a fire by striking sparks from lumps of iron ore, and they carved their sewing needles out of bones, each one with a tiny hole through which a thread could pass.

With these they sewed their clothes, using skins of animals. They made tiny cutting tools, half as long as a paper match, from flint, and glued them with resin into holes in handles made from wood or antlers.

They invented the spear thrower, which is a short shaft with a hook at one end that fits into the back end of a spear. It enables a hunter to throw a spear very hard. Some ancient artist carved the end of a spear thrower that was found in the Pyrenees Mountains in the shape of a fawn. Its head is facing backward, and it is looking at a little bird that is perched atop a lump of feces emerging from the fawn.

When someone died the early humans often left his necklaces of teeth and shells on his body, and food and tools beside it. They made a powder from the soft red stone called ocher, and sprinkled it on his body. So they clearly thought of death as meaningful and solemn. Perhaps they thought the one who died would have an afterlife where he or she would once again need tools and food, in a place where beauty mattered.

NOTHING THAT WE KNOW about the early humans is as awesome as what they painted in the depths of caves. Prehistorians first learned about these paintings in 1875, when an amateur archaeologist was hunting bones and tools in a cave at Altamira near the northern coast of Spain. His little daughter, whom he'd brought along for company, wandered into a nearby chamber. Holding up her candle, she saw paintings on the ceiling of two dozen nearly life-size bison, drawn in yellow, red, brown, and black. The paintings are so masterful that experts quickly—wrongly—called them modern fakes.

The greatest find of prehistoric paintings took place at Lascaux in southwest France soon after the start of World War II. Four teenaged boys were rambling on a hillside. In a place where a storm had uprooted a tree, the boys discovered that where the roots had been there was now a deep hole in the ground. A few days later they returned with a kerosene lamp, and one of them climbed down inside the hole. In the scanty light he clambered down a rocky slope and found that he was in a cavern.

The boy was stunned by what he saw. On the cavern walls were mural paintings of short and shaggy horses, bison, oxen, deer with spreading antlers, and that mythic beast the unicorn. Some of the animals were merely staring; others running for their lives. In a sloping gallery near the main one, other searchers later came on sketches of a stag swimming across a river. In another cave they found a drawing of a man with a horned head and an erect penis. He seems to be pursuing a reindeer and an animal that is part deer, part bison.

Since the Lascaux find, explorers in caves in southern France and northern Spain have discovered thousands of paintings and drawings. These were not just casual doodles; painting them required a lot of trouble. To prepare, the artist and his helpers would have gathered minerals and clays of different colors and prepared the paints. Then they would have carried the equipment down inside the caves, and someone would have built a scaffold to support the painter. Finally, with others lighting up the cave with torches, he would have set to work.

Painters often worked in chambers that are hard to get to. One such chamber can be reached only by wriggling through a narrow 200-foot-long tunnel. A portly priest who was an expert on these paintings once got stuck inside this tunnel. Others had to pull him out.

Some of these deep and scary chambers may have been the scenes of solemn rites. One can picture adults, holding flaming torches, leading children through the narrow tunnels and then, as torchlight flickered on the paintings, explaining what they meant. The caves hold proof that children then were much like children now. Deep inside them modern-day explorers sometimes come on footprints left by children running barefoot who made a point of splashing through the puddles.

Explorers found some stunning sculptures in the Pyrenees. Moving first by boat you enter a cave where a river flows out of a mountain. Then you walk for a mile through narrow passages, then through a kind of hall with long and twisted stalactites, then through other bending tunnels, till at last you reach a chamber where you have to stoop. Lying in the middle of it are two bison, two feet long, which someone sculpted out of yellow clay fifteen-thousand years ago.

No one can be certain what their art reveals about the culture of these ancient folk. The animals no doubt reflect a great concern with hunting, and perhaps with magic. By painting mammoths and bison, they may have hoped to master these fierce beasts and raise the odds of killing them. Some animals are pierced with spears, and one painting shows a mammoth trammeled in a pitfall.

These early men (and maybe women) were skillful hunters. They discovered when to wait at places where the big game passed on their migrations—for example, where reindeer always forded rivers. Or they camped beside the open ends of narrow valleys where cliffs closed off the sides and the farther end. These were natural traps where they could drive an animal or herd and kill it then or later when they needed meat.

Prehistorians in France have found the bones of between 10,000 and 100,000 horses in a giant heap at the bottom of a cliff. Over many years, no doubt, ancient hunters stampeded horses over the cliff or ambushed them in the narrow pass below. At a village in the Czech Republic prehistorians found a pile of bones of more than a hundred mammoths, and on a site in Russia searchers found remains of more than two hundred of them.

BY FIFTY THOUSAND YEARS AGO, we had spread through Africa, Asia, and Europe. Now we would spread into three other continents where no humans had set foot.

The most mystifying of these migrations is the one from Southeast Asia to Australia. It's hard to see how humans did it. Getting as far as the Indonesian islands wouldn't have been hard if they did it at a time when a "land bridge" connected the Asian mainland to Indonesia. (A land bridge is a crest of land that appears in a shallow sea during an ice age, when much of the earth's water

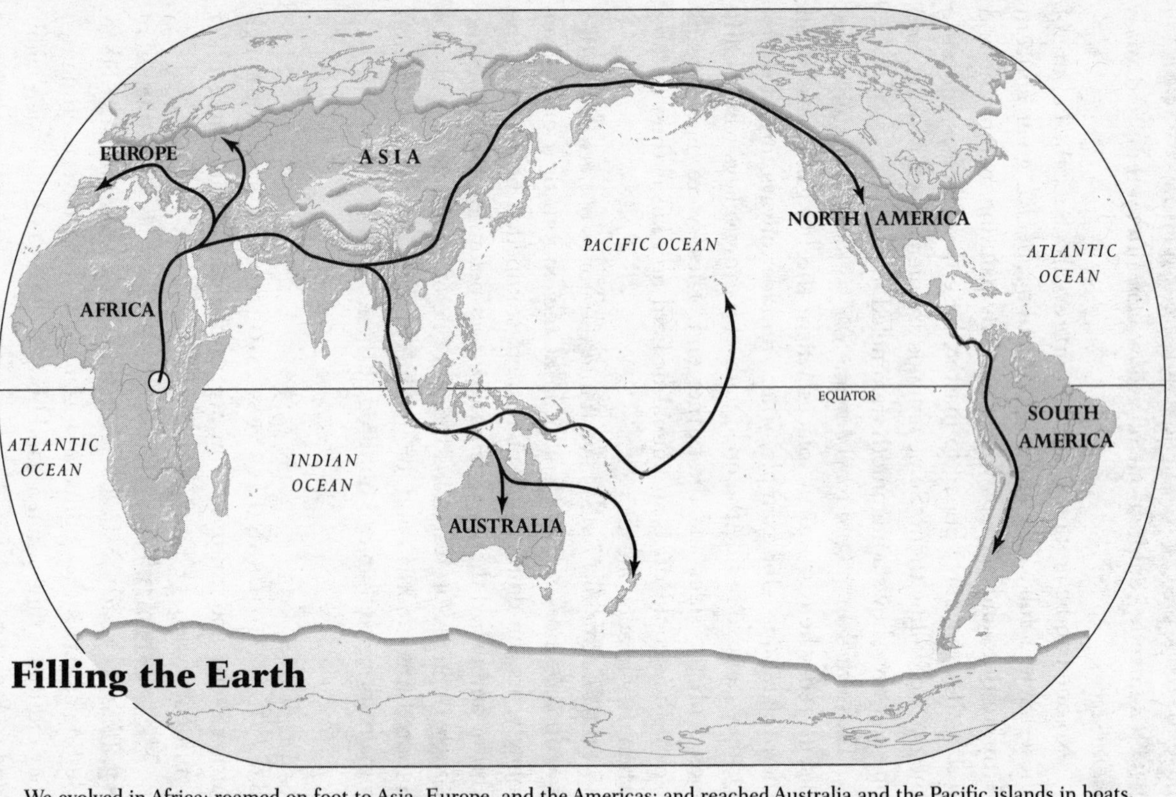

Filling the Earth

We evolved in Africa; roamed on foot to Asia, Europe, and the Americas; and reached Australia and the Pacific islands in boats.

freezes into glaciers, lowering the level of the sea.) From Indonesia they might have moved to the nearby island of New Guinea in boats made with skins, or rafts of bundled reeds.

To move from New Guinea to Australia, however, they would have had to sail or paddle over roughly sixty miles of open sea. It's hard to see why would they have done such a dangerous thing, since they had no way of knowing they would ever come to land. Most likely they made the crossing without meaning to, blown there during storms. These were the people whose descendants now are called the "Aborigines," although they were not in Australia "from the beginning," which is what *aborigine* means.

From the Australian mainland, some of these pioneers must have wandered down into Tasmania, which then was a peninsula jutting south from the continent. What happened to them in this place is interesting and revealing. Like the other Australian Aborigines, these people lived in Tasmania as simple hunters and gatherers. Then, more than ten thousand years ago, the oceans slowly rose, and stormy waters drowned the link to the mainland. Tasmania became an island.

Its people, who were isolated now from other Aborigines, clung to ancient ways for ten millennia. When the first Europeans came upon them two centuries ago, they were living specimens of life in the remote past. They had a rich social and ritual life, but they still used crude stone tools. In the early 1800s, British settlers nearly exterminated the Tasmanians in what was called "the Black War." They hunted them down with dogs, and moved the remnant to an offshore island, where they died of disease and civilization.

While the Tasmanians were changing not at all during 10,000 years, the Aborigines on the mainland evolved a somewhat more complex culture. They learned to tie stone points to wooden shafts, and they used spear throwers. Even today some Aborigines using spear throwers can hit a kangaroo three out of four times from more than a hundred feet, and kill it in one throw from thirty to fifty feet. And of course the Aborigines learned to make the boomerang, the well-known throwing stick.

• • •

OTHERS OF OUR subspecies moved to North and South America, two other continents where humans had never been. Prehistorians of today disagree about who made this move, and how and when.

The long-accepted story of the settlement of North America went like this. Humans lived in the extreme northeastern tip of Siberia, at the Arctic Circle, which reaches far to the east. Today, fifty-three miles of rough and icy water separate this tip of Asia from North America, but at the time in question the seas were low. So a wide land bridge of tundra and marshes connected Asia and North America. Bands of Siberian hunters moved back and forth along this land bridge (from continent to continent, but that they didn't know) following mammoths and wild horses.

Then some of them—from this point on we will refer to them as Indians—wandered away from the eastern (North American) end of the land bridge. From there the Indians probably followed game to the south and east. Since this was during one of earth's cold periods, glaciers covered two-thirds of North America, but these pioneers could have walked south along the ice-free Alaskan coast. Or they could have trod a narrow ice-free corridor that we know led south between the glaciers. As they trudged along this corridor mile-high walls of ice would have flanked them on both sides. When at last they were south of the ice, they would have found themselves in the northwest of what is now the United States.

That is the old story of the arrival of humans in the Americas. The newer variations on it are so many that we can only briefly list them. The Indians may have come at different times. They may have come from South Asia, Japan, even Australia. And they may have come by water, not by land, perhaps hugging the shore all the way from Asia, along the Bering land bridge, and down the western coasts of the Americas.

After scattering through North America, the Indians made their way to the other continent that lies to the south. We don't know when, but after decades, or centuries, or perhaps thousands of years they wandered down the isthmus between the continents, and then on through the jungles, mountains, and grasslands, all the way to the windy southern tip of South America.

By luck, we know just a little about the life of the Indians who lived on a site near the southern tip of South America nearly eleven thousand years ago. After the Indians abandoned it, a peat bog covered their village, and the acids in the partially decomposed plants preserved what would otherwise have rotted. Not only bones survived here, but also garbage, wood, and even the chewed leaves of a shrub that the villagers used as a drug. Surprisingly, these early Indians lived in parallel rows of huts, each hut covered with skins and floored with logs or planks.

Both of the newfound continents were full of game, a hunter's paradise. The archaeologists who found the evidence tell about an event that happened one day ten thousand years ago on the plains east of the central Rocky Mountains. Using some imagination, we believe it may have gone like this. A band of Indians saw a herd of bison. Approaching from the downwind side, so that the bison couldn't smell them, the hunters crept up close. Suddenly they shouted, and they threw their spears so as to panic the bison and drive them into a deep gully. The animals stampeded. They tried to jump the gully, but many fell short and landed at the bottom. In no time, writhing, bellowing animals filled the gully.

The hunters moved in and killed the bison that had not been crushed to death. When they had finished, 190 animals lay dead. Then the hunters butchered them. They had tons and tons of meat, enough for feasting and plenty to dry and eat later—probably much more meat than they could use.

HUMANS ALSO REACHED the islands of the Pacific Ocean. The earliest settlers in Oceania almost surely were seagoing farmers and traders from the islands off southeastern Asia. These people were used to moving in simple canoes or rafts from one inhabited island to another and trading pigs, pots, and yams. Then they moved to far-off, unpopulated islands, far to the east of the islands that they knew. First they went to Melanesia, the "Black Islands" northeast of Australia, and later to Micronesia, the "Little Islands" north of Melanesia. Much later still they showed up in the "Many Islands" of distant Polynesia, which are scattered in the mid-Pacific.

No one knows just why these islanders left their homes and sailed across the ocean to such far-off places. (New Zealand, Easter Island, and the Hawaiian Islands are 1,000 to 1,800 miles from the nearest inhabited land.) Some have guessed that they were fleeing from their home islands, where people were dying of hunger, or slaughtering each other in their wars. Some experts tell us these were skillful seamen, who could sail for hundreds of miles, guided only by the sun, the stars, and the trend of the ocean swells until they reached a coral or lava island they had somehow heard of.

In fact, however, there is a simpler but likely explanation for the voyages. It appears again and again in the writings of European travelers who visited Pacific islands in modern times. These travelers often heard of islanders who had been sailing in familiar waters when storms arose and blew their sailing canoes far out to sea. Ocean currents then carried them for hundreds of miles to another island. In 1696, for example, families sailing between two islands in the north Pacific were blown away by a storm. After seventy days they reached the Philippines, 1,000 miles away. Most likely many islands were occupied by people blown out to sea by storms, and lucky enough to land on far-off islands.

Seven men and women on a raft were blown away from the island of Mangareva in southern Polynesia. With great good luck, they landed on the lonely isle of Rapa, 600 miles to the southwest. The Rapans urged them to stay, but the Mangarevans decided to return home. They believed incorrectly that their island lay to the southeast, so they waited for a wind from the northwest, and then pushed off. What a terrible mistake! South of Rapa lies nothing but Antarctica.

WHEN THE PACIFIC ISLANDS had been reached, human beings had filled the earth.

CHAPTER 2

We gather by the rivers.

LONG AFTER WE had spread around the earth, humans still survived by hunting and by eating seeds and berries, insects, seaweed, lizards, eggs, and roots.

But about ten thousand years ago we began to make a basic change in the human way of life.

Imagine that a band of humans comes upon a flock of wild sheep. They kill and eat a sheep or two, but the animals are more than they can eat so they stalk the flock for weeks. From time to time they kill whatever sheep they need for food, mostly those that can't keep up.

One day the hunters find a valley, open only at one end, where they can hold the sheep. To take this step means settling down, for a while at least. Along the valley's open end they raise their tents or build some huts. Now the hunters are turning into herders. They can treat the sheep as their reserve, something to fall back on when they can't find other game. They train their dogs to herd the sheep.

As they cull and eat unwanted sheep, they slowly breed a species with more useful traits. The one-time skinny beasts evolve until they're fat, with thicker coats of wool. As the herders catch and pen other beasts, these too evolve in useful ways. Cows, which had been dangerous when living in the wild, grow docile, and they keep on giving milk even when their calves are weaned. Lean and agile boars turn into fat, nutritious pigs.

Of course, the herders we're discussing don't subsist on meat alone. All of them, but especially the women, gather other kinds of food. Most important are the grains—the barley, maize (or corn), and rice, and wheat—that grow untended on the plains and in the swamps around them. Harvesting these grains is easy: if you tap the stems the seeds will tumble in your basket.

They (probably the women) realize that the grains are seeds. So in a year when food is hard to find they scratch the soil and scatter seeds upon it, hoping they will have more cereal grasses. It's true that digging up the earth with sticks and planting seeds is much more work than simply gathering the foods that nature offers. But when they reap a harvest they have more grain to eat or to save for winter.

Now that they are planting crops, they have another reason (in addition to their herds) to stay in place. After all, it makes no sense to clear the fields and use them for only a year. So they build themselves some bigger huts, with space to store their tools and seeds. Around the huts they build a fence and then a wall to keep the cattle in and bandits out.

And that is roughly how farming and village life began. It didn't happen that a solitary genius thought up farming, and that news about this great invention raced around the earth. Humans must have made the shift to farming in something like the manner we described above. Little groups of people, bit by bit, scarcely thinking what they did, changed the way they dealt with animals and plants. (Incidentally, this change coincides with the beginning of what is called the "New Stone Age," when humans made finer tools from stones by polishing or grinding them.)

Farming took the place of food gathering, almost everywhere on earth, in a mere ten thousand years. That rapid widespread change almost demands a one-size-fits-all global explanation. Here is one:

Perhaps at a certain moment humans faced a problem of supply. A worldwide drought might have forced people everywhere to seek new ways to get more food. Such an explanation is appealing, but historians have found no proof of such a drought.

All right, perhaps the cause of change was not inadequate supply but rising demand: a sudden, global population increase, with so many mouths to feed that people everywhere were forced to drop the easy life of gathering food and learn to do the harder, more productive, work of farmers. This *may* have taken place, but evidence of such a population rise is hard—and probably impossible—to find.

In short, why farming appeared at almost the same time throughout much of the world remains a mystery.

Archaeologists have uncovered places where one can follow the transition from food gathering to farming. One such place is Jericho, an oasis near Jerusalem that lies more than 800 feet below the level of the sea. From the beginning, each generation in Jericho left behind a layer of earth and trash. Eventually the layers formed a mound that rises seventy feet above the plain. The lowest layers show that in the beginning wandering hunters used to camp at Jericho beside a spring. They hunted gazelles and camped in flimsy huts or tents and put up a shrine, perhaps devoted to the spring.

Later people slowly settled down and started farming. The early farmers raised wheat and barley and probably kept goats; later ones had dogs to help them tend their flocks. They took about a thousand years to carry through their farming revolution. By that time two or three thousand people lived here in a crowded village. They dwelt in small round mud-brick huts, but they built a massive wall around their village and a tower ten yards high. They were prosperous enough to have to fear the hunter-bandits in the nearby hills and desert.

THE INTRIGUING ICEMAN lived amid the snowy Alps in Europe, far from sun-baked Jericho. But like those people by the desert spring, he too was a creature of that moment when we humans turned from gathering to farming. He reappeared a dozen years ago when tourists walking in a mountain pass in Italy found his body, which a glacier had covered until then. His body was intact, and partly mummified by dehydration.

We will never know the reason he was in that mountain pass. He may have been hunting or tending sheep or, as we'll see, he may have climbed up there to fight. He was in his middle forties—pretty old for those hard times—but trim and fit except for some arthritis and worms in his intestines. He wore a coat of skins and on top of that a cape of woven grass. On his head he wore a fur cap, and on his feet leather shoes that he had stuffed with grass for warmth. Someone had cut his hair evenly to a length of three and a half inches. He carried all kinds of things—too many to list here—but among them were a rucksack, an axe with a copper blade, a bow and arrows, a flint dagger, two mushrooms, a sloe berry, and a tassel with a marble bead.

The Iceman lived when humans where he lived were settling down, beginning to farm, and learning the use of metals. In his stomach were bits of goat meat and cultivated wheat, the kind that farmers sow and later harvest. So apparently he lived where people farmed. If we consider when he lived, his axe's copper blade was a triumph of technology. Only recently had humans mastered the tricky task of roasting copper ore over a fire, and pouring molten metal in a mold.

On his body were charcoal tattoos that someone had pricked in his skin in places he could not have reached. He had lines on his lower spine, a cross behind his left knee, and stripes on his right ankle. Perhaps they tell us that the Iceman was a chief or wise man. But they may have been a therapy for pain, a kind of acupuncture.

If we don't know why the Iceman went where he was found, we do know how he died. The berry in his pack informs us that the month was August or September. He ate a meal, the last he ever had. Then someone shot an arrow into his left shoulder. He crawled into a natural basin in the rocks that was twice as deep as he was tall, and there he bled to death. His body froze. Snow covered his corpse before an animal could find it, and a glacier inched across the basin that he lay in. For more than fifty centuries he lay in peace in his frigid refuge underneath the glacier.

Two tourists found the Iceman when the glacier had melted just enough to expose his head and shoulders. Immediately, he became an international sensation. Scientists refroze his partially thawed body, and began to study it. Several women volunteered to bear his child, but that is still beyond the reach of science.

AND NOW, WITH trumpets fanfarading and a roll of drums, we turn to when we humans started to collect in "civilizations." By the word *civilization* we mean a place where people live in villages and towns, work at many trades, obey a government, worship a god or gods, and read and write. Of course humans didn't become "civilized" at some specific moment.

However, the place where this happened first was probably the south of what is now Iraq. The reader needs to visualize this place. In western Asia the Tigris and Euphrates rivers flow side by side for 750 miles from Turkey's lower edge through Iraq to the Persian Gulf. (See the map on p. 16.) The region's older name is Mesopotamia, which in Greek means "Land Between the Rivers." In their lower courses the rivers flow through level farmland, deserts, swamps, and marshes to the Gulf. The country is so flat that, as they say, if you step on a phone book you get a view.

Between 4500 and 4000 B.C., a farming people living in this lower part of Mesopotamia stood upon the brink of civilization. They drained some marshes so that they could farm them, and they dressed in woolen cloth and leather, made their pots from clay, and built their huts with bricks they made from mud.

After a millennium had passed another people came and occupied this place, probably absorbing the natives. Their skeletons and later paintings show that they were short with sturdy bones. Their neighbors farther up the rivers later called them the "Sumerians" and their region Sumer, but they called themselves "the dark-headed" and they knew their country as "the Land."

Water, precious water, shaped their lives. At first they may have lived among the marshes. A Sumerian myth relates that in primeval

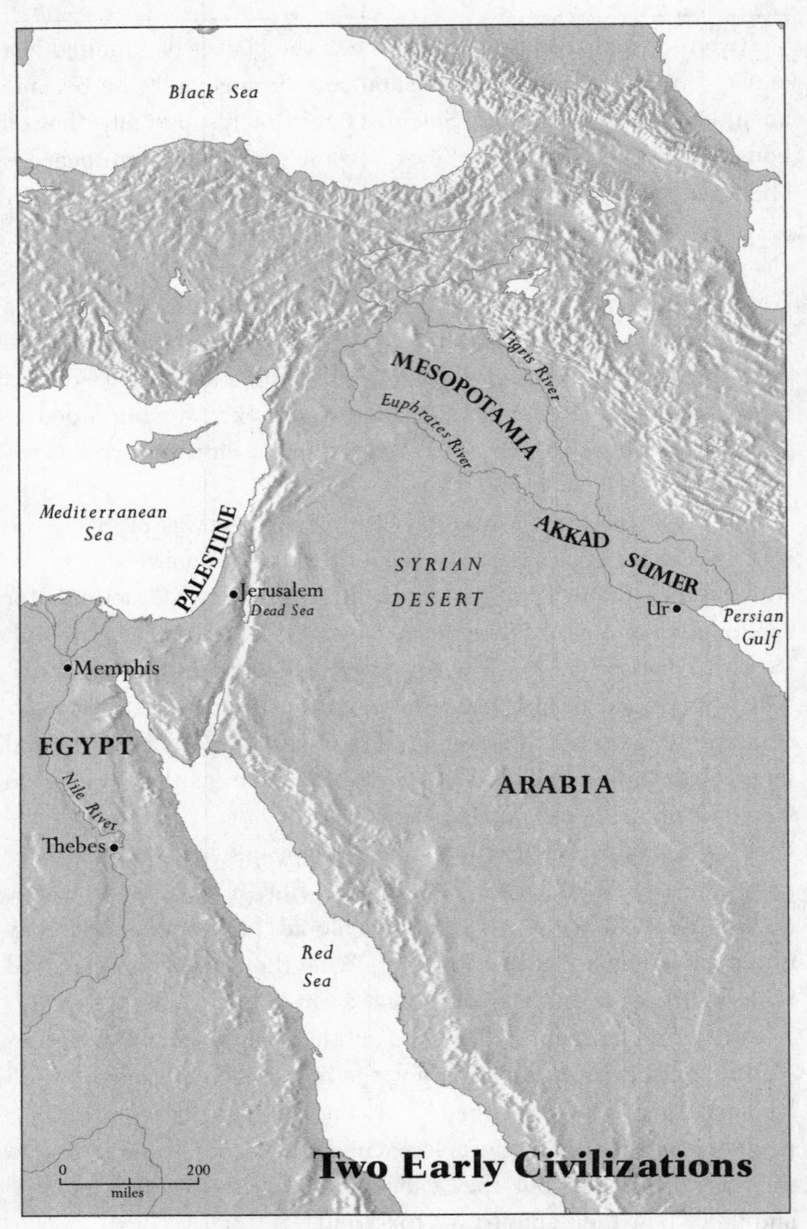

Sumer lay in lower Mesopotamia, between two rivers.
Egypt (then as now) stretched along the Nile.

times a god laid reeds on the face of the waters and, by pouring mud on the reeds, made the floor of a hut. To this day the people who live here in the swamps dwell in huts of reeds whose floors of reeds ooze mud at every step. Most of the Sumerians we're concerned with didn't live in marshes, but they did reside along the rivers.

The early farmers learned to dig canals so that when the rivers flooded they could first divert the water to the fields and later drain it off (so as to prevent salt from accumulating in the soil). Slowly they enlarged the network of canals in order to bring water to farmland farther from the rivers. Villages grew up and then developed into towns.

The soil was fertile, and with irrigation yielded barley, wheat, sesame, and dates. Farmers who had surpluses to sell might buy their neighbors' plots and thus become rich landlords. Meanwhile other Sumerians became merchants. Using donkey carts they carried barley, oil of sesame, dried fish, and cloth to the people in the mountains east of Sumer. They returned with things that Sumer lacked: precious metals, copper, cedar boards, and building stone for temples. Merchants sailed to other towns along the rivers using boats they made from reeds and skins. In this way, the growing towns became acquainted, and they built a common civilization.

Solitary peasants couldn't plan canals and dig them and keep them clear of silt. They needed chiefs to run the digging and to drive away the bandits from the desert. So in all the larger towns a chief emerged. He was called a "big man," and he had tax collectors, judges, and supervisors of canals. The merchants and the well-off landlords sometimes gave the big man their advice, but peasants, laborers, and slaves hadn't any say at all. According to a Sumerian maxim, "The poor do not have power."

Women, on the other hand, didn't fare so badly in this early civilization. Legally, they had more rights than women had in many other places for at least five thousand years. They could own a house or land, take part in businesses, and testify in trials.

However, women mostly did what women usually have done. One task of course was having babies. If a married woman failed to have them then her husband was entitled, under law, to take a second wife. The other "woman's work" was weaving, gardening, and cooking. In a Sumerian myth, the creator god Enki puts gods, that is, males, in charge of fishing, plowing, digging canals, making bricks, and building. But he puts goddesses in charge of growing grain and vegetables and weaving cloth, which the teller of the myth calls "woman's work." This is not to say that all the women did all they should. In a Sumerian joke a husband says, "My wife is at the shrine, my mother is down by the river, and here I am dying of hunger!"

Gods and goddesses helped to shape the look of Sumer's towns. Sumerians worshipped hundreds of them, and they had to please them all. In every town the "big man" was also the religious leader, and he and his priests would invariably build a tower on a lofty terrace. The bigger towers, known as ziggurats, looked like pyramids of boxes, each box being smaller than the one below it. Atop the highest was a shrine. On the level plains of Sumer one could see a ziggurat shimmering in the heat from miles away. A ziggurat figures in the Bible as the tower "with its top in the heavens" that the presumptuous Babylonians (not far from Sumer) tried to build until God took offense.

As the years went by the merchants and the bureaucrats discovered that they had to deal with ever-growing quantities of stuff: bricks for ziggurats, salaries for copper workers, dates and barley to be fed to slaves. They couldn't store such quantities of data in their heads. In about 3200 B.C., their scribes (or secretaries) solved this problem. With their hands they patted tablets out of what lay all around them, sodden clay, and in the clay they scratched their records of their bosses' dealings. They invented symbols representing donkeys, chisels, male or female slaves, jugs of beer, and so on.

Then the scribes found ways to write new words by joining symbols. In the diagram opposite, an upside-down triangle with a short line from the center to the bottom represents a vulva. It signified a woman. Three semicircles, two below and one above, meant "mountain." Sumer got its slave women from the mountains to the east, so if a scribe combined these symbols, the triangle and the semicircles, the result meant "slave woman."

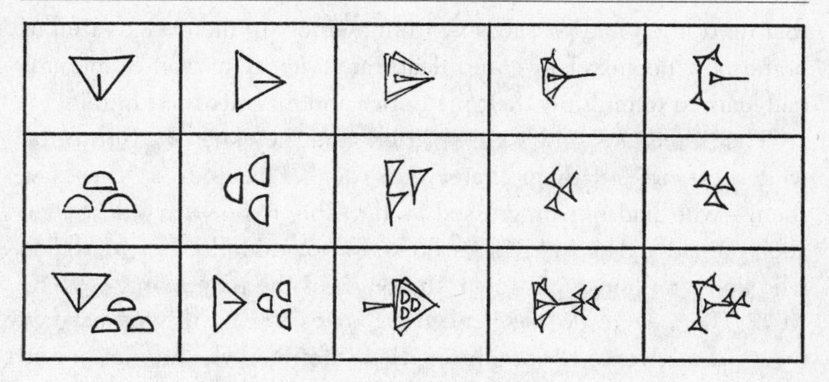

How writing arose in Sumer

The scribes incised the symbols using what was just about as common as the clay they wrote on: the reeds that grew in marshes. If you cut a reed not straight but at an angle and push it into clay, it makes a wedge-shaped mark, a slender triangle. After a while the scribes reduced the symbols, which had been recognizable sketches of things, to abstract sets of wedge shapes that the scribes could write faster but looked nothing like the things they stood for. (The diagram above shows how this happened.) Now the scribes could write with dazzling speed. A saying ran, "A scribe whose hand moves as fast as his mouth, that's a scribe for you."

The scribes became aware that many words had different meanings but sounded the same. Because of this they could often use one group of wedge shapes to represent two things. For example, *ha* meant "fish" and also "may." So they could use the same wedge shapes that meant "fish" to express "may." Sumerian writing now became a mixture of sets of shapes that stood for things and sets of shapes that stood for sounds.

At first, the scribes spent all their time recording business deals and inventories. But then Sumerians discovered other things that one could do with written words. A "big man" had his scribes record his boasts about his victories in war on his buildings. Priests had scribes write down hymns. Wealthy fathers dictated letters telling sons to mind their manners. Scholar-scribes recorded ancient myths that up to then had passed only from memory to memory, as well as brand-new poems of love and death and victory in war. Sumerians had learned to multiply their memories and magnify their minds.

The Sumerians now were civilized, and they knew it. But in the wilderness around them there were rougher people—so they saw them—who had not progressed as far. This is how they described such nomads: "The MAR.TU who know no grain. . . . The MAR.TU who know no house nor town, the boors of the mountains. . . . The MAR.TU . . . who do not bend their knees [to farm], who eat raw meat, who have no houses during their lifetime, who are not buried after their deaths."

Some towns grew to little cities whose people numbered as many as 35,000. These included Nippur, Uruk, Kish, and Ur. Since all of them depended on the rivers, they were fairly close together; Ur was visible from Eridu. As time went by, the larger cities conquered smaller ones so that they were no longer merely cities but city-states. Along their borders they dug ditches and planted markers.

Bigger states and warfare called for tougher rulers, so the "big men" gave way to iron-fisted warrior-kings who claimed that they had been chosen by the gods. The warrior-kings fought their neighbors over water rights and land, and if they beat them they removed the border markers. (A proverb ran: "You go and carry off the enemy's land; the enemy comes and carries off your land.")

The kings had courtiers, wives and concubines, many scribes, hordes of servants, companies of infantry, and troops of lancers who rode to battle in chariots, hurling spears. In Ur at least, when rulers died their subjects buried them in outsized graves. Then their court musicians, bodyguards, and concubines (in costly dresses) drank a painless poison and, their goblets still in hand, lay down to die beside their lords.

One such king was Gilgamesh, the ruler of Uruk. (Now we reach the point in history where we know some names of people, mostly kings.) We don't know much about his life, but Sumerians wrote poems and tales about him and transformed him from a kinglet to a godlike hero. In the stories Gilgamesh is a complicated man; not only does he do great deeds but he also asks big questions about life and death. In the first poem, Gilgamesh saves Uruk from devastation by the king of Kish. But when he sees men perish and he gazes on "dead bodies floating in the river's waters" he becomes aware with sorrow that he too will one day die.

He decides that he must make his name before he meets his fate, and he and his companion, Enkidu, leave home and have adventures. After one of these, Enkidu bravely descends to the world of the dead to recover a drum and drumstick that a goddess gave to Gilgamesh. But then, because he broke the netherworld's taboos, Enkidu can't return. Gilgamesh begs the god of wisdom to help him, and finally the spirit of Enkidu rises to the surface of the earth. The two embrace, and Enkidu tells Gilgamesh how dismal existence after death is. In the final poem, Gilgamesh dies.

About five centuries after the Sumerians had founded them, the age of independent city-states was near its end. The first Sumerian to govern more than several of them may have been King Etana of Kish. In about 2800 B.C. he "stabilized all the lands." Several centuries later a people from what is now Iran conquered all of Sumer and ruled it for about a century. Later, in about 2350 B.C., another Sumerian, King Lugalzaggesi of Umma, ruled all Sumer. He claimed that under his wise rule everybody "from the rising to the setting of the sun" lived in peace like cattle in a meadow.

But during Lugalzaggesi's reign another foreign conqueror appeared. Up the rivers from Sumer, near the site of modern Baghdad, was the kingdom of Akkad. King Sargon of Akkad stormed Lugalzaggesi's capital city, conquered it, and razed its walls. He clamped Lugalzaggesi's neck in stocks by a city gate where everyone could spit on him.

Sargon, "King of Battle," conquered all of Sumer, to his south, and Assyria, to his north, and more besides. "Now," he said, "[let's see] any king who wants to call himself my equal go wherever I went." His empire stretched from the Persian Gulf to the Mediterranean. For several generations, Sumer was merely a province.

Sumer wasn't finished yet, however, as a kingdom or as a cultural innovator. The Sumerians won their independence back, and King Ur-Nammu united the country. Among his other deeds, Ur-Nammu wrote a code of laws, probably the very first. No doubt he had the words inscribed on large stone tablets for everyone to see, but a damaged clay copy is all that now survives of them. In an introduction to the code, Ur-Nammu boasts that he has brought his people justice. He has rid them of the men who grabbed their oxen, sheep, and donkeys, and has seen to it that "the widow did not fall prey to the powerful" and "the man of one shekel did not fall prey to the man of [sixty shekels]."

Ur-Nammu's laws set money fines to compensate for damage. For example: "If a man has cut off . . . the nose of another man, he shall pay two-thirds of a mina of silver." And: "If a man violates a virgin slave girl without the owner's consent, that man shall pay five shekels of silver." He paid them to the owner, of course, not the slave. Ur-Nammu died in battle with invaders from the mountains to the east. According to a poem he was "abandoned on the field of battle like a broken pot."

Sumer now was all but finished. In about 1750 the kingdom of Babylon, farther up the Euphrates, conquered all of Mesopotamia. Henceforth Sumer would be merely an unimportant province of Babylon and later conquerors. The Sumerians abandoned the cities where they had invented civilization, and were absorbed by other peoples. The canals choked up with silt, and the cities crumbled until nothing was left of them but awesome hills of rubble, big bumps in the flat desert. Until the early 1900s, when archaeologists uncovered it, humankind forgot the place where civilization began.

Sumer lived on, however, in what it had taught the nearby peoples and what others learned from them. The most important thing that others learned from Sumer was how to read and write, but Sumer also taught its neighbors how to live in cities, how to shape their pots on wheels, how to make an inventory, how to put their vehicles on wheels, how to fight with chariots and axes, how to measure fields, and how to figure square and cube roots. Even peoples outside of Mesopotamia, especially the Hebrews, borrowed Sumer's myths and laws and made them their own.

ANCIENT EGYPT LAY a thousand miles southwest of Sumer, but these two civilizations were, in a way, sisters. Both arose at about the same time along the banks of rivers, and both showed humans who live in clusters how to organize themselves.

Although people who live beside the Amazon River in Brazil deny it, the Nile is probably the longest river on earth. It begins at Lake Victoria, deep in East Africa, and from there it flows north through other lakes, threads its way through highlands, deserts, and a vast papyrus swamp, and picks up the Blue Nile and other rivers. Only then does it enter what in ancient times was Egypt. At this point it is still 750 miles from the Mediterranean Sea. From here on, the river promenades along a course (see the map, p. 16) shaped like one of its own water lilies. It curves to the right, it curves to the left, and when it reaches Cairo it divides into several streams (forming the lily blossom) and empties into the Mediterranean.

Every year, heavy rains in the forests and hills far to the south swell the Nile. Until recent times the river flooded in late summer, and when it receded it left a coat of fertile mud along its banks. (Today a dam across the Nile prevents the floods.) In very ancient times, the narrow strips of fertile land on either side of the river held swamps and slender forests. Just beyond them was the barren desert. The contrast was astounding; one could stand with one foot on fertile soil and the other on gleaming sand.

Many thousands of years ago, hunters and gatherers lived along the Nile. They reaped wild grains, fished in swamps for perch and catfish, and hunted crocodiles and hippopotami. Then they made the shift to farming that we pictured in a general way at the start of this chapter. They began to keep goats, sheep, cattle, and pigs, and they planted wheat and barley. After centuries had passed, thousands of villages lined the Nile. Each year, just after the Nile had flooded and receded, the peasants would sow their seeds in the mud and later they would reap the grain.

Not long before 3000 B.C., a conqueror united Egypt. What we know about him isn't clear. Archaeologists have found a victory monument that seems to say that the man who united Egypt was a warrior with a chilling name: Scorpion. On the other hand, several ancient lists of Egypt's rulers start out with a man named Menes. Perhaps Scorpion and Menes were the same person. Menes is said to have made the northern town of Memphis his capital. He reigned for sixty-two years and was killed by a hippopotamus.

Not long before the time of Menes, the Egyptians invented a writing system. They may have borrowed the idea of writing from the Sumerians (with whom they traded), but recently archaeologists found pots and labels in an ancient Egyptian cemetery with inscriptions that seem to be older than the earliest Sumerian writing. At first the Egyptians painted pictures on wood or clay, and these are known today as hieroglyphs, a word derived from the Egyptian name for them, "the god's words." We see some hieroglyphs in this drawing of a tablet that records a victory of an early king. Experts disagree

King Narmer's victory tablet

about this, but early Egyptians may have read the hieroglyphs just to the right of the king's head as, "the falcon-god Horus [i.e., the king himself] has defeated the people of the papyrus country [i.e., "lower," or northern Egypt]."

Already in this victory tablet the Egyptians were using pictures to represent things that cannot be drawn, such as names. Often, though, the pictures stood for sounds, which could be combined to form words. Later the Egyptians took the next step, which advanced writing beyond what the Sumerians had done. They made up twenty-four symbols to represent the sounds of consonants. They did not make symbols for vowels, and got by without them. But the Egyptians did invent the principle of the alphabet, with symbols for the individual sounds our voices make.

Th nly trbl s tht th lck f vwls mks t hrd t knw hw thy prnncd sm wrds, r vn wht th wrds wr. ("Wr" cd mn *war, whore, where, ware,* nd s n.)

The surviving Egyptian hieroglyphs are hard to understand, and most of them do not tell us the things we really want to know. As a result, figuring out the history of Egypt during most of its three thousand years is like boating down the Nile on a moonless night, with no hint of the life on either bank except rare, mysterious glows of light from dimmed lanterns.

Egypt's unifying vision through three millennia was the pharaoh. What, you say, not the gods? But Egyptians believed the pharaoh *was* a god, a living god. This is why the king is shown on his victory tablet (opposite) as Horus, the ancient sky and falcon god. Pharaohs stood for power and for everlasting order and justice. In their statues and in paintings they are nearly always shown as strong and calm.

Some of the earliest pharaohs built the pyramids, which were really giant tombs. Since the ruler was a god he lived on after death, and his pyramid would be his house and tomb forever. In about 2680 B.C., Pharaoh Djoser set the example by building himself a step pyramid, which looks something like a Sumerian ziggurat. His successors in the centuries that followed had their architects design smooth-sided pyramids that were triumphs of geometry and art.

Historians know fairly well how the Egyptians built these tombs. Each year in the time of floods, when farming halted, the pharaoh's builders took on many thousand peasant-workers. Gangs of workers hacked great blocks of stone in quarries hundreds of miles away and then rafted them to the building site. Skillful workmen dressed the stones, and then more gangs of peasants dragged them up long earthen ramps and shoved them into place. What a feat! Pharaoh Khufu's pyramid at Giza measures 451 feet high. The men who built it had to move two million blocks of stone, and some of these weigh more than one large car. It took them twenty years to build it.

The pharaohs may have had some earthly motivations to construct these massive tombs. For one thing, since the pyramids were visible from far along the Nile, they reminded those who saw them of the ruler's might. What's more, building them must have pleased the ruling class and bonded them to the pharaoh. It gave the royal bureaucrats a goal that they believed in, since their own lives after death depended on the pharaoh's. They would have their own tombs near the ruler's great one, with their bodies mummified like his so that they as well as he would have an afterlife. So building his colossal pyramid may have helped to make them loyal to him.

As for all the peasant-workers, they depended on the government for what they ate throughout a quarter of the year. So being paid to work may have fostered loyalty. And working next to peasants who had come from other places may have made them all feel that they were a single people.

A single people yes, but of what race or color? Today (in the early 2000s) this question interests many people, black and white, who want to claim these gifted people as their own. The artists who painted scenes of daily life on the walls of tombs showed most of the men as having reddish skin, and the women yellow. However, the bodies of some people in these paintings, especially the slaves or servants, are black. Furthermore, to judge by the faces carved on their statues, a few of the pharaohs and queens and high officials were black Africans. Archaeologists have x-rayed the carefully preserved bodies of both royal and nonroyal Egyptians that are found in tombs. They reached the same conclusion that the paintings and sculpture point to: The ancient Egyptians were a multiracial people.

In about 2100 B.C., after it had thrived a thousand years, Egypt fell apart. Rival factions fought for power, and the country suffered turmoil for a century and a half. Then able rulers from southern Egypt gained control, and they moved the capital to Thebes, far south of Memphis. They gave the country two centuries of order.

During these two hundred years the pharaohs, so it seems, held themselves less distant from their people than the earlier pharaohs had. Their tomb inscriptions picture them as shepherds of their flocks. Their statues show them old and worn, nothing like the strong, sure rulers in both earlier and later Egyptian art. Perhaps the sculptors wanted to convey how much the pharaohs cared about the ordinary people.

Then an Asiatic people, riding horse-drawn chariots of war and wielding battle-axes, conquered Egypt. The Egyptians called them "Hyksos," which apparently means "rulers of foreign lands." They learned the Hyksos's ways of waging war, and a century later the Egyptians drove the Hyksos from the country. The leader of the victors boasted, "When the earth became light, I was upon him like a hawk. . . . I overthrew him, I razed his wall, I slew his people, and I caused his wife to go down to the riverbank. My soldiers were like lions with their prey, with serfs, cattle, milk, fat, and honey, dividing up their possessions."

In the next five hundred years, the pharaohs often went to war to expand their country and create an empire. We can guess why they were free to do this. The rulers no longer needed to employ their armies to keep peace at home, so they used them to extend their power. At times they ruled the little countries on the Mediterranean's eastern shore, Palestine and Syria, and they also drove far south along the Nile and conquered gold-rich Kush.

In about 1350 B.C., Pharaoh Amenhotep IV broke shockingly with beliefs that the Egyptians had held for two thousand years. Early in his reign he told his people to abandon all the old gods and to worship only Aten, a god of the sun. He pictured Aten as a loving world creator, whose brilliant rays brought life to humankind. The king originally had borne the name of an old god—Amenhotep means "Amen is content"—but now he changed it to Akhenaten ("One Useful to Aten") to demonstrate his love of the one true god. All through Egypt Akhenaten's allies hacked the name of Amen off inscriptions.

The pharaoh may have had to struggle with his priests and nobles as he rammed his reform through, but for about a decade Akhenaten had his way. Then he ran into trouble. Busy with reform, he had neglected his governors and his army. Enemies attacked, and Egypt lost a good part of its empire. Apparently he then recanted his beliefs and started to restore the old Egyptian gods.

After Akhenaten's death first one and then another of his daughters' husbands took his place. The second was a teenaged prince named Tutankhaten, who changed his name to Tutankhamen, thus honoring the god his father-in-law had deposed. He issued a decree admitting Akhenaten's errors and restored the old religion, and then he died when he was about eighteen.

Rightly or wrongly, later pharaohs deemed that Akhenaten and the first three pharaohs who succeeded him were all tainted by the same offenses to the gods. So they struck their names from the official list of kings. One result was that Egyptians forgot young Tutankhamen and, what's more, the location of his tomb. Because of that, and since officials had concealed his tomb so well, grave robbers didn't find it. But three millennia later, in 1922, English archaeologists uncovered Tutankhamen's tomb. As Howard Carter shone a flashlight through a hole they had pierced in the tomb, his colleague Lord Carnarvon hoarsely asked him, "What do you see?" Carter whispered, "Wonderful things!"

Wonderful indeed. Robbers long ago had looted almost all the tombs of Egypt's pharaohs; and the treasures that were in them are forever gone. But when Carter and Carnarvon found it, Tutankhamen's mummy, or embalmed cadaver, lay inside a set of "Chinese boxes": first two gold-and-wooden coffins, then another coffin made of solid gold. On his head he wore a golden portrait mask. In other chambers funerary beds and chests, statues, thrones, and chariots gleamed with gold. And this was the tomb of a minor pharaoh!

Pharaohs liked to claim (in inscriptions on the walls of tombs) that they gave their people justice. Egypt did have an effective legal system, and the proof of this survives in letters, wills, and wall inscriptions. On a wall in the tomb of a scribe archaeologists discovered the revealing story of a trial.

This is what had happened. Long before, a pharaoh had given fourteen acres of land to a navy captain to reward him for his service. Three centuries later, the captain's female descendants quarreled in court about who was the rightful owner of this land. Then, for a while, a male member of the family named Huy farmed it. After his death, his widow, Nubnofre, was driven off the land by a relative of her husband named Khay. She took the matter to court, but the judge decided against her.

Later Mose, the son of Huy and Nubnofre, appealed this verdict. When another judge examined the title deeds he realized that one of them must be forged, so he sent a court official with Khay to consult official records. Khay and the official schemed together, and returned with papers that appeared to show that Huy, Mose's father, had never had any right to the land. So the judge decided in favor of Khay.

Mose then gathered witnesses who swore that Mose's father Huy had farmed the land for years and paid the taxes on it. The end of the inscription is missing, but it seems likely that the final verdict was in Mose's favor. The tangled story suggests that courts worked carefully to provide justice.

Egypt was a power in the ancient world until about 1100 B.C., and even after that it sometimes showed its former strength. Later, one after another of the great empires of antiquity conquered it. But, as they did with Sumer, others would absorb Egyptian lore and learning. Egypt stands as the great example of a civilization that remained true to itself for thousands of years, almost as unchanging as her great river, which still

> *flows through old hushed Egypt and its sands,*
> *Like some grave mighty thought threading a dream,*
> *And times and things, as in that vision, seem*
> *Keeping along it their eternal stands.**

*Leigh Hunt, "A Thought of the Nile."

CHAPTER 3

The wanderers settle down.

ABOUT FOUR THOUSAND YEARS ago a group of Hebrews left their home in Sumer and wandered north along the Euphrates River. Wandering was nothing new for them, since they were seminomads whose name, Hapiru, means "wanderer." They led their flocks from pastureland to pastureland. Possibly they were like those desert nomads whom (you may recall) Sumerians looked down on as a people "who do not bend their knees [to farm], . . . who have no houses during their lifetime, who are not buried after their deaths. . . ." When these Hebrews reached Harran, just inside the southern edge of modern Turkey, they settled down awhile and then moved on again.

We know what happened to them next because the Jews, a thousand years later, wrote about the Hebrews in their book of history, myths, and laws called the Bible. We now have crossed a border in this story of mankind. Modern scholars wrenched the history that you have read so far out of stones and bones in digs, or scraps of writing left by folk who didn't care about their past. But now we meet a people whose history enthralled them. They told it and retold it, no doubt making many changes, and then they wrote it down.

Perhaps you wonder why a tale of errant shepherds should deserve a chapter in this book. Well, for reason one, their story illustrates how other wandering peoples on the borders of the settled places such as Sumer, Egypt, ancient China, and the Indus valley settled down. The Hebrews' story illustrates this civilizing process. What is equally important (reason two), the religion of this group of seminomads, after they had settled down, later influenced the creeds of several billion people.

ONE DAY, SAYS the Bible, God appeared in Harran to a Hebrew named Abram and gave him both an order and a promise. According to the Bible, "The Lord said to Abram, go from your country and your kindred and your father's house to the land that I will show you. And I will make of you a great nation, and I will bless you, and make your name great." So Abram and his kindred and their flocks and herds moved west, then south, skirting the Syrian desert, into Canaan.

The little city-states of Canaan took up much of Palestine, the strip of land along the Mediterranean's eastern shore. (See the map on p. 16.) They therefore lay astride the route from Egypt to Mesopotamia. Compared to the Hebrews, at least, the Canaanites were more civilized. Some were farmer-herders; others lived in market towns in houses built of sun-dried bricks. They wove and sold rich crimson/purple cloth of great prestige.

Once arrived in Canaan, the Hebrews who had come with Abram didn't settle down. Unlike the Canaanites, the newcomers were always on the move to fresher pasture for their flocks. For generations they would wander over Canaan's semiarid hills, the adults walking, the donkeys bearing tents and children, and the sheep and cattle grazing. From time to time, they parleyed with each other over grazing rights. At one point, for example, Abram and his nephew Lot agreed that Abram and his kin would pasture in the hilly country westward from the Jordan River, while Lot would graze his flocks along the river.

The Hebrews traded animals and cheese and wool for goods the Canaanites produced, goods the Hebrews hadn't learned to make. They learned and used the language of the Canaanites.

When Abram was ninety-nine years old (according to the Bible), God appeared again and made a pact with him. He promised Abram, "I will give to you, and to your descendants . . . all the land of Canaan, for an everlasting possession; and I will be their God." But Abram and his people had a duty in return. "This is my covenant . . . between me and you and your descendants after you: every male among you shall be circumcised. You shall be circumcised in the flesh of your foreskins, and it shall be a sign of the covenant between me and you." On that same day Abram, whom God now renamed Abraham, had himself, his son, his slaves, and no doubt all his tribesmen circumcised.

That is how the Bible explains why the Hebrews became a people with one god. Nowhere else, so far as we can tell, were there any other monotheists. In Mesopotamia, from which Abraham had come, the Sumerians and Babylonians and Assyrians worshipped many gods. So did the Egyptians. Even Pharaoh Akhenaten, when he ordered worship of the sun and scrapped the other gods, required that he be worshipped also, as a pharaoh-god.

Only Hebrews worshipped just one god. Even after they had made their pact, the Hebrews may have still believed that other gods *existed.* In his covenant with Abraham, God had not forbidden such belief. But whatever they believed regarding other gods, the Hebrews made a pact of loyalty to only one.

On one occasion, so the Bible tells us, God tested Abraham's commitment to him. He ordered Abraham to take his only son, whose name was Isaac, to a distant mountain and to offer him as a sacrifice, on a fire, to God. When they reached the mountain, Abraham put up an altar made of stones, prepared some sticks to make a fire, tied up Isaac, and laid him on the wood. But as he raised his knife to slay his son, the "angel of the Lord" stopped him, saying "now I know that you fear God." He told him that God promised that he would multiply Abraham's descendants "as the stars of heaven and the sand upon the seashore."

After Abraham, Isaac became the Hebrews' leader. God renewed to him the promise he had made his father: "Fear not, for I am with you and will bless you and your descendants for the sake of Abraham, my servant." A generation later Jacob, son of Isaac, took his father's place, and God renewed his promise to him. Once when Jacob traveled to the family's homeland at Harran, he dreamed he saw a ladder reaching up to heaven and angels climbing up and down it. God appeared and said, "The land on which you lie I shall give to you and your descendants; and your descendants shall be [many] like the dust of earth."

On another journey Jacob met a stranger, seemingly an angel. He wrestled with him all the night. Just before dawn the stranger changed Jacob's name to Israel ("God rules"), because "you have striven with God and with men, and have prevailed."

Perhaps these ancient leaders, Abraham, Isaac, and Jacob (Israel), really lived and led their families and flocks amid the hills of Canaan. The Bible tells so much about them that they seem like men of flesh and blood. If they were, they probably lived in about 1800 B.C.

But it seems more likely that when later generations told the Hebrews' story they used a family as a narrative device. They may have used, indeed invented, a father, son, and grandson to connect events involving many long-forgotten forebears. This would have made the story vivid, easy to remember. It may be too that during many tellings they added made-up stories of the ancient leaders' talks with God. Thus they may have made these Hebrew patriarchs the founders of a religion that others (as we'll see) perhaps developed many centuries later.

Their covenant with God didn't cause the Hebrews to abandon their nomadic ways. The Bible tells of an event that makes that point. A Canaanite ruler's son raped a daughter of Jacob, but he fell in love with the girl and wanted to marry her. His father held a meeting with the Hebrews, and he urged them to permit the marriage and to "give your daughters to us, and take our daughters for yourselves. You shall dwell with us; and the land shall be open to you; dwell and trade in it, and get property in it." With words like these he urged the wanderers to settle down.

The Hebrews hid their wrath about the rape. They pretended they'd permit the marriage if the Canaanites would meet just one condition: to be circumcised like Hebrews. The Canaanites agreed and did what they had promised. The demand for circumcision, though, was just a ruse. "On the third day, when [the Canaanites] were sore," Jacob's sons surprise-attacked the town. They slaughtered every male and "took their flocks and herds, their asses, and whatever was in the city and in the field; all their wealth, all their little ones and their wives." And they continued with their wandering life.

Slowly, though, the Hebrews did give in to civilization. No doubt they couldn't help seeing merit in the Canaanites' settled way of life. Here and there in early chapters of the Bible we read about the Hebrews buying land or houses—sometimes even farming.

IN ABOUT 1600 B.C., a group of Hebrews moved from Canaan down to Egypt. The Bible personalizes the story of the move like this. Israel, the former Jacob, had twelve sons. Joseph, next-to-youngest, was his father's favorite, but his brothers hated Joseph, especially after their father gave the boy a handsome robe. They seized their brother and sold him to a caravan of merchants taking balm and myrrh to Egypt. They killed a goat, dipped Joseph's robe in blood, and brought it home to Israel. They told their father that some animal had killed his beloved son.

In Egypt, Joseph had remarkable success. Although he was a foreigner and slave, the pharaoh summoned Joseph to interpret his strange dreams. In one of them, seven fat cows were eaten by seven poor and gaunt ones; in the other, seven full ears of grain were eaten by seven withered ones. Joseph told the pharaoh the meaning of his dreams: seven years of plenty lay ahead, then seven years of famine.

His explanation satisfied the ruler, who made the lowly slave his minister-in-chief. In the seven years that followed Joseph gathered great amounts of grain and stored it in the cities.

Later came the famine, as the pharaoh's dream had warned, and not to Egypt only but to all the lands around it. Up in Canaan, Joseph's family was hungry, so his brothers journeyed down to Egypt, where they had heard that they could purchase grain. Joseph, now a great official, met his brothers and forgave them. He told them to go home and then return with Israel (or Jacob) and all their wives and children. They settled in the delta of the Nile, near the sea, and took charge of the pharaoh's cattle.

This story is believable. Egyptian documents reveal that pharaohs did allow other peoples to trade in Egypt and to settle there in time of famine. And the pharaohs did own herds of cattle in the delta that the Hebrews might have tended. However, the Hebrews (not just a family but a people) may really have entered Egypt as warriors. They arrived there at about the time when the "Hyksos" conquered Egypt and ruled it for two hundred years. Documents suggest that these invaders included Hebrews.

Joseph died, and centuries may have passed, and the Hebrews in Egypt multiplied. But, the Bible tells us, "Now arose a new king over Egypt, who did not know Joseph." He enslaved the Hebrews and made them farm his fields and build his fortress towns. This biblical account fits what is known of Egypt in the 1200s B.C. Just at this time the pharaohs moved their capital northward to the delta, and they would have needed laborers to fortify the area against invaders coming down through Palestine. Egyptian texts from this time mention the use of "Habirus," perhaps Hebrews, for forced labor.

The Bible doesn't tell us much about the Hebrews in Egypt, but apparently there were twelve "tribes" or clans of them. Their names reveal a little. Some of those who soon would lead the Hebrews in their great adventure had Egyptian names. These included Moses, Phinehas, Puti-el, and maybe Hophni, Aaron, and Merari. These names suggest that their captivity had lasted long enough to partly Egyptianize the Hebrews.

Moses, who would later be the Hebrews' leader, was young when the oppression reached its peak. One day he witnessed an Egyptian overseer cruelly beat a Hebrew slave. Moses killed the man and fled from Egypt, eastward to the Sinai Desert. He tended sheep there for a priest named Jethro, whose tribe, significantly, worshipped one God.

One day as Moses roamed this barren country, seeking pasture for his flock, he saw a bush that burned and yet was not consumed by flames. Intrigued, he neared the bush, and from it God called out to him: "Moses! Moses!" God told Moses that he was "the God of your father, the God of Abraham, the God of Isaac, and the God of Jacob." He had heard his people's cries from Egypt, so he said. He ordered Moses to approach the pharaoh and to tell him he must free the Hebrews. Moses then should lead the Hebrews up to Canaan, a land that flowed "with milk and honey."

The Bible's rapid telling leaves one point unclear. Did Moses already believe in the God of Abraham when he slew an Egyptian to help the Hebrew worker? Or did he believe in a sole god only after he had learned from Jethro's tribe about their father-god? Or did he start to worship this god only when he spoke to Moses from the burning bush?

Moses now returned to Egypt and he tried, again and then again, to do as God had ordered and persuade the pharaoh to release the Hebrews. Finally the ruler gave consent, or else the Hebrews simply fled—the Bible story isn't clear. With several thousand Hebrews and some other captives Moses headed east. The pharaoh led an army in pursuit, and his soldiers trapped the Hebrews near a reed-choked lake. According to the Bible, God then sent a wind to drive the water back and let the Hebrews through. When the water rushed back it swamped the pharaoh's chariots.

For years, perhaps for decades, Moses and his people wandered in the Sinai Desert. Food and water were hard to find, and often they looked back with longing to their years as slaves, when at least they rarely suffered hunger.

They reached the barren land around Mount Sinai, which was "wrapped in smoke, because the Lord descended upon it in fire."

Moses climbed the mountain to the top, and God declared to him, "I am the Lord your God, who brought you out of the land of Egypt, out of the house of bondage." God renewed his pledge to help the Hebrews win the promised land of Canaan. "I will deliver the inhabitants of the land into your hand, and you shall drive them out before you."

The Bible says that he gave to Moses ten commandments, carved in stone. They ordered those who heard them to honor God and their parents, and forbade such crimes as murder and adultery. (Some modern scholars date the "Ten Commandments" earlier than Moses, others later.)

It wasn't easy, though, for Moses to convince the Hebrews and the other former slaves that they had a pact with God. Even while their leader stood atop Mount Sinai and conversed with God, they melted their earrings and made a golden calf, a symbol of fertility. When Moses came down from Mount Sinai he found them dancing by it. But he persisted, and he shaped his dozen Hebrew "tribes" (and presumably the others) into a fairly united group. The Bible now begins to call these tribes the "Israelites," meaning the children of Israel (or Jacob). They had a single purpose, to conquer Canaan.

One may well wonder why anyone would hanker much for Canaan. While some of Palestine, where Canaan was, is green and fertile, much of it is dry and bare, and rain is scarce. But the Israelites were trudging in the desert, where they may indeed have pictured Canaan as a land that flowed with "milk and honey." In any case, Canaan was the land of their forefathers, and God had promised it to them.

But another people occupied this promised land. As we said above, the Canaanites were more advanced than those who now were poised to take their home. They lived in towns and may have been the first people in the world to write with an alphabet of letters that stood only for sounds. They cooked and ate on handsome pots and plates and made their tools with iron, which the Israelites did not know how to do.

Just before the conquest started, Moses climbed Mount Pisgah, east of Canaan, so that he could view the land that God had promised. There he died. According to the Bible, God buried him and "no man knows the place."

THE BIBLE OFFERS two conflicting stories of the conquest. According to the first, Joshua, the Israelite leader after Moses's death, led his warriors across the Jordan River. As God directed, he circumcised his men with knives of flint. They stormed the town of Jericho and then they swiftly conquered all the promised land. They did this cruelly (says the first account), "utterly destroying" towns and "smiting," "wiping out," and "slaughtering" the Canaanites. Archaeologists have found remains of towns that were indeed destroyed and then rebuilt at just this time.

After this account, however, comes a different story. In this account, the taking of the land was slow and piecemeal. This does make sense, because the invaders were a loose confederation. According to this telling, some tribes fought and others simply settled where they could. Many of the Israelites probably began to farm the land, as they had done as slaves in Egypt. (No longer were they nomadic herdsmen, like their Hebrew forebears.)

Another people settled in the towns on Canaan's seacoast shortly after the Israelites had entered from the desert. The new arrivals were the Philistines, from whom Palestine would later get its name. Like the Canaanites, the Philistines were more advanced. For example, they could work with iron. Not the Israelites. The Bible tells us that "[No] smith was to be found [in] all the land of Israel . . . but every one of the Israelites went down to the Philistines [when they were at peace with them] to sharpen his plowshare, his mattock, his axe, or his sickle."

Until the Philistines became a danger to them, each tribe of Israelites had a leader, usually a warrior. (But one tribe had a female head who sat beneath a palm and settled their disputes.) However, when they had to fight the Philistines and other enemies, some Israelites concluded that the tribes must now unite beneath a single chief, a warrior king. Others didn't want one; God, they said, was king.

Just what happened next is not quite clear because again the Bible offers two conflicting stories. The end result, in any case, was that the Israelites chose a tall young farmer, Saul, as their first king. He quickly proved to be a first-rate fighter. One day he learned that Ammonites, a fighting people from the desert to the east, had invaded the lands of the Israelites. He chopped the oxen he was plowing with in pieces, and he sent a piece to every tribe with a message that "Whoever does not . . . [join me], so shall it be done to his oxen!" The Israelites gathered in large numbers, marched all night, surprise-attacked in the morning, and crushed the Ammonites.

The king next struck the greater enemy, the Philistines. He won some battles but his reign began to sour. For one thing, Saul could not quite crush the Philistines, who fought with chariots and iron weapons. Then his chief supporter, Samuel the prophet, turned against him after Saul began to act as if he were a priest as well as king. Worst of all, the king began to lose his mind. He fell into depressions and he lashed out savagely at all around him.

Another hero rose, a warrior named David. His father was a farmer near Jerusalem, and David as a boy had tended sheep. From what the Bible tells about him, one senses that David may at times have been a bandit leader, and only a marginal Israelite. As Saul's career collapsed, David's rose. He married a daughter of Saul, who permitted David to wed her after David had killed two hundred Philistines and given Saul their foreskins. David was also a close friend of Saul's son, Jonathan. Saul grew jealous of David and afraid as women chanted in the streets, "Saul has slain his thousands, and David his ten thousands." What more could David want, he asked himself, but to be king? He drove the warrior from his court, and for a while David commanded the bodyguard of a Philistine king.

Saul died as he had lived, in battle. The Philistines, allied possibly with David, overwhelmed the Israelites. They killed the king's three sons and badly wounded Saul with arrows. Rather than be murdered by his enemies, he held his sword point up and fell upon it.

Defeated by their enemies, and with their leader dead, the Israelites were deep in trouble. A bloody struggle for the crown began. If the Israelites could not agree upon a king, they risked becoming just a snarl of quarrelsome tribes, easy prey for enemies.

One man only had the prowess and respect to reunite them. This, of course, was David. Shortly before 1000 B.C., the elders of the tribes anointed him their second king. David made Jerusalem, among the hills, his capital, and he went from triumph on to triumph. He crushed the desert tribes and extended his kingdom north, and he beat the Philistines so badly that they never again were a threat to the Israelites.

Now that he was king, he had to bind the tribes together. One means to do this, which he often used from early manhood to old age, was to marry women from different tribes and factions. As we said, the first of his wives was Michal, daughter of Saul. Another was the lovely Abigail, whose husband handily dropped dead when David wanted to marry her. Marrying her strengthened his ties to a powerful clan in the south. Driven partly by politics and partly by desire, David gathered twenty wives and concubines.

Religion was another instrument with which to bond the tribes and build the monarchy. Until the time of David, the Israelites had viewed their pact with God as One-on-one. The pact did not involve the king of course, for they never had a king till Saul, and Saul ran into trouble when he tried to be a priest as well as king. But David partly changed the view that God and man did not require a king as go-between. Probably his victories alone had proved to many that their God approved of kings, and especially of David.

But David pushed the matter further, beyond merely having God's approval. As David saw it, he personified the covenant with God. He served not only as king but also as religious leader, and he made Jerusalem both capital and holy city. He may have written (experts disagree) many of the psalms, or sacred songs, that Jews and Christians still recite or sing today. (The most famous of them may reflect his boyhood task of tending sheep: "The Lord is my shepherd, I shall not want; he makes me lie down in green pastures. . . .")

David brought to Jerusalem the ancient "ark" that held the tablets with the Ten Commandments God had given Moses. The Bible says that as the ark arrived and a crowd looked on, David leaped and danced and gave to everyone "a cake of bread, a portion of meat, and a cake of raisins."

When he lay near death, David chose his second son to take his place. He said, "Let Zadok the priest and Nathan the prophet . . . anoint him king over Israel; then blow the trumpet and say, 'Long live King Solomon!' " And so it happened. Solomon was crowned before his father's death.

As often happens, the father built the business and the son enjoyed it. Solomon was fortunate to reign when the rulers of both Egypt and Mesopotamia happened to be weak. His little country thrived, and Solomon prospered with it. Whereas David had had 20 wives and concubines, the Bible tells us Solomon had 700 wives (one of them a pharaoh's daughter) and 300 concubines. Using craftsmen from abroad, he built a splendid temple, or "house of the Lord," of limestone, cedar, bronze, and gold.

While his reign was crowned with glory, underneath lay discontent. His people, mostly poor and frugal, found they had a king who sat on an ivory throne, drank from golden vessels, and collected apes and peacocks. He taxed them hard and drafted many men to quarry stone and cut the lumber for the temple.

After David and Solomon, the kingdom fell apart. The ten tribes north of Jerusalem seethed about the taxes and forced labor. Solomon's son and successor parleyed with them, but he bungled the discussions. When he tried to force the tribes to follow orders, they stoned his labor boss to death. The king himself was lucky to escape. The northern people formed a kingdom of their own, called Israel. The people in the southern remnant called their tiny kingdom Judah.

ONE MIGHT SUPPOSE the breakup of the Saul-David-Solomon kingdom meant disaster. But in the short run the tribes' division introduced their greatest age. Only after they had quarreled and divided did the tribes, teetering between two major civilizations, make their contribution to our fund of great ideas. Until this time the Israelites had seen their god as, yes, a father quick to help his children and concerned about their morals. But they also knew him as a tyrant who demanded sacrifices to him and was pitiless to those who crossed him.

In the eighth and seventh centuries B.C., men called "prophets" broadened and deepened the Israelites' religion. They claimed that they revealed the will of God, often in his very words. One of them was Amos, who called himself a shepherd. Amos had the courage to denounce the cruelty and greed of the king of Israel and the upper classes. "They sell the righteous for silver," he declared, "and the needy for a pair of shoes—they that trample the head of the poor into the dust of the earth." God is not concerned with worship but with decency and justice. Amos quotes him: "Take away from me the noise of your songs; to the melody of your harps I will not listen. But let justice roll down like waters, and righteousness like an ever-flowing stream." Amos predicted that both kingdoms, Israel and Judah, would fall.

Another prophet, Hosea, preached to the people of Israel when they were warring with Assyria, and when four of their kings had been murdered within fourteen years. He used his own experience, a tragic one. He had married a prostitute, and she betrayed and left him. In the same way, Israel had let down God by loving other gods—their version of adultery.

Israel, Hosea said, would surely feel the wrath of God. Ahead lay anarchy and defeat. "They shall fall by the sword, their little ones shall be dashed in pieces, and their pregnant women ripped open." But just as he, Hosea, forgave his wife and took her back, so God would one day pardon Israel. "They shall return and dwell beneath [God's] shadow, they shall flourish as a garden; they shall blossom as the vine."

Hosea, Amos, and the other prophets gave their listeners a new and different view of God. Yes, he does demand our rites and worship, said the prophets, but not from hypocrites and victimizers of the poor.

More than for religious rites he cares how humans live and how they treat each other. The prophet Micah says, "He has showed you, O man, what is good; and what does the Lord require of you but to do justice, and to love kindness, and to walk humbly with your God?"

As early as the time of David, learned Israelites began to write down certain poems and tales that people until then had learned only by word of mouth. And historians now wrote about events they witnessed, like the tragedy of Saul and the victories of David. For a thousand years others added to these writings. Israel's children couldn't forge a knife or shape a handsome pot, but some could write like Sophocles or Shakespeare.

Scholars gathered many of these treasured texts and slowly formed the Bible. Among the jewels were ancient myths, the stories that explained to people who they were and where they came from. All the peoples of the Middle East had myths, which probably originated not in Palestine but in the villages and towns of Mesopotamia, along the Tigris and Euphrates. Ancient tablets found in mounds of rubble near the rivers tell us myths much like the ones that introduce the Bible. They tell how Enlil, the creator, separated Earth from Heaven; and how Enki and his mother fashioned men from clay; and how the mother-goddess nearly killed the water god because he ate forbidden plants in paradise; and how the gods once caused a flood to kill "the seed of mankind" and (apparently, but part of the tablet is missing) warned a pious king to flee aboard an ark.

The Bible's writers tell these ancient tales to great effect, and they have stimulated many minds. Consider, for example, the myth of the creation. After God had made a man and woman, Adam and Eve, he gives us human beings our mission: "Be fruitful and multiply, and fill the earth and subdue it; and have dominion over . . . every living thing that moves upon the earth." During four millennia we humans have discovered in these words a validation of our (so-called) mastery of earth, namely: God ordained it.

In another Bible myth God punishes the first two humans when they disobey an order. Before he drives them from their paradise he tells them that from that day hence women will give birth in pain, and everyone will have to work to eat. "In the sweat of your face you shall eat bread." At the end of their lives Adam and Eve will die and return to the earth from which God had created Adam. "[Y]ou are dust, and to dust you shall return."

The editors included in the Bible age-old tales about the patriarchs and how they met with God, and how the Hebrews were enslaved in Egypt. They included tales about the flight, the wandering in the desert, the settling in Canaan, the three kings and the union of the tribes, and the breaking off of Israel from Judah. And they added other writings: prophecies and laws, the Ten Commandments, poems and proverbs, psalms and stories.

The Bible gave these former nomads, who now were needy peasants in Palestine, what Egyptians and the people by the Tigris and Euphrates lacked: a memory.

Among the most affecting writings in the Bible is the poem of Job. The Israelites did not invent the tale; Sumerians had told an older version. But some forgotten genius plucked his harp and made the legend sing. The poem concerns a mystery that troubles everyone: if there *is* a god or other topmost power, why does he or she permit the suffering of those who don't deserve it?

As the book begins, Job is wealthy and contented. He has a family (a wife, three daughters, seven sons) and "very many" servants, and possesses many camels, oxen, sheep, and donkeys. He is "blameless and upright," fears God, and turns away from evil. God decides to find out whether Job can keep his faith if he is stricken with disasters. So messenger after messenger approaches Job to give him awful news. They tell him first that nomads robbed his oxen and his donkeys, then that fire burned up his sheep, then that other nomads took his camels, then that wind blew down his house and killed his sons and daughters.

Job at first is stoic. "Naked I came from my mother's womb, and naked shall I return; the Lord gave, and the Lord has taken away." Even after God has Satan cover Job with loathsome sores from head to toe, his faith holds up. "Shall we receive good at the hand of God, and shall we not receive evil?"

Finally, however, Job despairs. When his friends arrive to comfort him, he curses the day he was born. He longs for death because "There the wicked cease from troubling, and there the weary are at rest." His friends and he debate the cause of evil. His friends uphold the customary view, that suffering is punishment for sin. God forgives the sinner who repents, and he emerges from his miseries morally stronger.

Job is not consoled. He lists the woes that have happened to him and denies that he did wrong. In despair he claims that God destroys a person at a whim, without mercy, while he lets the wicked go unpunished. Job appeals to God to show himself and justify to Job the sorrows in his life.

And God does appear, speaking from a whirlwind. He spends no time on Job's afflictions but reminds him of the size and splendor of the earth. He challenges poor Job, if he believes God's rule is wrong, to quell the force of evil. Does Job pretend to understand the power and purposes of God? At the end of their discussion Job confesses, "I have uttered what I did not understand, things too wonderful for me, which I did not know." In fact, God really hasn't answered Job, but the poet wants us to take some comfort in knowing of the immensity of earth and the mystery of God. We cannot limit God's transcendent purposes with human notions of what's just and good. As the story ends, God restores Job's blessings.

As the Bible started taking shape, the prophets' gloomy warnings to the kingdoms of the north and south came true. The northern kingdom, Israel, lasted until 721 B.C. In that year the Assyrians, from the Tigris and Euphrates region, defeated it. The conquerors, who were famous for impaling prisoners on stakes, merely scattered Israel's upper classes through their empire and totally absorbed them. History knows them as the Ten Lost Tribes.

Little Judah, in the south, survived by paying tribute, but in 586 B.C. the Babylonians, also from the land between the Tigris and

Euphrates, conquered it. They burned Jerusalem and drove off many people to Mesopotamia. When the Persians later conquered Babylon they freed these captives, and most returned to Judah. Since the northern tribes of Israel had disappeared, the Judeans—or in English, Jews—were now the only people left who worshipped the God of Abraham and Moses.

IN JERUSALEM THE Jews rebuilt the Temple, and their learned men completed the assembling of the Bible. But the troubles of their homeland didn't end. The Greek Seleucids conquered Judah in 198 B.C. They put a statue of the Greek god Zeus inside the Temple. Still later, in 63 B.C., the Romans conquered Judah and attached it to their empire. Under Rome the life of Jews was hard, and many of them left their homeland. In A.D. 66, Jerusalem revolted, but the Romans crushed the rising, burned the Temple, and all but razed the city. Two generations later Jews again rebelled, and the Romans crushed them once again. They carried some away as slaves.

Many other Jews left their homeland. They no longer had a land they called their own, and they scattered all around the Mediterranean and beyond. For nearly two millennia they lived in ghettos, far from other Jews, often badly treated.

Though scattered, they possessed the book that told their story and explained their faith. This treasure-house of myths and facts, laws and psalms, and poems and prophecies was the product of their former lives as tribal people on the fringes of great civilizations. Now that they were scattered it would help them to retain the learning and the feeling for their past that made them unlike any other folk on earth.

But that's not all the Bible would accomplish, since other humans too, not Jews alone, would read it. It added to the human fund of great ideas about the cause of suffering, our duty to the poor, the morality of war, and the purposes of history. And two far-spreading religions would partially adopt the belief, once held by seminomads and lowly farmers, in a single, strict, and watchful god.

CHAPTER 4

Two ancient cities follow diverse paths.

ABOUT 2,500 YEARS ago two city-states of Greece flared like stars exploding. In different ways they both excelled. For a century and a half they gave mankind an opportunity to see how wide is the range of things that we can do.

Three peninsulas hang like udders from Europe's underbelly. On the west is Iberia (home of modern Spain and Portugal), which almost reaches Africa. In the middle is Italy, looking like a boot that's poised to kick. And on the east is the Balkan Peninsula, full of rocky hills and quarrelsome peoples.

Greece is at the bottom of "the Balkans," where they dwindle to an end. It is formed of juts of land and 1,400 islands strewn with low but rugged mountains. Before our story starts, Greeks who needed wood for fuel or lumber to construct their ships had largely stripped the hills of trees. Erosion followed. Plato, the Greek philosopher, once wrote of his homeland that "what remains . . . is like the bony body of a sick man, with all the rich and fertile earth fallen away, and what is left is only the scraggy skeleton of the land."

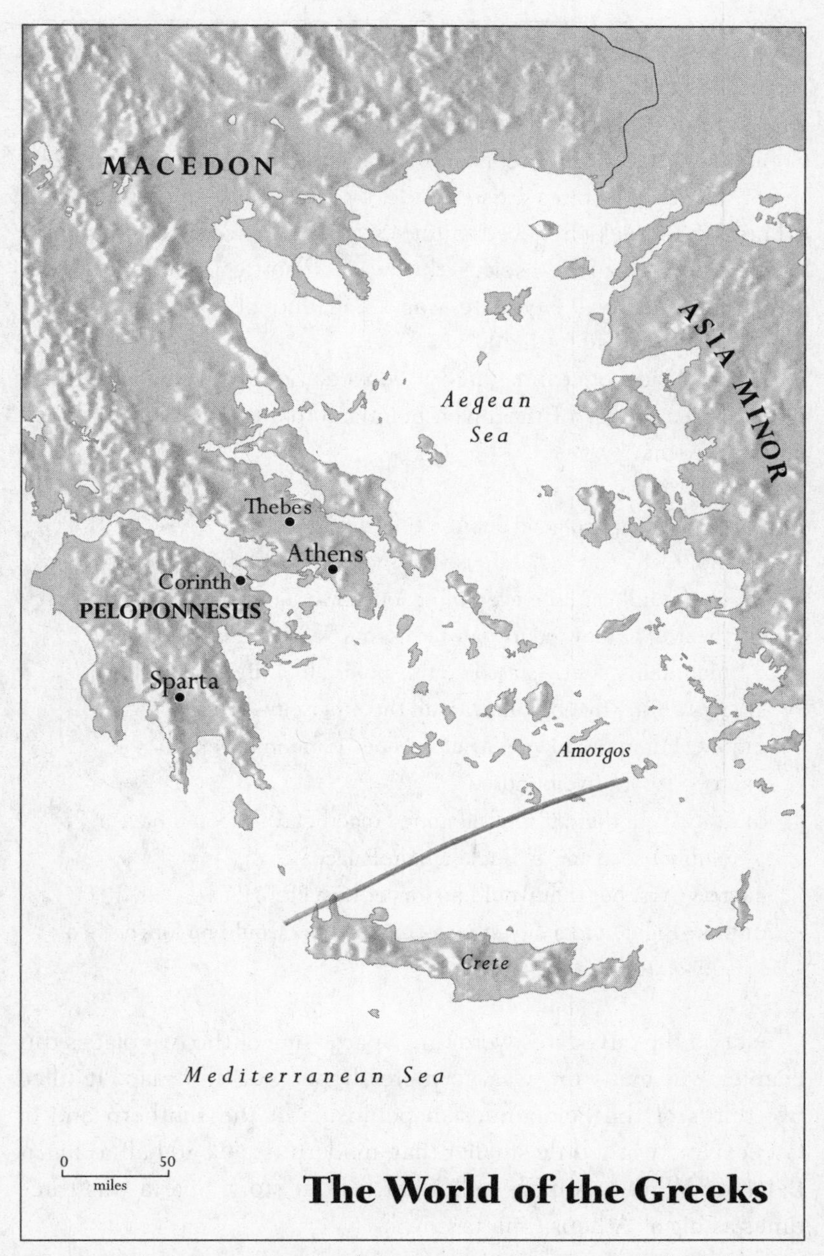

City-states (too many to show here) crowded the Balkan Peninsula (left), the west coast of Asia Minor, and islands in between.

Because the villages and towns were by the sea or walled by mountains, they lived in partial isolation. Many towns developed therefore into small and independent city-states, about 200 of them. Some were tiny. On the island of Amorgos, which covered fifty miles, three city-states existed side by side, each one running its affairs. The Greeks believed in little states; they were certain people ought to organize themselves this way. Aristotle (another thinker) held that the ideal city-state was small enough so that everyone knew everyone else by sight.

An historian once imagined a conversation between an ancient Greek and a modern Briton who belonged to a London club. It went about like this.

BRITON: What's so good about a tiny state?

GREEK: Can't you see? Life is vivid. You can stand by our fortress on the hill, and see everything: our town, our temples, our theatre, our law court, our farms, our seashore, our ships, our mountains. A city-state like this holds all of life's possibilities.

BRITON: But why not join with all the other city-states?

GREEK: Hmmm. How many clubs does London have?

BRITON: About five hundred.

GREEK: Well, then, if they all joined together, they could have a clubhouse as big as Buckingham Palace.

BRITON: Yes, but that would no longer be a club.

GREEK: Right, and a city-state as big as yours would no longer be a city-state.*

Not all the city-states were tiny. Sparta, one of the two places this chapter will focus on, was no mere flyspeck on the map. It filled two-thirds of the Peloponnesian peninsula at the southern end of Greece, and was a little smaller than modern Israel and half as big as Belgium. What was more important for our story, Sparta was three times as big as Athens, Sparta's rival.

*Freely adapted from H. D. F. Kitto, *The Greeks* (1951), p. 79.

In most places, governments develop slowly, with trials and errors and the shedding of some blood. But if the legend is correct, Sparta missed all that. Instead, a Spartan named Lycurgus, in the 600s B.C., devised quite quickly the system that the Spartans used for centuries. Lycurgus saw that Sparta needed basic rules to run the government and for the Spartan way of life. He wrote them singlehanded and won approval of them from the god Apollo, whose cave was in the mountains of central Greece. The Spartans agreed to use his rules, but Lycurgus feared they might later change their minds. So he prepared to leave Sparta, and just before departing he made the Spartans swear to leave the constitution unaltered until he returned. Then he went away and starved himself to death.

The government Lycurgus is supposed to have designed was run entirely by the Spartan upper class of warriors and landowners. At the top were not one but two kings, who, in time of war, served as generals. (This arrangement sometimes worked as badly as one might guess.) Sparta also had a council of twenty-eight men aged sixty or more; and a board of five overseers with very broad powers; and an assembly of all the warriors in the heavy-armed infantry. More unusual than their form of government was the way the Spartans lived. According to the legend, Lycurgus also planned this, but no doubt it simply evolved. For the Spartans, these were the facts of life: they numbered only about 25,000, while the conquered people whom they ruled were perhaps twenty times as many. They had to somehow manage all these people. The Spartans decided to leave some of them free, but with no political rights. The rest they made into publicly owned farm laborers—in a word, serfs. As badges of their lowly status, the serfs were made to wear dog-skin caps, and each year (the philosopher Aristotle tells us) the Spartans formally declared war on them. They may have done this so that their secret police could legally murder any troublemakers.

It was their determination to oppress these others who outnumbered them that shaped the ruling class's way of life, the customs that we have in mind when we use the word *Spartan*. They had to make themselves a weapon that could crush revolts, so they invented a social system designed to make every Spartan male a keen and loyal soldier. A committee of elders inspected newborn baby boys and approved only babies who were fit. They threw the others into a mountain ravine.

After the age of twelve a boy lived in a camp with other boys. He ate wretched food and slept on a bed of reeds that he gathered by a river. He was taught to read and was trained in marching and gymnastics and in the kind of music that was played in drills and battles. The older boys were whipped once a year to see who could stand it longest. They also went through a "time of hiding," in which they had to live alone in the wild and emerge at night and kill any serfs they found.

At the age of twenty a young man joined an army squad of fifteen men, and he trained and ate with them. The food was so dreary that a visitor who ate in a Spartan mess is said to have exclaimed, "Now I understand why the Spartans have no fear of death!" So that he could pay for his food, each man was allotted income from one of the state farms worked by serfs. And in this way most Spartan men passed their lives, training and fighting, living off the labor of an underclass, and always under supervision of the state.

The Spartans were known to all as warriors. Among the Greeks, not just the Spartans, it was disgraceful to run from battle and to throw away your shield. Nevertheless, a lighthearted poet (not a Spartan) wrote:

Some lucky Thracian has my noble shield:
I had to run; I dropped it in a wood.
But I got clear away, thank God! So hang
*The shield! I'll get another, just as good.**

Compare that with the story of the Spartan mother who is said to have told her son, as he was going out to battle, either to triumph or be killed. "Come back with your shield—or on it."

*Kitto, *Greeks,* p. 88.

Women's lives in Sparta were not much easier than men's. Girl babies were less likely to be thrown in a ravine, but girls shared some of the same training as the boys. Their main duty was to produce more warriors for the state, but Sparta made even sex a challenge. In the first years of marriage their husbands could visit them only at night and by stealth; this was intended to be a test of the men's cunning and drive. Spartan couples, it was said, had children before they ever saw each other's faces by daylight.

To outsiders, it seemed that Spartans had sold their souls. Not only did they oppress their serfs, but they had surrendered much of their freedom to their war machine. Culturally their lives were barren. Long before, the Spartans had been known for their poets, their music, and their bowls and weapons cast in bronze. Poets from all over Greece had competed every autumn at a Spartan festival. But in its heyday Sparta was a cultural wasteland. Their capital city was a mere collection of villages, with hardly any handsome temples. Music, for the Spartans, was only war songs and the piping of the flutes as they marched to battle.

Grim as their lives were, the Spartans took pride in giving them to their state. According to a Spartan custom, when one of their kings went to war he always had beside him a Spartan who had been crowned as victor in the all-Greece Olympic games. One year a Spartan at the games was offered a bribe to lose a wrestling match, but he refused it. After he had thrown his opponent, someone asked him whether it would not have been even better to take the bribe. "No," he answered with Spartan brevity, "I shall fight the enemy by the side of my king."

ATHENS, SPARTA'S RIVAL, lay to Sparta's northeast, on a peninsula nearly girdled by the sea. It contained only a city and about a thousand square miles of hills, villages and little seaports, olive groves, and tiny fields of grain. But so small were the other city-states that Athens was second in size only to Sparta.

Before the fifth century B.C., Athens' history was like that of many other city-states. It was ruled first by a line of kings, then by landlords who had made their fortunes raising grapes and olives, then by ordinary people (a limited democracy), and then by three dictators in succession. When the second of these tyrants was murdered, his brother succeeded him. But he proved to be harsh and vengeful, so a clique of wealthy men overthrew him. For the next two years, these men struggled for power.

This competition led to a basic change in the way Athens was ruled. Among the struggling politicians was the able Cleisthenes. When he found he needed help, this politician took a fateful step: he turned to the common people. He promised the ordinary Athenians that in return for their help he would bring about reforms.

When he had won, in 508 B.C., Cleisthenes actually did what he had promised. All free men living in the Athenian state at that time became citizens with full rights. Those over thirty years old were eligible to take part in the ruling council, whose members would be picked by lot. Without having planned to do so when he started, Cleisthenes had taken Athens a long way toward democracy.

Later, however, the Athenians went farther. They paid their government officials for their service, which at least in theory meant that even the poor could afford to serve. And they set up law courts in which juries of hundreds of citizens, without any management by judges, decided cases by majority votes. Every citizen might hope to sit on the council at some time, or hold an office, or serve on a jury. Athenians were certain of the wisdom of the common man.

True, Athenian democracy was far from pure. For one thing, only men took part, and, of the men, only those who were citizens. The rest were excluded because they were slaves or had been born somewhere else. And although officeholders were paid, the pay was low, so that most men could not afford to serve again and again. In the posts that dealt with crucial matters such as taxes, the navy, and foreign affairs, a handful of rich men served time after time.

Meanwhile, as we saw, Sparta, the other leading city-state, was ruled autocratically by a military caste. Here, then, were two sharply different systems. And now both systems would be tested in war.

This is the background. To the east, across the Aegean Sea, was Asia Minor (modern Turkey). Centuries earlier, many Greeks had crossed the Aegean and settled here along the coast. These Ionians, as they were called, stayed in touch by sea with the motherland Greeks; they were all one civilization. Politically, however, the Ionian Greeks were subjects of the Persian Empire, whose capital was far away in what is now Iran. The Ionian city-states had a grievance against their overlord. Persia often interfered with them in order to ensure that they were ruled not by Athenian-style democrats but by tyrants obedient to Persia.

In 499 B.C. the Ionians rebelled. Athens helped them by sending twenty ships, and another Greek city-state (Eretria) sent five, but Persia nevertheless stifled the revolt. The Persian emperor Darius was furious with the impudent Greeks for helping the Ionians. He vowed to punish them and is said to have ordered a slave to repeat three times at dinner every day, "Master, remember the Athenians." Darius did remember, and seven years after the revolt he sent an army to punish Athens and Eretria. The campaign, however, was a disaster. His fleet of 300 ships, bearing 20,000 men, was wrecked in a storm off northern Greece, and the surviving Persians went home.

Despite this failure, Persia now sent envoys to many Greek city-states demanding tokens of submission: earth and water. The Spartans threw the envoys down a well, telling them that there they would find plenty of both. In 490 B.C. the Persians struck again. This time they sailed straight across the Aegean and captured Eretria and sacked it. Then they sailed on to the bay of Marathon, about twenty-five miles north of Athens, and landed an army on the coast. The Athenians debated what to do, and they considered using a very cautious strategy: simply to stay at home and defend their city when the attack came. While they were debating, they sent a messenger to Sparta asking for help.

Soon, however, the Athenians decided to get in the first blow at the Persians. (This daring was typical of them.) Along with a small force from the nearby town of Plataia, they marched to the plain of Marathon. There they boldly attacked the much larger Persian army and drove the Persians back to their ships. Then, just as the Spartans finally arrived, the Athenians rushed back to their own city before the Persians could sail around the coast and attack it. The frustrated Persians once again gave up and returned to Asia.

As a mighty power, the Persians could not put up with such humiliation. Ten years later, a huge Persian army marched along the northern shore of the Aegean and down into Greece. On the sea, not far away, the Persian fleet sailed with it. Athens, Sparta, and other Greek cities joined forces to oppose the Persians. Their ships fought them well, off Greece's eastern coast, but then they had to retreat to the waters near Athens. Meanwhile the allied Greek army tried to block the Persian one on the eastern coast of Greece at Thermopylae, where a narrow pass ran between the cliffs and the sea. Here, the allies hoped, their small force could hold off the Persians' large army.

On the first and second days the Greeks, fighting in shifts on the narrow shoreline, blocked the enemy. But on the night before the third day, a Greek traitor showed the Persians a mountain path that led to the far end of the pass. When the Greek army learned of this betrayal, they retreated, leaving a rearguard of only three hundred Spartans and some others to delay the Persians. These men fought hard, and all were killed. The epitaph on their mass grave reads, "Passerby, tell them in Sparta that here in obedience to their laws we lie."

The Persians soon won most of Greece—all the way down to the slender isthmus that links the main body of Greece to the Peloponnese. The Athenians had abandoned their city, and the Persians sacked it. But the Athenians still had their fleet. They lured the Persian navy into a bay near Athens that was so narrow that the lightweight Persian ships could not outmaneuver the heavier Greek ones. While the Persian ruler Xerxes, Darius's son, watched in horror from a throne on a hill overlooking the bay, the Athenians sank most of his ships.

With no navy, Xerxes had no supply line, so he withdrew the bulk of his army from Greece. In a later battle, a force of Spartans and other Peloponnesians beat the remnant he had left behind. On that very same day, it is said, on the faraway coast of Asia Minor, Greeks stormed a beach and destroyed what was left of the Persian fleet. At a sacred shrine on Mount Parnassus the Greeks erected a bronze column simply inscribed: "These fought in the war." Below are the names of thirty-one city-states, beginning with Sparta and Athens.

No one could be sure that Persia would not attack yet again. So the Athenians organized a league of the Greek towns that lay on the coasts and islands of the Aegean Sea. They swore oaths of allegiance that were to last, they pledged, until lumps of iron, which they threw into the sea, should rise again. Each of them contributed a sum of money or an agreed-on number of ships and men. While many of the members were assessed at only one ship, Athens supplied its fleet of two hundred.

Naturally, Athens dominated the league. After a while, Athenians began to shape the foreign policies of the league without consulting the other members. And then, when the island of Naxos tried to secede from the league, Athens crushed this "revolt" and forced the Naxians to pay the league what it demanded. Other protests had the same result, and so what had begun as a defensive league turned into an Athenian empire. Many Greeks now viewed Athens with fear.

IN THESE EXCITING decades the Athenians were not only fighting a great empire, along with other Greeks, and building a small empire of their own. They, and other Greeks as well, also had an adventure of the mind. They probed the nature and the meaning of life itself, and told what they had learned in buildings, statues, and above all words that still can move us.

This chapter has already shown the Athenian mind at work in one arena. Our narration of the Persian wars drew on the writings of Herodotus, who is sometimes called the "father of history." Herodotus was born in a Greek town in Asia Minor, but he later lived in Athens. About a generation after the Persian wars, Herodotus set out to write the history of that struggle between the empire and the city-states. This was not an easy task, for he had scarcely any public documents or memoirs to work with. He had to rely largely on the memories of aged army veterans, and what their sons and daughters still recalled of tales their fathers told them.

Herodotus wrote mainly to enthrall and entertain the Greeks who paid to hear him give public readings from his *History.* He tells many anecdotes such as this one about the homeward voyage of Xerxes after the Greeks had beaten him. A storm blew up, and the captain told Xerxes that the ship would sink " 'unless we can rid ourselves of this crowd [of Persian noblemen] on deck.' "

Herodotus writes, "On hearing this Xerxes is supposed to have said, 'Gentlemen, now is the moment for each of you to prove his concern for the king; for my safety, it seems, is in your hands.' The Persian noblemen bowed low, and then, without more ado, jumped overboard; and the ship, lightened of her load, came safely to her port on the Asian coast." As soon as he was on shore, Xerxes gave the ship's captain a gold crown as a reward for saving his life, and then, "to punish him for causing the death of a number of Persians, cut off his head."*

Herodotus was not just a storyteller; he was also an artist and historian. Running through his *History* like a crimson thread is a majestic theme: how a mighty despotism fell victim to its pride and failed to overcome a people who were poor but free and gifted. Even his anecdote (above) about Xerxes helps to make that point, though Herodotus comments skeptically about that story that it seems unlikely that Xerxes would have sacrificed his noblemen when he might simply have flung overboard an equal number of rowers. The point of the anecdote is that the Greeks valued life more than did the Persians.

Herodotus wanted to show how sharply customs sometimes differ. On one occasion, he says, the Persian emperor Darius summoned some Greeks who happened to be at his court and "asked them what they would take to eat the dead bodies of their fathers.

*Herodotus, *The Histories,* trans. Aubrey de Selincourt (1954), pp. 539–40.

They replied that they would not do it for all the money in the world. Later, in the presence of the Greeks . . . he asked some Indians [from India], of the tribe called Callatiae, who do in fact eat their parents' dead bodies, what they would take to burn them. They uttered a cry of horror and forbade him to mention such a thing. One can see by this what custom can do, and [the poet] Pindar, in my opinion, was right when he called it 'king of all'."*

History, someone said, surveys the past in order to find what it is that makes us human. The Greeks also invented another way of exploring our humanity: the tragic drama. Greek tragedies were short, tense plays that explored the meaning behind the suffering caused by some unintended harmful act. Even for the poorest Greek, these tragedies became a passion. At a festival held in Athens once a year, audiences of many thousands sat on stone in open-air theaters for as much as ten hours a day, on four consecutive days. Each day they watched four or five dramas.

The most famous of all Greek tragedies is Sophocles' *Oedipus the King*. Like almost all these dramas, it is based on an appalling tale. The audience would already have known this story before attending the play, but that was quite all right. They expected not to hear a new story but an old one, retold in such a way as to move them to pity and understanding.

This is what had happened before the action starts in *Oedipus the King*. Young Oedipus is raised in Corinth as the son of its king, until one day he learns from an oracle that he is fated to kill his father and marry his mother. Since these would be horrible offenses to the gods, to avoid committing them Oedipus flees from Corinth and heads for the city-state of Thebes. At a remote crossroads, he meets a stranger, who arrogantly orders him out of the way. Oedipus kills this man. After he arrives at Thebes, he rids the town of a man-eating monster by solving a riddle. He is rewarded with the throne and also with the hand in marriage of the queen, whose husband, Laios, has recently been murdered.

*Herodotus, *Histories,* pp. 190–91.

When the play begins, Thebes is suffering. Cattle are dying, crops are failing, and a dangerous disease is spreading. Oedipus, who is now a respected king and the father of four children, learns the cause of the troubles: the gods are angry because the killer or killers of the late King Laios live, unpunished, in Thebes. Oedipus decrees a terrible punishment for the killer or killers unless they give themselves up, and then he sets out to solve the murder. "I must not hear," he says, "of not discovering the whole truth."

Piece by piece Oedipus unearths the truth: It was he himself who killed King Laios, who was the stranger he fought with on the road to Thebes. What is more (the horror mounts) he is himself the son of Laios and of Laios's widow, who is *his own wife,* Jocasta. When he was born, it turns out, an oracle had told Laios that his son was fated to kill his father. To avoid this, Laios had the child abandoned on a hill to die. But the child was rescued and ended up in Corinth, where he was raised believing himself the son of the Corinthian king. That son of course was Oedipus.

When he learns all this, the horrified Oedipus rushes into the palace and learns that Jocasta, his mother, his wife, and the mother of his children, has just realized the truth and hanged herself. He seizes her brooch and gouges out his own eyes. As he is led away, the onlookers wail: "Everyone said, 'That is a fortunate man'; and now what storms are beating on his head! Call no man fortunate that is not dead. The dead are free from pain."

What is Sophocles telling us in this haunting tragedy? Perhaps that fate is far, far stronger than we are. And also that Oedipus trusted too much in his own reason and his own senses. He successfully used them to solve a riddle and to learn the truth, but with what a terrible result! Of course, Sophocles was a Greek, and he certainly approved of the use of reason. He admired the way his fellow Athenians used their minds to plan wars, run a government, make money, design temples, and write history and philosophy. But he didn't want to see them follow reason too far, and devalue the gods, and delve into mysteries better left unsolved. They must not imagine that they were stronger than the fate that ruled them.

Among the crowds in the theater, and on the streets, in the marketplace, and at the sports parks was a fat, bald, snub-nosed man in shabby clothes who loved to talk. This was Socrates, a teacher who charged nothing for his lessons, and a philosopher who wrote nothing at all. He liked simply to chat with friends and ask them questions that made them reconsider their ideas. We would know very little about him if his best pupil, Plato (of whom more below), had not recorded, and perhaps partially invented, these discussions.

In one of the dialogues, Socrates runs into a professional seer named Euthyphro, who is about to prosecute his aged father for murdering a man who was himself a murderer. Euthyphro thinks that prosecuting his father for this crime is the pious or holy thing to do, and the gods demand it. Socrates is not so sure.

> SOCR. By Zeus, Euthyphro, do you think you yourself know accurately how matters stand respecting divine law, and things holy and unholy, that . . . you can prosecute your own father without fear that it is you, on the contrary, who are doing an unholy thing?
>
> EUTH. I would not be much use [as a seer], Socrates . . . , if I did not know all such things as this with strict accuracy. . . .
>
> SOCR. Then tell me, what do you say the holy is? And what is the unholy?
>
> EUTH. Well, I say that the holy is what I am doing now, prosecuting murder and temple theft and everything of the sort, whether one's father or mother or anyone else is guilty of it.

Socrates, however, doesn't want mere examples of holiness. He wants to know what is the *nature* of holiness so that he may decide for himself what things are holy. Euthyphro tries one definition after another, but each time he does a few questions from Socrates make it clear that he has not thought the matter through. Come on, says Socrates, you must know what holiness is or you would not be prosecuting your father. Tell me!

EUTH. Some other time, Socrates. Right now I must hurry somewhere and I am already late.*

That is how the dialogue ends.

Sophocles, Herodotus, and Plato. These are only three of the many Athenians in the fifth century B.C. who reflected on the human past, explored the meaning of life, and built temples and meeting places worthy of human beings. While they were doing these things, other Athenians directed the city-state and the empire, fought battles, and of course ran their farms and shops. They were a busy people, many of them with several interests. Sophocles, for example, not only wrote about 120 plays; he was also a dancer, a lyre player, and a wrestler, and he served Athens as a diplomat and general. This was the Athenian ideal (so different from the Spartan one): to seize and savor all of life.

ITS BRILLIANCE, THOUGH, could not save Athens from the fear and envy of the other Greeks. This was the price it paid for forming a league against the Persians and then turning the league into an empire. A leader among its enemies was Corinth (where Oedipus grew up), which lies on the northeast end of the Peloponnesus. Corinth was known for its produce (the seedless raisins called currants are named for it), and for its trade, its thinkers, and its beautiful pottery. It also gave the name Corinthian to an elegant building style that architects have used for thousands of years. The Corinthians saw empire-building Athens as a threat to their own trade and wealth. It was time, they said, to go to war, so they allied themselves with other Peloponnesian city-states. Then they urged the Spartans to join them.

The Spartans were not sure what to do. A war might realize their worst nightmare by leaving their serfs unwatched, thus giving them a chance to rebel. For some time, however, they too had been alarmed

*Plato, *The Dialogues of Plato,* trans. R. F. Allen (1984), pp. 41–58.

at Athens's growing power. A couple of decades earlier they had even fought a brief war against Athens to block its expansion. And they had this to think about: if they stayed out of the coming war they might lose their place as the leaders of the Peloponnesus. After some hesitation, Sparta joined the alliance, and war broke out in 431 B.C.

We know a great deal about this war because one of the Athenian generals, Thucydides, wrote a history of it. At the beginning, Thucydides warns his readers that in his history they may miss "storytelling." This is probably a gentle dig at that fine storytelling historian, Herodotus. Thucydides too could tell a story, but above all he wanted to explain *why* things happened. "I shall be content," he wrote, "if the future student of these events, or of similar events, which are likely, given human nature, to occur in later times, finds my account of them useful. I have written not for immediate applause but for posterity." He succeeded in this aim; readers still call his *History of the Peloponnesian War* a masterpiece.

At first, Thucydides tells us, the Athenians fought with care. They used a strategy that made sense for a sea power battling a land power. Every year they allowed the Spartans to invade the land around Athens and lay it waste, knowing that the Spartans could not get over their high walls and into the city itself. Meanwhile, the Athenians took to their ships and sailed around the Peloponnesus, raiding towns along the coast.

Thucydides shows us how the war brought out the basic natures of the city-states. After the first campaigning season of the war, the Athenians held a public funeral for its men who had been killed in battle. The whole town gathered, and their wartime leader, Pericles, addressed them. He aimed to inspire them by contrasting Athens with Sparta and other cities. "Our laws," he reminded them, "provide equal justice for all in their private disputes. Our public opinion welcomes and honors talent in every field of achievement . . . ours is no mere work-a-day city. No other provides so many recreations for the spirit—contests and sacrifices all the year round, and beauty in our public buildings to cheer the heart and delight the eye day by day. . . .

"[The Spartans] toil from early boyhood in a grinding pursuit of courage, while we, free to live and wander as we please, march out none the less to face the very same dangers. . . . Our citizens attend both to public and private duties, and do not allow absorption in their own various affairs to interfere with their knowledge of the city's. We differ from other states in considering the man who holds aloof from public life not as 'quiet' but as useless."*

Eventually, the Athenians dropped their strategy of sticking to what they did best, naval warfare. Instead, they set out to conquer the large Greek-settled city of Syracuse on the island of Sicily. Their rationale was that Syracuse was a wealthy friend of Sparta, and therefore a danger to Athens. The attack on Syracuse was a disaster, and most of the Athenian invaders were killed. But Athens wasn't finished, and it battled on. A decade later though, with the help of gold from Persia, Sparta won the war. In 404 B.C. the Spartans sailed into Athens's harbor at Piraeus and tore down the walls that for so long had kept them out of Athens.

Its victory left Sparta the strongest city-state, but it used its power unwisely. During the war Sparta had promised all the city-states that it would restore freedom from Athens's domination. Now, however, Sparta turned over the Greek Ionian cities on Asia Minor to Persia, the old enemy of Greek liberty. Sparta bullied the others, and for about thirty years it tried to force them into an empire of its own, harsher than the shattered one of Athens.

The truth, however, is that Sparta was itself declining, and had been for decades. Yes, it had won the war, but largely because of Athens's blunders. Sparta's military system was breaking down, both at home and abroad. Many Spartan soldiers lost the farms that fed them, and when this happened they also lost their status as full members of the ruling class. When there were no longer enough of them to fill the army, Sparta refused to recruit new citizens. Instead it armed freed serfs, and social outcasts, and even men who fought

*Thucydides, *The History of the Peloponnesian War* (1960), pp. 110–17. The translation is by Sir Alfred Zimmern. I have made minor changes in the wording.

for pay. Meanwhile, Spartans forgot the sacrificing spirit they had learned as boys. They now took bribes, and their kings took the biggest ones.

Athens pulled itself together fairly quickly. A decade after the war had ended, Athens, Corinth, Thebes, and Argos and some other towns allied to overthrow the rule of Sparta, but nothing came of this. A generation later it was Thebes, not Athens, that defeated Sparta. Thebes did this in a way that up till then had seemed impossible: its army beat the Spartans in 371 B.C. in a straight-out battle. It was the first time in three centuries that Sparta had lost in such a fight. The bad news reached Sparta during a festival, while the men's chorus was performing. The overseers let the chorus finish, and then they informed the families of the dead. As Spartans would, they told the women not to mourn in public but to bear the loss in silence.

Athens, meanwhile, defeated its old rival in a way that counted even more than victory in war: it remained true to its vibrant self. In the fourth century B.C. Athens was still a prosperous business center, and able men were still proud to serve in its government. Athens was still the school of Greece, rich in artists, writers, and especially teachers and philosophers.

The most famous of its thinkers were Plato and Aristotle. Plato was a pupil of Socrates, the teacher and gadfly. He wrote down his teacher's dialogues and may also have composed some of them, as well as other books of his own. His mind ranged very far, and he touched on all the topics that a thinking person ponders at some point in life. Someone once observed that the history of philosophy is only a series of footnotes on what Plato wrote about the big questions.

Perhaps the most important of these questions was this: how does one produce good men and a good state? In his most famous book, *The Republic,* Plato gives his answer. The important thing is to select a few talented people and educate them well. Sparta, he decided, was good at choosing future rulers and preparing them; the only trouble was that the Spartans did not teach the right things. Plato wanted to instruct his future rulers so that they could understand the goals that they should strive for. The basic test that Plato wanted rulers to apply to any public act was this: will it make us better humans than we were before?

Aristotle was at one time a pupil of Plato, but he became a different kind of thinker. Whereas Plato dealt with great ideas, Aristotle surveyed fields of knowledge. His rules were: gather facts, analyze them, classify, discuss. This may sound dull, but, after all, facts are interesting. And so are Aristotle's opinions. He believed, as Plato did, that the aim of each of us should be to understand what constitutes a good life, and then to live it. The purpose of a government is to help its members do this. However, he thought Plato's ideas about philosopher-kings would not work. After studying the constitutions of many states, he decided that the best thing a city-state could do was to leave power in the hands of the middle classes, by which he seems to have meant the smaller landowners. They had the leisure time needed for government and were likely to oppose violent changes.

Aristotle thought that human beings are at their best in city-states. "Man is an animal," he wrote, "whose nature is to live in a city-state." If you did not live in one, if you were unlucky enough to live outside Greece under a different kind of government, then you were something less than man at his best.

Even these philosophers could get things wrong. Often, when they were describing an ideal government, both of these Athenians used Sparta as their model. The reason is clear: they admired Sparta's order and discipline and its skill in shaping minds. They seem to have forgotten that in Sparta they, with their inquiring minds, would never have been encouraged to think freely on a range of topics, much less given the freedom to teach about them.

IN THE FOURTH century B.C., the days of the Greek city-states were nearing an end. They had fought each other for too long, and maybe they were tired. More than that, many Greeks (but not the Athenians) seemed to have lost interest in their cities; no longer would they dedicate their lives to public service. Perhaps this happened because their activities were now more specialized. As trade in the Mediterranean developed, business grew more complicated and it took more time. Meanwhile, the art of war now called for full-time soldiers; part-time wouldn't do. So a citizen could not easily be a farmer or trader and at the same time a politician and a soldier, as his grandfather might have been. The active citizen had been the strength of the city-state. Now he was gone, and the city-state was bound to follow.

Just north of the city-states in the Balkan Peninsula lay Macedon, a kingdom that was partly Greek. In the middle of the 300s B.C., King Philip II transformed this backwoods country into a military engine with a well-trained army. He then began to conquer Greece. Striking first to one side, then the other, making peace with one enemy so that he could strike the next, he conquered neighbors to the east and south. The Athenians tried several times to stop him, but in the autumn of 338, Philip and his brilliant son Alexander totally defeated Athens and its allies. Soon all of Greece was under his control.

Lively Athens, martial Sparta, and all the other city-states of Greece that showed how much the human species can achieve had sunk to insignificance.

CHAPTER 5

China excels and endures.

LIKE THE OTHER ancient civilizations, China rose beside a river. The Huang Ho or Yellow River (see the map opposite) starts its journey near the "Roof of the World" in Tibet. From there it rushes eastward, cutting through the treeless mountain ranges, picking up the yellow silt from which the river takes its name. Then it makes a northward loop through a desert, bends south and east again, ambles over the North China Plain, forms a swampy delta, and empties in the Yellow Sea.

One might suppose the Chinese first begin to farm and build a civilization on the North China Plain, which is famous for its endless fertile fields of cotton, rice, and wheat. But that's not how it happened. Chinese civilization began farther inland, at places where the river runs through highlands, and where natural terraces lie between the hills and the swampy land beside the river. In these places the soil was so light that early farmers could loosen it with sticks, and the terraces were high enough to escape the floods that make the Yellow River "China's sorrow." Here the farmers could safely plant their cabbages and millet.

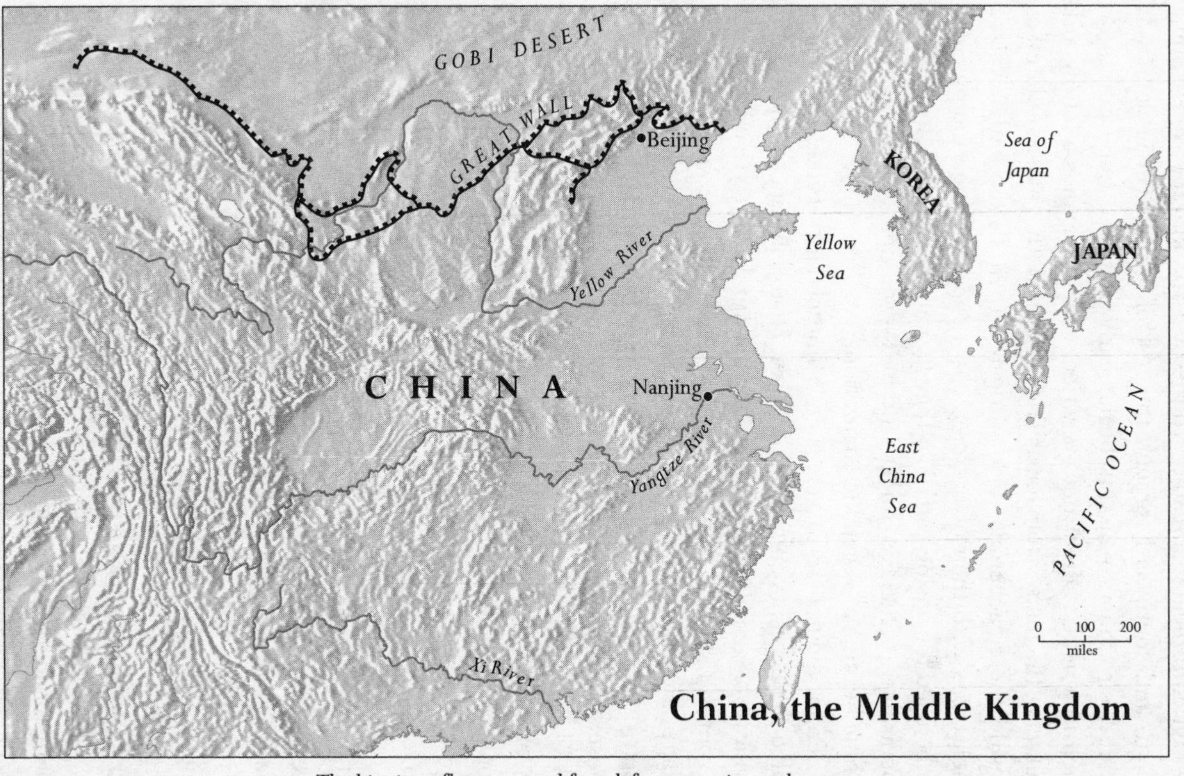

The big rivers flow eastward from lofty mountains to the seas.

Agriculture later spread down the river and into the North China Plain, and southward through the rest of China. Fertile land was scarce, so China's peasants began early to spend huge efforts on their tiny fields. Often they grew rice, which demands a lot of work but can fill many stomachs. They terraced the hills, built dikes beside the treacherous rivers, irrigated their fields, and fertilized their crops with their own excrement. Land was much too precious to be used for raising cattle, so they kept animals that scavenged for themselves—chickens, ducks, and pigs—and not too many of them. China became a "vegetable civilization."

From about 1800 to 1100 B.C., a line of kings called the Shang ruled the peasants on the Yellow River. Historians once believed this dynasty a legend, concocted out of ancient tales of mythic kings, but a century ago that view began to change. Chinese scholars heard that pharmacies in Peking were grinding strange old bones to make their pills. Scratched on the bones were picture figures much like those still used in modern Chinese writing.

The scholars traced these bones to a place beside the Yellow River, and began to dig. Soon they came on large deposits of the bones, and realized that these were ancient royal archives. In time, the scholars learned to match the figures or characters on these bones with their modern Chinese equivalents.

Then the scholars figured out the purpose of the bones. Fortune-tellers working for the ancient kings used to pose a question about the future, then split apart a turtle shell or an animal's shoulder blade. The way it cracked gave the answer to the question—usually a yes or no. Sometimes the seer then scratched on the bone the question he had asked. Some typical questions were: "Will the king's child be a son?" "Will tomorrow be good for hunting?" "If we raise an army of three thousand men, will we drive away the enemy?"

The bones assure us that the Shang existed, since the writings on them name specific rulers once considered legendary. They show that kings and warriors spent much of their time fighting cruder peoples on their borders. Other digs have told us even more about these ancient people. The farmers dwelt in villages in thatch-roofed huts, while ruling families, servants, craftsmen, priests, and certain soldiers lived in thick-walled towns. Their artisans were masters of the art of casting stately bronze goblets, steamers, and cauldrons. Some of these the Shang employed in sacred rites; others they filled with food and wine and placed in tombs to feed the spirits of the dead.

Another point about the writings on the ancient bones: The Chinese have always felt a oneness with the ancient people of their land. The writings scratched on bones prove that they are right to do so, since the characters so nearly match the modern Chinese characters. The path of Chinese culture leads directly, for four thousand years, from the Shang to now.

After several hundred years had passed, a semi-barbarous people called the Zhou (pronounced "joe") settled on the borders of the little kingdom. The Shang ruling class looked down on the Zhou as country bumpkins, but these neighbors had confidence and frontier energy. The Shang people, meanwhile, were fed up with their cruel and dissipated king. In 1122 B.C., a Zhou king, Wu the Warlike, routed a dispirited Shang army fourteen times as large as his. As Wu took over the kingdom, the Shang ruler fled to his palace, set it on fire, and died in the flames.

The Zhou king taught the conquered Chinese a view of human events that runs like this: Heaven, the supreme being, rules the cosmos, and no one rules on earth unless he has the Mandate, or consent, of Heaven. The Zhou, had not fought the Shang kings to win power or glory, but because Heaven had commanded them to punish the Shang for ruling badly. The Chinese therefore should accept the Mandate of Heaven and obey their new rulers. The Zhou also pointed out that rulers were obliged to rule justly in order to remain in the good graces of Heaven. Otherwise they would themselves lose the Mandate of Heaven to another king. And how would this new king prove that he had the Mandate of Heaven? Simply by succeeding in overthrowing the bad king.

The Zhou rulers enjoyed the Mandate of Heaven for many centuries, and they conquered most of northern China. Finally, however, they too fell. A time came when their governors in the provinces would no longer obey the king, and semi-barbarians began to invade the country. Tradition says that the last Zhou ruler was infatuated with a concubine and used to amuse her by lighting beacon fires to summon his army to fight invaders. His officers got tired of this game, and when real enemies appeared, they ignored the signals. The king had lost the Mandate of Heaven, and the invaders sacked his capital and slew him.

For the Chinese, the Mandate of Heaven became an organizing concept. From Zhou times onward, intellectuals viewed history as the story of the rise and fall of dynasties of kings, each of which began with the Mandate of Heaven and then lost it and fell from power.

In the decade from 230 to 221 B.C., the rulers of a western kingdom conquered all the other Chinese states. For the first time ever, all of China was united under a single ruling family, the Qin. (Pronounced "chin." From *Qin* comes the Western name for the country: China.) The newly unified country was as big as Europe—so big that it took about a month for horsemen to carry a ruler's orders to the borders. Compared to the rest of the world, it was also populous; the Chinese were a quarter of the people on the earth.

The conqueror of China thought big. He decided that the ruler of so much land was more than a king, so he adopted a new title that suggested he was both a god and hero. Westerners usually mistranslate it as "emperor" and call him the "First Emperor." He moved the noblemen of states he conquered into mansions on one side of the Wei River, from which they could gaze across the river at his enormous palace on the other side.

Earlier rulers of the northern border states had built long walls to keep out raiding horsemen from the Asian grasslands. The First Emperor joined these into a single Great Wall that stretched fourteen hundred miles from the Pacific Ocean into central Asia. This awesome wall, which others rebuilt many centuries later, is so big that astronauts could spot it as they circled earth. When the emperor died, officials buried him beneath a massive mound of earth with an army of eight thousand life-sized terra-cotta soldiers and horses.

Chinese emperors of later centuries sometimes lost control of this vast land, but others always put it back together. Even when China was divided and weak, it was the greatest force in eastern Asia. Its neighbors borrowed China's picture writing and its thoughts on politics, religion, and art.

Although they exported their silk westward, the Chinese knew little and cared less about the peoples to the west, beyond the Himalayas, the Gobi Desert, and the Asian steppes. As a result, they viewed their country as the center of civilization, surrounded on all sides by "barbarians" (they viewed the Japanese as shoeless, tattooed "dwarfs") and so they named their country Zhonghua, which is usually translated as "Middle Kingdom."

China's self-satisfaction must have infuriated its neighbors. On one occasion an emperor was angry with the shogun of Japan because he would not curb the Japanese pirates who raided the ports of the Middle Kingdom. "You stupid eastern barbarians!" he wrote. "Living so far across the sea . . . you are haughty and disloyal; you permit your subjects to do evil." The shogun coolly answered, "Heaven and Earth are vast; one ruler does not monopolize them."

Not many civilizations can boast, as China can, that the leading figure in their history was a man who taught others how to live. Such a man was Kongfuzi, whom English-speakers call Confucius. He was born in the small state of Lu in 551 B.C., while the Zhou dynasty was breaking up. Confucius is said to have been very tall, and he had big ears, a flat nose, and buck teeth. His family were minor gentry, and for a time he held a minor government position. However, he was chiefly a professional teacher, who collected fees from students and taught them his opinions on government and life.

Some of his students rose to high official posts, but their teacher didn't. In his middle years, Confucius spent a dozen years traveling through northern China hoping to persuade some ruler to give him a position in which he could put into practice his views on government. But while everyone respected Confucius, no one hired him, and at the end of his life he viewed himself as a failure. He resented, he said, being treated "like a gourd that is fit only to be hung on the wall and never put to use."

As a source of lasting inspiration, though, Confucius proved to be a huge success. This might surprise anyone who reads the compilation of his sayings made by his disciples after he had died. Many of their teacher's sayings are, quite frankly, dull and flat. They have only the merit of conviction.

He was wedded to the values of the past. He was certain that the troubles of his time would vanish if his countrymen would only return to the way of life that the first Zhou king (who taught that rulers had the Mandate of Heaven) had shaped centuries earlier. To do this, said Confucius, men must play the roles that life had given them. "Let the ruler be a ruler and the subject a subject," he said. "Let the father be a father and the son a son."

The ruler's major duty is to be a model for his people and promote their good. Confucius was convinced that people would obey and imitate a moral ruler, but not an evil despot. "If one leads them with virtue . . . they will have a sense of shame and will rise to expectations." As for men of means, they should strive to be "gentlemen," that is, people of moral worth. All people, but especially gentlemen, should constantly ask themselves, "What is the right thing to do?" and then try to do it. Such men should help the king to rule by example rather than with harsh and rigid laws. "The gentleman's essential quality is like wind; the common people's like grass; and when the wind is on it the grass always bends."

Confucius hadn't much to say about the role of women, but a female Confucian named Pan Chao later filled that gap. In a manual called *Lessons for Women,* she made their humble place in life quite clear. "Let a woman modestly yield to others," she says, ". . . let her put others first, herself last." She must serve her husband because if she does not, "the proper relationship and the natural order of things are neglected and destroyed." She must also obey his parents: a woman should "not act contrary to the wishes and the opinions of parents-in-law about right and wrong; let her not dispute with them what is straight and what is crooked."

Not everyone agreed with Confucius and his followers, and rulers didn't always listen to them. The First Emperor despised Confucians; once he urinated in a scholar's hat. A well-known Confucian teacher asked an irresponsible Chinese ruler what should be done "if within the state there was no good government." The ruler "turned his head to the left and right and spoke of other things."

A group of practical politicians and philosophers known as Legalists objected to Confucius's do-goodism. It will not work, they said, since bureaucrats, who Confucius wanted to be gentlemen and moral models for the people, are in fact self-seeking and untrustworthy. Perhaps the common people yearn for order, but in fact they are stupid and selfish. So rulers must use rigid laws and heavy punishments. What the ruler wants is right.

In 81 B.C. some Confucians and Legalists had an argument that illustrates their opposite positions. This was the background: The emperor's finance director had set up offices throughout the country to buy surplus grain wherever the crop was good in order to prevent the market price from dropping. The government then sold this grain in other districts where the crops were poor and grain was scarce, so as to keep the price from rising to the point where people starved. Meanwhile, the government also had monopolized the smelting of iron and the evaporating of seawater to get salt. It made nice profits from the grain deals and monopolies.

Businessmen complained about the state's invading the marketplace, so the finance minister called together sixty Confucian scholars from all over China to discuss the government's policies with its officials. The very fact that the minister called *scholars* to such a meeting shows the respect in which the Chinese held learned men. Notes taken at the meeting by a Confucian have survived. They show that the earnest Confucians found themselves in a tough debate with hard-boiled Legalists.

The Confucians argued that the government competed unequally with ordinary people, and that this encouraged official greed. The

Legalists answered that the government needed income from monopolies to pay the troops who guarded China's borders from attacking nomads. To this the scholars gave the classic Confucian answer: military strength must be rooted in morality. If rulers and officials would develop moral virtue and reflect it in their lives, the barbarians would beg for peace and war would end forever. The scholar who recorded this debate believed his side had won it, but the policy didn't change.

Legalists were not the only Chinese who disagreed with Confucius's ideas. A stream of thought called Daoism (or Taoism) held that the only reality was Nature (Dao), a cosmic force. Humans shouldn't try to make things better, as Confucians did, or they would smash the harmony of Nature. The best thing they could do was to adapt to Nature's mold. Let Nature take its course! The Daoists' ideal was a city-state in which people heard the barking dogs and crowing roosters of a nearby city-state but were so content that none of them had ever gone to see it.

It was not the Legalists or Daoists but Confucius and his devotees who altered China. His thought appealed especially to bureaucrats and scholars, as was natural since he had aimed his teachings at them. But emperors too, even hard-boiled ones with Legalist opinions, saw merit in Confucianism because they did need good administrators. They started putting Confucian scholars on their staffs.

Then the government began to give examinations based on the Confucian writings to the empire's would-be bureaucrats. Ambitious boys prepared for these exams at schools all over China. It's said that the examination graders rewarded graceful composition more than logic. But examinations served the rulers' purpose, which was finding bright and well-Confucianized young men. The rulers also had a system for promoting the most able after they began to serve. So Confucianism helped to shape a thoughtful, loyal ruling class.

WHILE CENTURIES PASSED and dynasties rose and fell, China's scholar-bureaucrats and wealthy gentry shaped a "high culture" like no other in the world. Many of these cultivated men—even certain emperors—were themselves fine poets, scholars, painters, and masters of the art of writing Chinese characters with style. Those wealthy men who were not artists themselves valued and supported those who were.

Chinese painting had its own aesthetic. Artists often painted the same subjects that their teachers had painted before them, but they did it with charm and freshness. A painter waited till he saw his painting in his mind. Then (a Chinese once explained), "with his brush sweeping across the paper, he pursues [the vision] as a falcon chases an elusive rabbit. If he slackens his pace, his vision escapes and may never return." A Chinese landscape painting on a yards-long scroll can have great appeal. Tiny people walk on paths through villages, across wild rivers on hump-backed bridges, through forests where they're dwarfed by tall and crooked pines, and up the slopes of rugged hills whose peaks are lost in mist.

Unlike painters, Chinese poets wrote about anything at all. Two much-admired poets of the 700s, Li Po and Du Fu, were officials at the court of an emperor who was an eager patron of the arts. Li Po considered himself "an immortal banished from heaven." He was a swordsman, drinker, wanderer, and friend of "singsong girls" and hermits, but he also found the time to write some 20,000 poems that glow with charm. Usually he was cheerful, as in "The Girl of Yue":

A girl picking lotuses beside the stream—
At the sound of my oars she turns about.
Giggling, she vanishes among the flowers,
And, all pretenses, declines to come out.

Sometimes, however, Li Po was thoughtful and nostalgic. Every child in China can recite his "Quiet Night Thoughts":

Beside my bed the bright moonbeams bound
Almost as if there were frost on the ground.
Rising up, I gaze at the mountain moon;
Lying back, I think of my old town.

Li Po is said to have drowned while drinking and boating with some friends. He had reached across the water to seize the moon's reflection.

Li Po's friend Du Fu had a reverence for the family that marks him as a Confucian. He wrote "Chiang Village" when he was reunited with his family after a separation caused by a rebellion:

Westward beyond the high purple clouds
The sun strides down to the level earth.
My rickety gate is loud with sparrows;
The stranger comes home from a thousand leagues.

My wife and children can't believe I've survived;
Recovered from shock, they still wipe at tears.
In times of great troubles I've been buffeted about,
And returning alive is nothing but luck.

Neighbors crowd along the top of the wall,
Groaning out sighs, or sniffling and sobbing.
Late in the night we light a new candle
And confront one another as if in a dream.

Not only were the Chinese splendid artists; they also were the most inventive folk on earth. They may have been the first to use umbrellas, wheelbarrows, matches, toothbrushes, and playing cards. Their magicians magnetized needles and used them to choose lucky places for houses and graves; later, Chinese sailors made compasses with them to find their way across the seas. More than a thousand years before anybody else, the Chinese were making paper, and later they discovered how to print words upon it. (It's said that a blacksmith named Pi Sheng made clay copies of Chinese characters, baked them hard, glued them on an iron plate, inked them, and pressed papers on the ink, thus making many copies.) Five centuries before Johannes Gutenberg printed the first European book (a Bible, in about 1450), the Chinese government sponsored the printing of a 130-volume set of the Confucian classics.

At about the same time, a Chinese inventor combined honey, sulfur, and saltpeter, and heated the mix. He was searching for a potion that would make one live forever, but he found one with the opposite effect. Later others dropped the honey and added charcoal. Sad to say, they used the resulting gunpowder not only for firecrackers, as historians have claimed, but also to make hand grenades and mines.

China's silk was such a major export that the route that caravans took through central Asia to the west was called the Silk Road. Chinese potters made porcelain cups, bowls, and vases that were thin as eggshells, lovely enough for the gentry and maybe cheap enough for the poor. Some of it was so hard (it's said) that steel couldn't scratch it.

IN THE 1200S, disaster struck. Mongol nomads from the grasslands to the north swept down and, shedding lots of blood, conquered all of China, as well as much of the rest of Asia and some of Europe. (We will have more to say about them in the next chapter.) The Mongols' ways were not like those of civilized Chinese. Later Chinese chroniclers pictured them as savages who washed in urine, if at all, and knew only the ways to rob and kill.

In fact, the Mongols governed China well for about a hundred years. They were wise enough to use some Chinese scholar-bureaucrats to help them rule their huge domain. As we'll see in chapter 8, Italian merchants who came to China while the Mongols ruled it were favorably impressed. Eventually, however, the Mongols met the fate of all the dynasties before them; they lost the Mandate of Heaven. The country fell apart, and by the 1350s it was wracked by civil war, and famine and disease were killing people by the tens of millions.

Let the story of one family stand for what happened to those millions. In the 1330s, Chu Shih-chen and his wife were hungry peasants in the North China Plain. When their landlord evicted them from their tiny farm they moved to another. But they found it so hard to feed their six children that they arranged for others to adopt two of their sons, and as soon as they could they married off their two daughters. Not long after this, their crops shriveled during a drought and were eaten by locusts. Shih-chen, his wife, and their oldest son all starved to death.

Only their youngest son, seventeen-year-old Chu Yüan-chang, was left, and he was so poor that he could not afford to bury his family until a neighbor gave him a scrap of land for their graves. In order to survive, he became a novice in a Buddhist monastery, which sent him out to beg. Soon, however, the monastery found itself so poor that the abbot disbanded it. Yüan-chang faced starvation.

We know that dismal story only because Chu Yüan-chang later became so prominent that someone thought it worth recording. At the age of twenty-five, the young beggar became a soldier, and soon he was a rebel warlord. In the 1360s he defeated other warlords and won control of the Yangtze valley in the center of China, and in 1368 he drove the last Mongol emperor back into the grasslands from which his people had come. Chu Yüan-chang then proclaimed himself emperor and founder of a new dynasty, the Ming or "Brilliant." As emperor, he took the name Hung-wu, which means "Swelling military power."

Hung-wu had his faults. After his first few years he felt a paranoid fear of being ridiculed because of his peasant origins and his pockmarked, pig-snout face. In a series of purges he killed tens of thousands of the very men who had helped him triumph.

However, he also undid the blunders of the Mongols and restored and sometimes improved on the best of the old Chinese ways. He built up a large standing army, which he kept divided so that no general could become the mighty warlord he had been. He seized the estates of big landowners and rented the land to the poor, perhaps because his own peasant family had suffered so much. What was perhaps most important, he brought back the old Confucian schools and government examinations. Hung-wu hated educated men—he often had officials thrashed with bamboo rods—but he knew the value of a good bureaucracy.

A few years after Hung-wu's death, Emperor Yong-lo began a project that could have shaped the history of the world. History is full of might-have-beens, and this is big. In 1405 Yong-lo sent a fleet of junks (flat-bottomed, square-sailed ships) on a voyage to impress other rulers and gather tribute from them. An able eunuch named Zheng He commanded the fleet. (Many eunuchs served in Chinese courts.) With many stops at royal courts, they journeyed south along the Chinese coast, around the Malay peninsula, west to India, and then back home. During the next quarter century Yong-lo and his successors sent six other fleets to courts as far away as the Persian Gulf, Arabia, and east Africa. They brought home tribute payments, ostriches, giraffes, and sometimes rulers who refused to pay.

These cruises were impressive, as they were meant to be. Usually about 30,000 men sailed on as many as 317 oceangoing junks. Some of the ships were four decks high and longer than a football field. With their compasses, the Chinese sailed across the open seas, something Europeans hadn't yet learned to do.

After the seventh voyage, in 1433, the voyages stopped, never to resume. What's more, the emperor decreed that henceforth Chinese ships must not sail outside of coastal waters, and no Chinese could go abroad. Why did the Chinese, when they seemed to have begun on the road to far-flung power and influence, suddenly stop the voyages? One reason was simply that the emperors could not afford to keep them up. Just then they needed all the money they could get to build their capital, Peking, and drive away the always-looming Mongols. The tribute money gathered on the voyages didn't match the cost of building junks and manning them.

The Chinese had another cause to halt the journeys, and it may have counted more. The emperors and their advisers saw their country as the center of the world. Probably they asked themselves, why should the Middle Kingdom bother any longer with those distant barbarians? They were satisfied to dominate their own large quarter of the world.

Halting the expeditions was a fateful move. If the journeys had continued, they might, first, have made Ming China a major naval power. Then its merchants might have followed up with trading voyages of their own, making China a colossus in the world economy.

China might have spread around the earth its inventiveness, its energy, its taste in art, and the teaching of Confucius that, "If one leads [people] with virtue . . . they will have a sense of shame and will come up to expectations."

Of course, after it had stopped the voyages Ming China didn't shrink to insignificance. Hardly! It was still the largest realm on earth, with more people than all of Europe. Its artisans made china, silk, and books so fine that no one else could match them. Its rulers, guided by the teachings of Confucius, ran the country well. Its people usually had enough to eat.

In the 1500s and early 1600s, however, troubles started. Several emperors grew bored and let officials and the palace eunuchs rule the empire. Hungry workers in the cities rioted because their pay was low and taxes high. A potter threw himself inside a blazing kiln to dramatize the workers' plight. Famine and disease were common. When the rulers didn't pay the troops on time, the men deserted. Former soldiers and hungry peasants took to the roads as bandits. Japanese pirates ravaged the coast, even after an angry emperor captured some and steamed them to death in giant pots, like lobsters. Court officials accused the finest Chinese general of treason and had him cut to pieces in the Peking marketplace.

The reader knows what was happening. After ruling for nearly three centuries, the Ming dynasty was losing the Mandate of Heaven.

Rebellions broke out in northern China, and in 1644 a rebel leader seized Peking. The empress killed herself. The emperor rang a bell to summon his ministers and ask for their advice, but none appeared. He walked to the imperial garden and hanged himself from a tree. After ruling China for nearly three centuries, the Ming dynasty had ended.

One of the Ming generals fighting the rebels asked for help from the Manchus. These were war-loving semi-barbarians who had already conquered China's northeast corner, right down to the Great Wall. Happy to oblige, the Manchus hurried south and drove away the rebels. They took Peking and declared their five-year-old king the emperor of China. They named his dynasty the Qing (pronounced "ching") or "Pure." It would take a generation, but they ended by subduing all of China.

The Qing emperors faced a dilemma. Should the Manchus adopt the ways of the Chinese? Or should they make the Chinese adopt the ways of the Manchus? The first option was not an option. For every two Manchus there were ninety-eight Chinese, and the Manchus had no wish to drown in that ocean. So the Manchus forbade their own people to marry Chinese, or to do manual labor, or to become businessmen. And for a while these measures worked.

As for forcing the Chinese to become Manchus, that too was out of the question. One could hardly make the vast majority accept the customs of a tiny semi-barbarian minority. However, the Manchus did order Chinese men to shave the front halves of their scalps and braid the remaining hair in pigtails, Manchu style, to show submission. The Chinese did this. "Keep your hair and lose your head," they summed it up, "or lose your hair and keep your head."

The Manchus understood that since they were so few they needed the help of the Chinese gentry, with their Confucian ideals of service. So they revived the civil service examinations and used Chinese scholar-bureaucrats to help run the country. Some scholars refused to cooperate, out of loyalty to the memory of the Ming, but by using great tact the ablest of the Manchu rulers, Kangxi, won over a number of them. He announced a special examination only for men of outstanding talent, and then announced their names. Those who were thus singled out could scarcely refuse to compete, so they took the examination. Kangxi put the winners to work on an official history of the Ming dynasty.

Kangxi was himself a genuine Confucian and therefore a "sage ruler." He published a series of maxims that summed up the Confucian virtues—generosity, obedience, hard work, thrift, duty, and so on—written in a down-to-earth style that the man in the street could understand. He also read Confucian classics, and debated the knotty points in them with a team of Manchu and Chinese scholars. Naturally, he made sure that word of his studies leaked out.

Here is an example of Confucian ethics at work under the

Manchus: In the late 1600s a scholar-official named Huang Liu-hung served as magistrate in a poor, bandit-ridden county in north-eastern China. He had qualified for his job by passing a Confucian examination. Years later, after he had retired, Huang wrote a hand-book about the office of magistrate, and he made his points by giving examples from his own experience.

One of his cases, he says, involved a woman named Wang and her husband, Jen, who lived in a one-room hut beside a forest. Wang ran away from Jen with another man, but when her lover abandoned her she went back home. Jen was murderously angry. One night he waited until she fell asleep, and then he put his knee on her belly and his hands around her neck and began to strangle her. She strug-gled so hard that her legs tore their sleeping mat, and her bowels opened. Finally she died.

Jen carried her body through the forest toward the house of a man named Kao. This man had struck him during a recent quarrel, and Jen wanted to fix the blame for the murder on him. However, when he heard some watchmen nearby, he dropped the body and went home. The next day he filed a murder complaint against Kao.

Huang, the magistrate, might easily have decided the case against Kao, but he began to suspect Jen. He rode out to the village, eight miles from his home, and examined Jen's hut. He noticed its poverty, the rips in the almost new mat, and the pile of dried excrement beside it. His assistants told him this must be the dung of an ox or a donkey, which poor people often burned for fuel, but Huang was not con-vinced. He ordered that water be boiled and poured it on the dung. The resulting odor decided it: the source was human. Taken together with the ripped mat, the excrement strongly suggested that Wang had been killed here, in her own home. Huang finally got Jen to confess.

By law, Huang might have executed Jen, but he decided not to. His reasons reflect Confucian values, especially the stress on duty to one's family. Jen was his father's only son, and the loss would be too much for the aged man to bear. Then too, Jen had no chil-dren, and if he was executed his family would die out, a serious con-cern. Furthermore, Wang had not been a dutiful wife, and since she had betrayed her husband she deserved to die. And finally, Jen had indeed been provoked by Kao, who should not have struck him. So Huang sentenced Jen to a beating with bamboo rods, and he made him wear a wooden collar around his neck.

But Huang also had to settle another matter. Who would pay for Wang's burial? Jen could not afford to, so Huang ordered Kao to pay the expenses. That would teach him not to strike people in the face when he lost his temper.*

From the late 1600s through most of the 1700s, Qing China enjoyed a Golden Age. In part this happened because the country had a succession of able emperors who kept the land at peace and guarded the borders. Two of them had very long reigns. Kangxi, whom we mentioned as the reviver of the Confucian system, reigned for sixty-one years. So did his grandson, who retired a few years before his death in order not to appear disrespectful by reigning longer than his grandfather.

At the same time, Chinese arts and letters thrived. A team of scholars produced what is perhaps the largest series of books in the history of the world, an anthology of the most famous Chinese literary and historical works of the past. The series took ten years to complete and filled 36,000 volumes. Meanwhile an aristocrat named Cao Xueqin wrote China's best-known novel, *The Dream of the Red Chamber.* It tells the story of a family like his own: rich and charming people who were headed downward.

China's greatest triumph in this Golden Age resulted from its earlier disasters. All those horrors of the mid-1600s—civil wars, invasions, bandits, floods, diseases, famines—had reduced the population by perhaps a third. Now that so many had died, good farmland was abundant, and therefore poor peasants could afford to rent some land, marry, and have children. At the same time, the Chinese were starting to grow plants that had reached them from the Americas: corn (maize), potatoes, and sweet potatoes. Peasants could grow these plants even on poor land, where it was hard to raise wheat or rice.

*Jonathan D. Spence, *The Death of Woman Wang* (1978), pp. 116–39.

The Chinese thrived like mice in a breadbasket, and their numbers swiftly rose. By the early 1700s, they had made good the earlier losses, and their numbers kept on rising. Growth like this can bring boom times for all. Everybody—peasants, potters, merchants, weavers, builders—profited by selling to each other.

They prospered until well into the 1700s, but then they ran into trouble. The population kept on rising, and by the end of the 1700s the Chinese numbered 300 or 400 million—about twice their number at the start of the century. Now they had too many mouths to feed, and hunger, the normal lot of many, started to return.

Meanwhile, as will not surprise the reader, the Qing began to lose the Mandate of Heaven. Bungling and crooked officials took over the bureaucracy, and a handsome young army officer became an emperor's favorite and scooped up huge amounts of money for himself and his friends. Sections of the Grand Canal, which ran for 850 miles through the North China Plain, silted up. The government allowed the grain supplies reserved for famines to run out. Chinese armies struggled with insurgent groups whose names—White Lotus, and Heaven and Earth Society—were charming but whose deeds were not. Border wars with Burma and Vietnam went badly.

As the century neared its end, so did China's Golden Age.

CHAPTER 6

Some attempt to rule us all.

EMPIRES. THE WORLD has known some big ones. Seven centuries ago the Malinke people built the empire of Mali, which nearly filled the western half of the grasslands below the Sahara Desert. Caravan trails across it ran many hundred miles from the salt fields of Taghaza and the markets of Gao and Timbuktu to the gold sands of Bambuk and Bondu. At about the same time, the Incas in South America won and ruled a swath of land that stretched along the Andes Mountains for 2,500 miles. These were no petty polities, the empires of the Incas and Malinkes.

This chapter, however, deals with even grander conquerors and empire builders. These are people who long ago fought for mastery in the biggest arena: the wide band of densely lived-in land that stretches across Europe, North Africa, and southern Asia. They conquered lands that stretched from continent to continent; they yoked together civilizations. Why, we will ask, did they *want* these giant pools of humans, how did they conquer them, and why did their empires finally collapse?

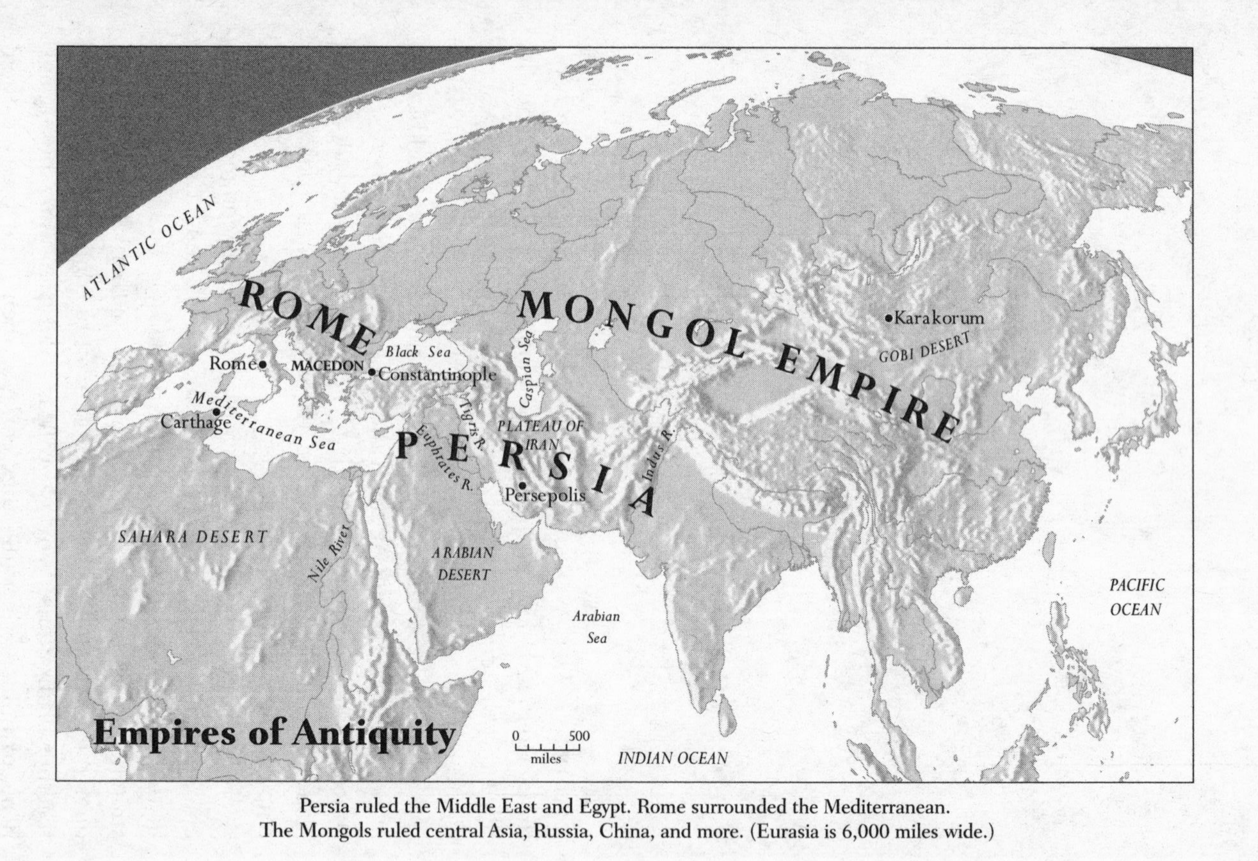

Persia ruled the Middle East and Egypt. Rome surrounded the Mediterranean. The Mongols ruled central Asia, Russia, China, and more. (Eurasia is 6,000 miles wide.)

. . .

In about 1000 b.c., groups of Aryans from somewhere to the north drifted downward to the parched plateau east of the Tigris and Euphrates Rivers and west of ancient India. The new arrivals rode on horses and were armed with short, strong bows and quivers full of arrows. Their speed and weapons gave them an advantage, and they pushed aside the natives. They named their conquered home Iran, "Land of Aryans."

Half a millennium later, Cyrus II ruled one of these Iranian peoples, the Persians. In 550 b.c. he led an army of horsemen northwest from their mountain home against their neighbors, the Medes, who were also Aryans and had themselves already begun to build an empire. With the help of two Median generals who defected to him, Cyrus beat the Medes and carried off their treasures. Then he rode farther to the north and west and conquered Asia Minor (most of modern Turkey).

Herodotus tells a tale about Cyrus's chief enemy in western Asia Minor. This was wealthy Croesus, king of Lydia, who gave us the expression "rich as Croesus." Croesus made the fatal error of sending his army across the Halys River to attack Cyrus. He had been encouraged to do so by an oracle who prophesied that "if Croesus crosses the Halys he will destroy a mighty empire." Croesus didn't understand the prophecy: the mighty empire was his own.

With the wealth of Croesus in his coffers, Cyrus now attacked in the opposite direction, to the east. He subdued most of the tribes and peoples of his own Iranian plateau and what is now Afghanistan. Next he turned to the west again and, seemingly unbeatable, captured Babylon and all the region of the Tigris and Euphrates. Then he pushed down into Syria and Palestine, at the eastern end of the Mediterranean. "I am Cyrus," he proclaimed, "king of the world, great king, legitimate king, king of Babylon, king of Sumer and Akkad, king of the four rims [of the world]. . . . All the kings of the entire world . . . brought their tributes and kissed my feet in Babylon."

After two dazzling decades of conquest, Cyrus turned again toward central Asia, and he fought some nomads east of the Caspian Sea. In their final clash, the Persians and the nomads fought each other first with bows and arrows, till they had no arrows, then with spears and daggers. The simple tribesmen beat the Persians and killed the mighty Cyrus.

The Persian Empire, though, continued to grow. Cyrus's son Cambyses conquered Egypt, and his successor Darius won the Indus Valley on the western edge of India. Darius and his two successors tried to conquer Greece, but, as we saw in chapter 4, they failed. So Persia ended where Europe began. How big this empire was! It held most of what is now the Middle East: southwest Asia and northeast Africa.

One has to wonder why the Persians wished to rule so much. For Cyrus the motive was probably nothing more than love of war and conquest. His courtiers once suggested to him that he move the Persian people from rugged Iran to some flat and fertile land, but Cyrus told them that they missed the point. If the Persians moved, he said, they would become slaves, not rulers, because "Soft countries breed soft men." As for the motive of his successors, they may have believed simply that it was their duty to keep expanding the empire, as Cyrus had done. Or perhaps they felt endangered, as conquerors always do, by still-unbeaten peoples past their borders, and thought they'd better conquer them as well.

The Persians proved to be judicious rulers who did the things they must to hold their lands intact. For example, to improve communications they built a Royal Road that ran 1,600 miles from near the Persian Gulf to near the Aegean Sea. Messengers in relays rode the length of it in less than seven days. Herodotus reports, "Not snow, nor rain, nor heat, nor gloom of night stays these couriers from the swift completion of their appointed rounds."

What was most distinct about the Persian rulers was that they let others do the ruling for them. They split the empire into many provinces and let their governors do as they thought best as long as they sent in the things required of them: recruits for the army; gold, silver, ivory, and ebony; wheat; frankincense; boys and girls; and eunuchs. Governors allowed the conquered peoples to preserve their laws and customs. For the ruling of so many different peoples no other policy was as sensible as tolerance.

AFTER TWO CENTURIES, an able foe appeared on Persia's western flank: Alexander of Macedon, known to history as Alexander the Great. We saw in chapter 4 how Philip of Macedon conquered Greece and how he then prepared to go to war with Persia. Probably he aimed to win no more than Asia Minor's western coast, a fraction of the Persian Empire. Before he could begin, a Macedonian aristocrat murdered him.

Philip left a son named Alexander, who had grown up fast. At sixteen he had served as regent of Macedon while his father was away at war. He is said to have fretted that his father would leave him nothing to conquer. At eighteen he led the cavalry of Macedon in a battle that gave his father mastery of Greece. At twenty, when Philip died, Alexander succeeded him and eagerly took over Philip's project, war with Persia.

At about the age of twenty-two, in the spring of 334 B.C., Alexander crossed from Europe into Asia Minor with an army of 35,000. He battled south along the western coast of Asia Minor; then he zigzagged eastward through it, winning battle after battle till he held it all. The Persian emperor Darius III sent a message: if Alexander would agree to peace, the emperor would give him money, all his lands west of the Euphrates, and his daughter in marriage. A grizzled Macedonian general named Parmenion told the young conqueror that if he were Alexander he'd accept. Alexander gibed that if he were Parmenion he too would probably accept. At about this point Alexander decided to go on and conquer all the Persian Empire.

Others generals might have asked themselves if so few men, so far from home, conceivably could win so much, but Alexander had no doubts. Already he had proved, and he would prove again and then again, how good he was at making war. His rugged body, often wounded, stood up well in hard campaigns. He moved his army swiftly through terrain that he had never seen, outthinking and outdaring enemies. Most important, he simply would not lose.

With Asia Minor in his hands, Alexander led his army south to Syria and Palestine, conquered them, and then took Egypt. He now had taken all of Persia's western provinces. Next he led his army eastward to the Tigris and Euphrates region, where he routed Darius. He marched past Babylon and eastward to Iran, the heart of Persia. (How easily a historian can write "he marched" and "took," as if no effort and no hardships were involved.)

At the royal city of Persepolis he burned the palace that Xerxes had built a century and a half before. This, said Alexander, was his revenge for Xerxes' burning of Athens. Darius fled before him until one of his own governors murdered him. Alexander gave the emperor a royal burial and put the murderer to death. Then he headed even farther east, skirting deserts, threading through the barren mountains. At times his men could reach the forts of enemies only by scaling snow-clad rocks with ropes and iron pegs.

And then he plunged down into India. What a daring venture, to attack that strange and densely populated land with some 30,000 men! (Of course, he had no maps and no idea how large India really was.) We can only speculate about his reasons. For one thing, Persia once had ruled the northwest fringe of India, along the Indus River, and Alexander must have felt he had to conquer that at least. He was always curious, and probably he wanted to explore this exotic land. And of course he was a conqueror; conquering was what he did. So he led his army to the Indus, and near the river he defeated a king who had attacked him with 200 trumpeting, terrifying elephants.

Legend has it that while he was in India, Alexander met some holy men, living in a forest, who told him that his conquests had no meaning. He owned only what they did: the land he stood on. He applauded what they said.

The conqueror now told his men that they would push through India. (He probably believed that he was near the eastern edge of the earth. The Greeks believed the earth to be a disk surrounded by a river called Ocean.) When he revealed his plans, his troops were stunned. They were now 3,000 miles from home, depleted by the heat, exhausted after years of war. They asked, would he never halt?

The unthinkable occurred: Alexander's loyal army told him they would march no farther. He argued with them, and for days he stayed inside his tent. Finally he yielded and agreed to go no farther, and they turned around. Suffering from heat and lack of food and water, Alexander and his army struggled rearward to Iran.

What did Alexander plan to do with all that he had conquered? His actions when he reached Iran provide some clues. He wed the daughter of the murdered Darius, and at his order 80 of his officers and 10,000 of his soldiers married Asian women. Alexander saw these weddings as a measure to unite the peoples he had conquered. With that same intention he had told his generals long before to unify his army. He was now delighted to review a force of 30,000 Persians trained to fight beside his men. At a banquet with the representatives of all the peoples that he ruled, Alexander told them what—perhaps—he really wanted. Apparently he prayed for peace, and that the peoples of his empire might be partners (and not merely subjects), and that all the peoples of the world might live in harmony and unity.

From Iran he marched his army west to Babylon, which he planned to make his capital, but here he sickened with a fever. His resistance had been weakened during years of hardships and by many wounds and much hard drinking. He was near to death. When they heard how ill he was, the king's devoted troops insisted on seeing him. He couldn't speak, but as his soldiers—every one—filed by in silence Alexander's eyes uttered his farewells. He died in June 323 B.C., at the ripe old age of thirty-two.

In a decade, never losing any battle, he had won an empire that spread over all the Middle East and Egypt. But the truth is that he had not done much with it. He may have dreamed of unity and global peace, but he had not yet achieved them. He lacked the time, perhaps the skill, and probably the will.

He left no heir to sit upon his throne. When someone asked him on his deathbed who should take his place he simply said, "the strongest." Not surprisingly his generals could not agree on who that was. So three of them cut his empire into smaller states, based in Egypt, Mesopotamia, and Greece, and each took one. For the next two hundred years, these men and their heirs waged war with one another, and with any of their provinces that tried to win its freedom. So much for Alexander's dream of unity and peace.

EVEN AS HIS empire fell apart, another Mediterranean people built a greater one. But their story really begins back in the late 500s B.C., just when the Persians were finishing their conquest of the Middle East. Far to the west of Persia, on the Tiber River in central Italy, was the farming city of Rome. The rulers of the Romans were a people north of them, the busy, luxury-loving Etruscans.

According to a legend, in 510 B.C. the son of the king of Rome raped a Roman woman. The angry Romans took up arms, expelled their rulers, and began to rule themselves. Since they were determined to have no more kings, they founded a government run by wealthy landowners who stayed aloof from ordinary Romans. Then, however, "plebes," who farmed the fields and formed the backbone of the tiny army, threatened to depart and form a rival city-state. So the "patricians" then allowed the plebes to choose their own officials who had the right to veto the decisions of the patricians.

The Romans fought one enemy after another. Close to Rome they mastered local "tribes" of semicivilized peasants and shepherds. In the north they absorbed the former masters of the region, the Etruscans, and in the south of Italy they conquered colonies of Greeks. Usually they went to war reluctantly, to save their skins, but they also wanted land. By about 265 B.C. they commanded all of Italy but the far north, and the nearby islands of Sicily, Corsica, and Sardinia.

Now came the turning point, when Rome strode out of Italy and began to make itself the greatest empire that the world has ever known. At this moment two big powers dominated the western end of the Mediterranean. One of course was Rome, which was strong on land in Italy but nowhere else. The other power was the Romans' fated enemy, seagoing Carthage. Its leaders were not farmers, like the Romans, but businessmen who dealt in ivory, pottery, gold, and cloth. Their capital, on the Mediterranean's African coast, occupied a point of land that juts toward Italy. With their potent navy they had all but made the western half of the sea a Carthaginian lake.

Here then was the classic matchup: a land power faced by a sea power, like Sparta and Athens two centuries before. The inevitable began when Carthage conquered land on Sicily that gave it control of the strait that separates the island from the toe of the Italian boot. The Romans claimed that now they had no choice; they had to fight to stay alive. In 264 B.C. they went to war.

In the early fighting on the island the Romans battled hard and well. But they had only a meager navy, and Carthage proved itself the master of the sea. So Rome decided to transform itself. The Romans built a fleet of galleys—ships that one could sail or row with oars—bigger than the Carthaginians'. In a naval battle north of Sicily, the Romans latched their ships to the Carthaginians' with grappling irons. Then they swarmed across on boarding bridges, and they beat the Carthaginians in the brutal melées that their foes had always been so good at. After much more fighting, the Romans won the war and took the islands, Sicily, Sardinia, and Corsica.

The Carthaginians, however, were not finished. In a second war, they showed that if the Romans could fight on the sea, they could fight on land. Carthage's commander was the brilliant Hannibal, twenty-five years old, whose father, when Hannibal was a little boy, had made him swear an oath of everlasting hate of Rome. Hannibal's army won the southern half of harsh Iberia and added half-wild swordsmen to their ranks. Then, complete with elephants from Africa, they marched through southern Gaul, across the Alps in winter, and down to Italy. They ravaged the peninsula. When the Romans tried to stop them, they shattered them three times.

For Rome the hour was dark. After one defeat the Senate ordered grieving women to remain indoors. It ordered human sacrifices, a savage rite the Romans barely knew of. Slaves were drafted to guard the capital, armed with sacred weapons from the temples. The Romans raised more troops to take the places of the fallen, and staggered on. Despite his triumphs, Hannibal couldn't win the war.

The Romans found the leader whom they needed. Publius Cornelius Scipio was strikingly like Hannibal: he too was a general's son; he too took command when twenty-five years old; he too was supremely confident. He first defeated the Carthaginians in Iberia, and then he crossed the sea and brought the war to Carthage. When he learned of this, Hannibal and his army hurried home from Italy, but in 202 B.C. Scipio defeated them, and the Carthaginians surrendered. The Romans gave them easy terms but joined Iberia to their growing empire. Hannibal had fled, but they pursued him till he swallowed poison rather than be caught.

However, half a century after its defeat in the second war, Carthage once again was rich and, some Romans claimed, dangerous. In the senate the fiery Marcus Porcius Cato constantly demanded action, thundering, "Carthage must be destroyed!" On at least one occasion he flourished figs from Carthage as proof of the enemy's revival. He overstated the danger, but the Romans chose to think him right. They sent another army to Carthage and in 146 B.C. destroyed the city. Much of North Africa became yet another province of the growing Roman Empire.

The conquest of the western Mediterranean was just the start. To the east of Rome were the three big chunks of Alexander's empire that his generals had taken for themselves. One by one, the Romans struck them. Even before the third war with Carthage, they seized Macedon and Greece. Then they took what had been the western end of Persia's empire, and then Egypt.

By now the Romans' aims had changed. No longer were they backing more or less reluctantly into wars and conquests. More and more they sent their armies into other lands because they had acquired an empire-building impetus. They fought to help their allies, fortify their borders, conquer markets for their goods, provide themselves with jobs, keep their armies busy, and, indeed, to slake their thirst for triumphs.

Two and a half centuries after the Carthaginian wars began, the Romans ruled the whole Mediterranean basin, all the lands that bordered on the sea. This had been a huge achievement. The Mediterranean is 2,300 miles long; in ancient times it took two months to sail the length of it.

And yet the Romans were not satisfied. In northern Europe, past the Alps, were tribes that often raided the borders. The Romans had to deal with them. In the first century B.C. and the first century A.D. the Romans battled long and hard against these tribesmen in their fens and forests. When these wars were over, Rome had won a strip of Europe that stretched from Britain to the Black Sea.

One place that the Romans couldn't conquer marred their map of Europe. This was Germany, which a Roman writer called "a region hideous and rude . . . dismal to behold or cultivate." In the year A.D. 9, tribesmen in a northern German forest slaughtered a Roman army. The 17th, 18th, and 19th Legions never again appeared in the Roman army list because, given their bad luck, no one would have served in them. Rome gave up on ruling most of Germany.

In the meantime, Rome endured an episodic civil war among its mighty generals. In Rome's early centuries a general had been a gifted amateur like the semilegendary Cincinnatus. When the country needed him, he left his plow to lead an army. He beat the enemy in a single day and went home to his bean fields. Now, however, when the Romans had an empire, generals commanded troops in distant provinces for many years. On the far frontiers they ruled like kings and often formed some big ideas, and when they journeyed back to Rome they brought their armies and ambitions with them. For decades men like these led coups and civil wars. General after general seized power in Rome, pushed the Senate to the side, and made himself the empire's soldier-despot—till another general grabbed his place.

It was one such army chief, the rich Augustus (meaning Blessed), who ended this long struggle. Not only did he push aside or crush his rivals, rising in this way to power as other generals had, he also kept that power, because Augustus wasn't just an able soldier but an able politician. Rather than destroy the Senate, as he could have done with ease, he stroked it. Rather than offend the Romans by eliminating ancient offices, he simply filled them all himself. Augustus was simultaneously commander of the armies, consul, censor, tribune, and chief priest. But far from wanting anyone to bow to him, he was easily approached and liked to be known simply as First Citizen.

Augustus ruled for forty-five years, and this was long enough to accustom Romans to one-man government. When he died, his power passed quite smoothly to his stepson. After *his* death, the next emperor was picked, in effect, by the prefect of the palace guards. Such an event could only lead to trouble. The man the prefect chose, Caligula (meaning Little Boot), clearly was a dreadful choice; he soon became a vicious madman. And yet Caligula took power as emperor with no trouble. For better or for worse, the form of government had changed, and everyone accepted that. From Augustus on, an emperor ran the empire.

Even when the rulers were incompetent, Rome gave its people what they wanted: order and civilization. Everybody profited from Roman laws, which were clear and firm. "Let justice be done," ran a Roman maxim, "though the heavens fall." Everybody also gained from Rome's amenities, such as its famous roads. These ran from Scotland to Arabia, dead straight wherever the terrain was flat, and built to last. Along them towns sprang up by the thousands. These towns boasted not only forts and victory arches but also law courts, schools, libraries, theaters, temples, piped-in water, public baths, and fountains—even sewers and central heating.

In the A.D. 200s, the empire began to face some worrisome problems. One of these resulted from the dreadful system, which really wasn't any system, for choosing emperors. In earlier centuries emperors had risen to power in various ways; usually a ruler chose his son or someone else to take his place. But now it often happened that a striving general seized the throne, just as generals had done before Augustus. Typically, his soldiers would acclaim him, raising him on his shield, and he would then assassinate the ruling emperor and force the Senate to approve the change.

Most of these usurpers came from Roman provinces: the Balkan Peninsula perhaps, North Africa, even distant Syria. They knew no other life outside the army, and they hadn't any notion how to rule an empire. Often that didn't matter much, for after they had taken power they rarely stayed alive long enough to make things worse than they had found them. Out of twenty-six emperors during one half century (235 to 284) all but one was killed or took his own life. The problem wasn't what these power grabbers did but that they had no chance to do anything.

Another problem came to light during the 200s: disloyal armies. Formerly the government had raised its troops at home in Italy, but now it drew them from the very peoples that the army was supposed to fight: tribesmen from North Africa, the Danube valley, even Germany. It was men like these who stared out from the wall the Romans built across the narrow waist of Britain, or from their forts in Europe, or from the lonely guard posts on the sands of Syria and Egypt.

Stationed far from Italy, which they never saw, these men cared nothing for the welfare of the empire, though they sometimes helped their generals to interfere in politics. They often fought each other rather than the enemies who pressed against the borders. Those enemies, especially Germanic tribes who swept down from the north, were almost more than Rome could handle. Because of their attacks, defenses on the borders crumbled. Undefended farmers left their lands, and these reverted back to swamp or sand. Lack of confidence in borderland defenses reached the point that Emperor Aurelian, known as "Hand on Hilt," built walls around the city of Rome itself.

Around A.D. 300, the tough and able emperor Diocletian tried to bring back order. To judge by statues and his portrait on his coins, he was a tall and hard-mouthed man. He had risen to the throne the usual way, through the army, but above all Diocletian was an energetic manager. He decided that the empire had become too big for one man to govern and defend. The biggest military problems were in the east, but it took three weeks to sail from Italy to Egypt or the Black Sea.

So Diocletian split the Roman Empire. He chose another man to rule the west and took the eastern half himself. Later each of them would choose another man to govern part of his domain, resulting in four "princes of the world." They might have quarreled with each other but Diocletian made them work together.

As Diocletian saw it, to stave off anarchy he had these tasks: he had to build a larger bureaucracy to help him govern, and he had to double the size of the army and pay the soldiers more. In this way he could keep the army out of politics and help it to keep the peace and fend off invaders. But to pay the added troops and bureaucrats would cost a lot. So Diocletian did the only thing he could; he raised the already heavy taxes.

These taxes caused such hardships that in order to avoid them many farmers, businessmen, and workmen fled their farms and shops. The response of Diocletian was a set of laws that tied the ordinary subject to his place of work. This measure probably was hardest on the farmers. Throughout the empire peasants found themselves forbidden to desert the land they rented, and their children might not leave after their parents' deaths. So in effect, the laws turned farmers into serfs, whom one purchased when he bought the land they worked on.

It was the army that enforced these laws. The army now wasn't just the empire's protector but its tool of oppression. In this grim era, common punishments were breaking legs and gouging eyes. So the empire was preserved, but only by grinding down its people and making them machines for paying taxes.

After twenty busy years Diocletian, of his own free will, gave up the throne. He retired to a palace on the Adriatic Sea, and it's said he spent his final years raising vegetables. A younger general named Constantine now fought his way to power and made himself the ruler of not only the eastern half but all the empire. He quickly made two big decisions: he would live in the eastern half of the empire, and he would build a new, eastern capital city, one that was splendid to behold, centrally placed, and easy to defend.

He chose a good location for his capital: the ancient town of Byzantium, near the entrance to the Black Sea. Strategically this location is important since whoever holds it commands the only route by water from the Black Sea to the Mediterranean. Since it filled a spit of land nearly ringed by water, it was splendidly defensible. Within a few years Constantine built walls that made it almost unconquerable. He filled the town with palaces and churches and adorned the streets with statues grabbed from other cities. He modestly renamed it for himself—Constantinople, meaning Constantine City.

The Roman Empire's older, western European half survived another hundred years. Then Germanic tribes broke through the border. Like ants to sugar they were drawn to cities full of goods to loot. They ravaged all the land, and in the year 410 and again in 455 they sacked the city of Rome itself. They set up kingdoms of their own, which would later be dynamic European nations. But the Roman Empire in the west was gone.

Historians sometimes say the eastern half survived another thousand years. But what survived was not always an empire. Neighbors nibbled on it, until nothing remained but Constantinople itself. And in 1453 a Turkish army blasted through its walls with cannons.

Not before, not again, have we seen the like of Rome. Its immensity and amenities, its laws and length of life make the Roman Empire one of our great achievements.

THE MEN WHO built the Roman Empire knew little or nothing about the Mongol tribes who roamed on Asia's far-off grasslands. Before the Mongols began their conquests in the 1200s, many centuries after the breakup of the Roman Empire, they were simple nomads. On sturdy little horses they wandered on their treeless plains in search of grass for horses, sheep, and yaks to graze on. They also lived by pillaging from other Mongol tribes. They lived in yurts—round tents made with sticks and felt—and fed on bits of meat and curdled mare's milk. Because they held that water was divine and must not be polluted, they never washed. Their filthy leather jackets shone with grease, and their southern neighbors, the Chinese, said the Mongols smelled so awful that no one could come near them.

In the late 1100s a tall young Mongol named Temüjin mastered his clan in a bloody struggle, and he then began to conquer other Mongols. He ruled from Karakorum, in north central Mongolia, but was usually on the move. By the year 1206, when he was in his forties, he called himself the ruler "of all tribes who live in yurts." At a meeting of Mongol chiefs that year he was given the title Genghis Khan, which apparently means "ruler vast as the ocean." (But had the Mongols ever seen an ocean?) He planned to rule the world.

Rule the world! That anyone should dream of such a goal, much less a nomad in this rarely heard-of corner of the earth! He had to justify this project to his people, since it would require enormous effort and great loss of life. He had his shaman (his go-between with heaven) tell the Mongols that he, the shaman, had soared to heaven on a dappled horse and spoken to the Mongols' god. Eternal Blue Heaven had decreed that Genghis Khan should rule the world.

So the mandate came from heaven, and the Mongols must obey their chief. Later Genghis Khan would many times remind the world that Eternal Blue Heaven had ordered him to rule all men. It followed that any person, anywhere, who opposed him also opposed Eternal Blue Heaven, and could only count on death.

On one occasion, though, Genghis Khan explained the reason for his endless warring differently. He didn't say he fought to satisfy Blue Heaven. It was simply in our nature to delight in mastering our fellow humans. "Man's highest joy," he said, "is in victory: to conquer one's enemies, to pursue them, to take what is theirs, to make their loved ones weep, to ride on their horses, and to embrace [his word] their wives and daughters."

Once he had the Mongols under his control, Genghis Khan rode south and struck his neighbor, China. In an early battle he used a ploy that many Mongol victims later on would fall for. Pretending to retreat, he left behind a small detachment in a valley. The Chinese general could not resist this decoy; he attacked it. Suddenly the horde of Mongols charged the Chinese from behind the hilltops, racing down from every side at once, shooting storms of arrows, screaming, slashing with their swords. The stink of Mongol men and horses shocked their Chinese enemies. Victory was quick and the slaughter that ensued was dreadful.

Other victories followed this until the nomad horsemen conquered all of northern China. But how could so few defeat so many? Of course, the Mongols had a military genius in their khan. He never lost a battle, and selected other generals who knew the way to win. Another reason for success: the Mongols traveled very far very fast. A Mongol army once rode 270 miles across Hungary in just three days. If they were in a hurry, the Mongols slept while riding on their rugged horses. If they hadn't eaten they would cut their horses' jugular veins, suck the blood, close the wounds, and journey on. When they struck, they gave their enemies no time for preparations, and they quickly overwhelmed them.

The Mongols had another edge in warfare: they weren't at all averse to slaughter. When they overcame a town they often murdered every woman, man, and child. Sometimes they would spare them, only to drive them out in front when they attacked another town, forcing them to fight their enemies and in this way getting rid of captives and saving Mongol lives. To the Mongols, massacres made sense. Their number wasn't big, so they couldn't spare the troops to pin down peoples they had beaten. It was better just to kill them. And it didn't hurt to earn a name as savages; future enemies might give up quickly in the hope, most likely vain, of being spared.

After they had conquered northern China, Genghis Khan turned west. The Mongols conquered what is now Afghanistan and Iran. Then they raided Russia, the first partly European country to confront them. North of the Black Sea, Mongol raiders overwhelmed an army led by Russian princes. They massacred the ordinary Russian soldiers, but they killed the princes as they always slew aristocrats, avoiding bloodshed. They buried them alive beneath a floor of planks, on which they held a party while the princes suffocated.

Genghis Khan died in 1227 while making war in China. The Mongols bore him northward to his homeland, killing everyone they met along the way to prevent the news of his death from reaching enemies. They buried him in a nest of coffins in a range of sacred hills, and permitted no one to approach the site. No one knows today on which of the hills his people buried Genghis Khan.

The khans who followed him were able men, even if they couldn't match his genius. They had a vaster store of human lives to work with, since, if they didn't choose to kill them, they could draft more troops among the many conquered peoples. With larger armies, they could fight on several fronts.

Southern China fell. The Chinese heir apparent fled and drowned near what is now Hong Kong. Baghdad, capital of the Mesopotamia region, fell. Mongol horses trampled on the caliph. A Persian poet wrote, "The world appeared as tangled as the hair of an Ethiopian. Men were like wolves." Much of Russia fell. According to a Russian epic, "No one remained to weep for the dead . . . for all without exception lay lifeless."

The Mongol empire now was twice as big as Rome's great empire at its biggest. It covered most of Asia from the Black Sea to the coast of China.

As Emperor Diocletian had split the Roman Empire in four parts, the Mongols too divided theirs in four, each one governed by a khan. One of them, the Great Khan, not only governed China, Korea, Mongolia, and Tibet but was also overlord of all the empire. He kept in contact with the other khans by using a vast system of couriers. (The Persians had done the same, and so would the Incas, in their South American empire, two centuries later.) Frequently the khans permitted conquered kings to keep on ruling, provided they took orders and paid tribute. The awful notion of a second visit by the Mongols was enough to keep these kings in line.

Now the khans pushed westward into Europe. Their armies ravaged Hungary and Poland and they raided Austria. One of the Great Khans wrote the Pope in Rome that he and all the European kings must come to China to do him homage. Unless they did, he'd consider them the enemies of Eternal Blue Heaven. And we know what that meant.

Finally, however, the gigantic empire fell apart. That was bound to happen; like the Roman Empire, the Mongols' khanates stretched too far. After the reign of Kublai, Genghis Khan's grandson, the Great Khans focused mainly on their base in China, letting the other three khans go their ways. The Mongols also made the fatal error of the victor: they learned the customs of the conquered and abandoned theirs. They lost their taste for war and slaughter, let their little horses fatten, dressed in silken clothes they'd looted, and began to learn the use of soap and water.

The worst thing that could happen happened: in every khanate the Mongols lost their skill in battle. In Palestine they lost to the Egyptians. The khanate of Iran and Mesopotamia fell apart. A Mongol invasion of Japan failed, and so did a venture into Southeast Asia. As we saw in chapter 5, in the late 1360s their Chinese subjects beat the Mongols, and they rode back humbly to their homeland in the north. And a decade later the grand prince of Moscow beat the Mongol "Golden Horde" in Russia; they didn't go away but they started on a centuries-long decline. The Mongols' brilliant day was done.

CERTAIN THINGS STAND OUT about these conquerors: the Persians, Alexander and his Macedonians, the Romans, and the Mongols. One is "frontier energy," a term we used before about the border tribes who crushed the Shang in China. The Persian horsemen, in their felt boots and leather breeches; Alexander's Macedonian shepherds (who ever heard of them before or since?); the plain-living Roman farmers, just moved into town from the nearby hills; the filthy Mongols with their rugged horses and their curdled mare's milk—all these frontier peoples had the same advantages. They traveled light and fast, regarded hardships as mere annoyances, and hadn't anything to lose.

Another point: the men who led these conquering peoples had iron wills. They astound us with their readiness to squash a million souls like bugs. Genghis Khan is said to have said, "I have committed many acts of cruelty and had an incalculable number of men killed, never knowing what I did was right. But I am indifferent to what people may think of me."

As well as vast amounts of evil, the conquerors did some good. The Persian Empire gave the Middle East two centuries of peace and tolerance. Alexander seems at least to have dreamed that he might bring unity and harmony to his conquered lands. Rome gave the Mediterranean world, and beyond, centuries of Pax Romana, "Roman Peace." Even the Mongols, cruel as they were, brought order to the biggest continent on earth. During Genghis Khan's lifetime a Persian wrote that one could safely walk from Persia to central Asia balancing a golden platter on his head.

CHAPTER 7

We found the worldwide faiths.

HOW MOVING IS the thought of early humans deep in caves in southern France and Spain! Solemn men and women, holding high their flaring torches. Artists, splashed with black and red, painting stags and dancing shamans on the walls. Children, scared and awed. Even then, it seems, we humans felt the mystery in life, the need to fathom who we are and what will come of us.

Nowhere on earth have people sought the answers to those mysteries so long and so deeply as they have in India. Our tale begins in towns that rose along the Indus River, which moseys southward through the hot, flat plains of what today isn't India at all but modern Pakistan. Humans began to live in towns here later than they did in Egypt and Mesopotamia, and a little earlier than in China. From early on they lived in homes made out of kiln-baked bricks, and they had a sewer system many cities of today might envy. They left behind them many soapstone seals on which they carved both words and pictures. Scholars still can't fully understand the words, so many mysteries about these folk persist.

Two things we especially would like to know about these river people are: what were their religious beliefs, and did these beliefs live on in the Hinduism of later times in India? We have very little evidence about the religion of the Indus people, but consider this: carved on several of their seals is a god who is seated like the Indian holy men of later times, with his legs drawn close to his body and his heels touching. On the largest of the seals he appears to have other faces on either side of his head, and on all of them, this god is naked and his penis is erect.

This is the intriguing thing: this figure on the Indus seals looks very much like Shiva, the god whom Indians of later times would worship as a destroyer and a re-creator of life. What is more, archaeologists found many cone-shaped objects in the Indus valley digs, and these are almost surely phalluses, abstract sculptures of the penis. In the Hinduism of later times a phallus is often a Shiva symbol.

In about 1750 B.C., town life on the Indus River ended. River mud then buried the remnants of the towns so deeply that historians knew nothing of the Indus people till the 1900s, when archaeologists found the ruins. It looks as if the civilization, at least in places, ended violently. In one town diggers found the bones of many people who were slaughtered in their houses, in the streets, and at a public well.

At about the time the Indus towns died out, light-skinned immigrants were filing southward through the mountain passes into northern India. They called themselves the Aryans, which means "the noble." They tended sheep and cattle, and were also warriors who rode to battle on chariots, wielding swords and axes. The Aryans were yet another of those energetic peoples who were moving out of central Asia's grasslands. They were pushing not only south into India but also westward to Iran (we have met the Medes and Persians), Mesopotamia, Asia Minor, and Greece. Their family of languages, called "Indo-European," would father many modern tongues.

Unlike the Indus River people, with their homes of brick, the Aryans lived in flimsy huts of wood and reeds. Long ago their homes decayed and disappeared. Whereas the Indus people left us bricks and seals but scarcely any words that we can read, the Aryans left us little else *but* words. What survives of them, apart from pots and tools, are many of their hymns. When they first arrived in India, the Aryans had no writing system. Their poets composed hymns in their heads and taught them to others, who taught them to others. But after several centuries the Aryans wrote them down.

Although the meaning of the hymns is often puzzling, they do suggest what the early Aryans did. It appears that after they had entered India, they conquered a native people whom the hymns describe as dark and ugly, bull-lipped and snub-nosed, and worshippers of phalluses. These natives may have been the people of the Indus valley. If so, it appears that the Aryans played a part in wiping out that earliest Indian civilization. The Aryans had a rigid social system, which is still alive in India. Every Aryan was born a member of a caste of priests, or warriors, or merchants and farmers, or artisans, laborers, and servants. Most other Indians were "untouchables."

At the heart of the Aryans' religion were their rites for sacrificing animals to gods. Only priests knew how to call upon the gods to come and join the worshippers and drink the sacred and intoxicating juice called soma. The gods were kindly and cheerful and usually granted what worshippers had asked for, such as children, victories in battle, or long life.

As centuries went by the Aryans came to believe that sacrifices were more than just a way of winning favors from the gods. They were reenactments of what had happened to a primeval man named Prajapati, who had existed before the founding of the universe. The gods, who seem to have been his children, had sacrificed him. When Aryan priests reenacted that original sacrifice, if they did it perfectly, the world was born anew. But if they made mistakes chaos would result.

As centuries went by, many thoughtful Indians turned against the priests and sacrificial magic. In poems and essays called Upanishads ("Sittings Near a Teacher") their learned men discussed the fundamental questions about our existence. Their biggest contribution was what they had to say about an afterlife, an existence after ours on earth. The older view, set forth in Aryan hymns, was that one who lived a full and upright life might hope to have an afterlife in heaven among the gods. If not, then he or she might finish up in hell.

The writers of the Upanishads, on the other hand, believed that most of us will have no afterlife at all. They saw existence here on earth as never-ending rounds of life and death. For every living thing—animals, plants, humans, or gods—they said, the normal fate is to die and then to be reborn, perhaps as a member of a higher or a lower caste, or perhaps an animal or a plant. Then one dies again, only to be reborn again in another form. These cycles of death and rebirth go on forever, and this everlastingness is a terrifying prospect. What to do about it is a problem that concerned Upanishad philosophers, and also, as we soon shall see, the Buddha.

The religion known today as Hinduism emerged from the beliefs of the Aryans and the Upanishad sages and others, and partly also from the religion of the earlier Indus people, which had never wholly disappeared.

It is often said that nothing can be said about Hinduism that cannot also be denied. It's the total of what a billion very different Indians believe and do. A Hindu may call for help on Shiva, Vishnu, or another of the many gods and goddesses, and perhaps on more than one of them. If possible a Hindu also reads the sacred poems and essays, and meditates on Brahman, the Self of every living thing. This spirit, Brahman, is so impersonal and all-embracing that it is "neither this nor that." "Verily, this whole world is Brahman," wrote one of the sages. "Let one worship it tranquilly as that from which he came forth, as that into which he will be dissolved, as that in which he breathes."

HINDUISM WAS ALREADY ancient when the man we call the Buddha taught in villages of northeast India. We have no writings about him that date from his own time, and so we know very little about him. Siddhartha Gautama was his name, and his father was a chieftain in the foothills of the Himalayan Mountains. While still a young man Siddhartha left his home, after which he lived for decades as an ascetic and then a wandering teacher. He died in the 480s or 470s B.C. at the age of eighty.

Those are the bare facts, of no great interest. But the legends that his followers told about him in later times are vivid and moving. Whether true or false, they influenced many million lives. In the legends, Siddhartha's father was not a mountain chieftain but a wealthy rajah, who reared his son in a charming palace from which every glimpse of misery was carefully screened out. Siddhartha had it all; he was handsome, rich, and brilliant, and when he was still very young he had been married to a lovely girl. But at the age of twenty-nine, he secretly left the palace several times with a servant and went to nearby villages to discover for himself what life was really like.

On these expeditions he came upon Four Signs that would forever shape his view of life. The first three were a feeble old man, a man who was hopelessly sick, and a corpse surrounded by weeping mourners. When, in shock and wonder, Siddhartha turned to his servant Channa, the man could only tell him, "Yes, my prince, these things must come to all." But how, Siddhartha asked himself then and later, did human beings endure their lot: the misery, the physical decay, the certainty of death? If this was life, then why be born?

One day, however, on another secret foray, he came upon the Fourth Sign, a holy man. This man had chosen, Channa told him, to wander homeless and to beg for food. When Siddhartha saw the calm and peace in this man's face he hoped that he had glimpsed a way to look for answers to his questions. But his inner angst persisted, and when it reached its worst he decided to leave home and find the cause and cure for evil, pain, and death. In the middle of the night, he stole away from his beloved wife and son and slipped outside the palace. He cut his flowing locks of hair, the badge of gentlemen, and flung them in the air, and swapped his rich man's raiment for a poor man's rags.

For several years, the legends say, he studied with some Hindu teachers, but these gurus could not meet his needs. Their lofty thoughts and endless rites were not what he was looking for. And so, for seven years he lived alone, a hermit in a forest. He meditated; he used yoga to tame his wants and guide his mind; and he nearly starved himself to death. When he had almost lost all hope of finding answers, he had a flash of wisdom. Why, Siddhartha asked himself, did he mortify his flesh? His body, after all, was all he had with which to seek the answers to his questions. It was time, he thought, to return to something nearer to the life of ordinary humans. So he left the forest.

A village girl offered him a bowl of milk curds. In an act that symbolized his transformation, he accepted and he ate them. For a man whose self-denial was a local legend, this simple act was stunning. It so upset five hermits who lived nearby that they rose and went away. Siddhartha bathed and changed his clothes, and he sat beneath a fig tree in the thinker's pose. He began again to meditate, determined not to leave until he had the answers.

Siddhartha knew, he knew for certain, that he himself had triumphed over what had so consumed the writers of the Upanishads: the endless rounds of birth and death and birth again. The chains that once had bound him to an earthly life were broken. Triumphantly he said,

Many a house of life
Has held me—seeking ever him who built
These prisons of the senses. . . .
Sore was my ceaseless strife!
But now You builder of this dwelling—You!
I know You! Never will You build again
These walls of pain. . . .
*Your house is broken, and the ridge pole split.**

Now that he had risen over every earthly care, he might simply have ignored the world around him. Why should he concern himself

*Nancy Wilson Ross, *Three Ways of Ancient Wisdom* (1966), p. 89. (I have simplified the language.)

about the rest of us? But Siddhartha knew how much he had to teach the world. He had found the answers to his questions about the meaning of our suffering and death. Could he not help his fellow humans by teaching them what he had learned? He decided to turn back from the bliss of inner freedom and return to troubled daily life. Leaving his place under the tree, he said, "I will beat the drum of the Immortal in the darkness of the world."

His first converts, tradition says, were those hermits who had left in horror when he ate the bowl of curds. He followed them to the holy city of Benares, and when they saw him there, in a deer park, they realized that he had been somehow glorified. He was now a buddha, a word that at that time meant simply an enlightened person. But in the hermits' eyes, and the eyes of the many millions who would later be his followers, Siddhartha was *the* Buddha.

The hermits bowed to him and sat down to be taught, and from the start Siddhartha proved to be a vivid and engaging teacher. By placing grains of rice on the ground he drew a picture of a wheel. This, he told them, was a symbol of the round of births and deaths that keeps on cycling because of our desires. The causes of our suffering and despair are our greed and self-absorption. But we *can* root out these deadly faults, which spring from our confusion.

To reach an understanding of the meaning of our lives we can do certain things. We must see clearly what is wrong with us, resolve to heal ourselves, and mind the basic moral laws. Then—but this is very hard—we must think constantly about our goal, which is knowledge of life's meaning. We must contemplate it "with the deep mind." Doing these things will lead us to a tranquil freedom from our drives and wants. If we escape from them, we also will escape the "wheel"—the endless rounds of birth and death—and reach a state of perfect peace.

That is what the Buddha taught the hermits, and later many others, as he wandered from one village to the next for the rest of his long life. No one knows just when he died—probably between 500 and 350 B.C. He had attracted many followers. No fewer than five hundred Buddhist monks convened soon after he had died to dumb his teachings down to formulas—precisely what the Buddha, who never was dogmatic, would have urged them not to do.

Buddhism spread much faster later, in the 200s B.C., when Emperor Ashoka governed most of India. Early in his reign, Ashoka ("without sorrow") was noteworthy mainly for the pointless slaughter of many thousand Indians. But later he became a Buddhist, and he spent his final decades spreading the teachings of the Buddha in India and in other lands. When the ruler of the island of Sri Lanka sent a gift of splendid pearls, Ashoka thanked him by sending him his only son, a monk, to bring "the Jewel of Truth."

Later though, Buddhism faded out in India, the land where it was born. Hinduism gave it birth, and Hinduism reabsorbed it. But even as it vanished in its motherland Buddhism spread through Asia like a rising tide. In Sri Lanka, Burma, Cambodia, Thailand, Vietnam, Laos, Nepal, Sikkim, Tibet, Mongolia, China, Korea, and Japan it would shape the lives of countless millions.

WE KNOW MUCH more concerning Jesus, who inspired the Christian faith, than we know about the Buddha.

Jesus came from Galilee, a little princedom in what long before had been a kingdom: Israel. In the time of Jesus, Galilee was just a puppet state within the Roman Empire. Jesus's father—that is, the man who raised him—was a carpenter and of course a Jew. The Gospels, or early stories of the life of Jesus, suggest that Jesus's mother was pregnant with Jesus when she married, and that his father was God. Jesus was not born in A.D. 1, as one might suppose (since A.D. means *anno Domini,* the year of the Lord, that is, Jesus), but at least four years earlier.

When he was about thirty-five years old, Jesus became a holy man who, something like the Buddha, walked from place to place and preached to little groups, often in the open air. He was warm and forceful; speaking just a word or two to someone who approached him, he sometimes changed the person's life. As he went, he gathered followers, including a tax collector and fishermen from the shores of the Sea of Galilee. Some of them would stay with Jesus till his death and then carry on his teaching.

Four people who had known him or had learned about him from his followers wrote short accounts of Jesus's life. These "gospels" (or "good news") tell us nearly all we know about him, but what they tell us isn't always clear. Jesus himself caused some of the confusion. He liked to teach with parables, or little tales that make a point, usually about God and us. But often the point was not apparent to his hearers (Jesus was aware of this), nor is it clear to readers now. Take, for instance, the well-known parable of the younger son who wanders off, wastes his money, nearly starves to death, and finally returns to his family. One might suppose that Jesus wanted to teach fathers to forgive their erring children. But he probably meant to teach that God accepts the person who rejects and then returns to him.

Did Jesus see himself as teacher, rebel, or the son of God? Did he intend above all to beseech us to be meek and kind, or did he want to warn us to get ready for the ending of the world? Careful students of the Bible hotly disagree about these matters.

This writer thinks that Jesus meant to bring us shocking news. The world was ending, he declared, and the "Kingdom of God" was at hand. It wasn't going to happen sometime in the future. No, the Kingdom of God was happening at that moment. You could enter it right then, right there. (He was not talking about a life after death, which is something he talked about very little.)

This means, he said, that the old moral laws were not enough. Yes, you must love each other (and he spoke movingly of that), but—and here he seems to contradict himself—you must also do things that will upset relationships. You must love not only your neighbor but also your enemy. You must leave your land and your family and follow Jesus. "Do not think that I have come to bring peace on earth," he said. "I have come to set a man against his father, and a daughter against her mother, and a daughter-in-law against her mother-in-law . . . he who does not take his cross and follow me is not worthy of me."

He probably did not believe he was the son of God, or a rabbi or a rebel, but how he saw himself is not apparent. Much depends on whether the Gospels give us his authentic words, and what those words meant in his time. He presented himself as one who somehow actualized the Kingdom of God and made it happen even as he spoke.

During the two or three years when Jesus spread his message, Jewish leaders opposed him more and more. This is no surprise, since he sometimes seared them using burning words. For example: "You are like tombs covered with whitewash; they look fine from outside, but inside they are full of dead men's bones and all kinds of filth."

At the end of his life, when he was still in his thirties, Jesus and his followers went to Jerusalem for Passover, the annual celebration of the Israelites' escape from Egypt. He knew the trip was dangerous, and not only because he had Jewish enemies in Jerusalem. The city lay inside the Roman Empire, and its Roman chiefs would never tolerate a man who talked about the coming of a rival kingdom.

Just the same, he told his friends, he had to go to carry out God's will. What might happen in Jerusalem was in the hands of God. His death could motivate his fellow Jews to make the change of heart he felt they needed for admission to the realm of God. "I must be on my way today and tomorrow and the next day," he said, "because it is unthinkable for a prophet to meet his death anywhere but in Jerusalem, the city that murders the prophets, and stones the messengers sent to her."

When they reached the city Jesus and his followers ate a Passover supper, which was to be their last meal together. The next day a follower betrayed him by pointing him out to a crowd of his enemies. A scuffle followed, and Jesus was led away. He appeared before a Jewish court, which concluded that he was blasphemously claiming to be the son of God. They turned him over to the Roman commissioner, telling him that Jesus claimed to be the king of the Jews. The commissioner asked Jesus if this was true, and Jesus answered strangely, "You have said so." The Roman hesitated for a moment, but he ordered the troublemaker put to death.

Roman soldiers took him outside the city and crucified him, probably by nailing him to a crossbeam fastened to an upright shaft. Someone fastened a sign on it that read "This is Jesus, the King of the Jews," and as he writhed in pain priests and others taunted him. At this point Jesus seemingly decided that his mission had been betrayed by God himself, who did nothing to prevent his suffering such pain. He wailed a sentence from a psalm, "My God, my God, why have you forsaken me?" and soon he died.

His followers were shattered. They had lost their leader and did not know what to do. But then, the Gospel writers tell us, shocking things occurred. A well-off friend of Jesus had laid his body in a tomb cut in the rocky hillside, and rolled a disk-shaped stone across the front. But when his followers came near the tomb, they found the stone was rolled aside and Jesus's body gone. What had happened to it? More amazing yet, Jesus then began appearing to his followers, briefly, here and there, like a ghost, but a ghost that one could speak to and touch. The appearances convinced them that Jesus had risen from the dead and that this confirmed the truth of all that he had taught.

His followers began to form religious groups. They spoke of Jesus as the Christ (meaning "the anointed") and of themselves as Christians. Not surprisingly, most Christians in Jerusalem, who were surrounded by Jews, saw themselves mainly as a special group of Jews. They stuck with Jewish dietary laws and circumcised their baby boys, as God had ordered Abraham to do long before.

But some of them denounced the usual worship in the Temple and started to reject or reinterpret Jewish laws and customs. What counted, so they said, was the spirit, not the letter, of the laws—that, and believing what Jesus had taught. To say such things was dangerous. Angry vigilantes stoned to death a Christian preacher, Stephen, who had called their leaders "stiff-necked people, uncircumcised in heart and ears."

A leading enemy of Christians was a pious Jew named Saul. Short, balding, and bowlegged, he was a man of destiny. Although he had been trained as a rabbi, he earned his living making tents. After

Jesus's crucifixion he persecuted the early Christians, arresting them and sending them to prison.

One day when Saul was trudging down a road in Syria, he heard a voice that said, "Saul, Saul, why do you persecute me?" The voice was that of Jesus. Saul immediately became a Christian believer. He was more than merely converted; he felt that God had called on him to bring a message of salvation "to the ends of the earth," and he set out to do so. For some reason he dropped the name Saul and began to call himself Paul.

As a missionary, Paul had the benefit of living in the peaceful Roman Empire, where travel was possible, though not always safe. For the rest of his life, three busy decades, he sailed in ships along the Mediterranean and hiked the Roman roads to towns in Syria, Asia Minor (modern Turkey), Greece, and even to the great capital, Rome itself. In all these places he talked to little groups and converted a few of his hearers. After he had left them, he kept in touch with converts by writing them encouraging letters. He took terrible abuse from men and nature: he was jailed, pelted with stones, bitten by a viper, three times beaten with rods, five times whipped, and three times shipwrecked.

In the meantime, Paul transformed the message of the man who hailed him on the desert road. Jesus had aimed his words at Jews, but Paul was preaching not to Jews but pagans. He was therefore free to found what was almost a new religion, greatly aided by the fact that he knew very little about what Jesus had taught. (After Jesus spoke to him on the road in Syria, Paul waited three years before going to Jerusalem and meeting Jesus's leading follower, Peter, and Jesus's brother, James.)

Paul made Christianity more palatable to pagans by simply dropping much of Jewish "law." For instance, he told the men who listened to him that in order to be Christians they need not be circumcised, like Jews. This teaching may have relieved the minds of many men.

What was most important to Paul was not what Jesus taught but who he was, and what his execution and his rising from the grave implied. Whereas Jesus had been almost guarded about who he was, Paul taught boldly that Jesus was the son of God, who had ushered in the soon-to-happen end of the world. Jesus's death made amends for the sins of everyone, thus making it possible for everyone to be pardoned by God and admitted among the people of God. Paul's teaching that Jesus had sacrificed his life *on purpose* was Paul's major contribution to Christianity. Jesus never made that claim.

But how did Paul *know* that Jesus was the son of God, who died to atone for our sins? The evidence was the fact that after his death Jesus had risen from the grave and then appeared in several places to his followers and then to Paul himself. Those who heard this thrilling news, and who truly believed that Christ (that is, Jesus) had died for them, would be saved. As the world came to an end, a judgment day would follow, and the believers would join God.

Near the end of his life Paul was arrested in Jerusalem when Jews charged him with bringing a non-Jew into the inner courts of the Temple. The Roman rulers probably considered him a troublemaker, like Jesus, but unlike Jesus, Paul was legally a Roman citizen. When he claimed his right to appeal to the emperor, they sent him to Rome. Our source for Paul's life, the Bible's book of Acts, does not tell what happened next, but according to tradition he was put under house arrest for two years, then tried, convicted, and beheaded.

And what had Paul achieved in three busy decades? Well, he had taken what he had heard about the death of an obscure holy man and shaped it into a religion for everyone. Then he had scattered the seeds of this new faith over as much of the world as he could reach. If Jesus was the most influential figure in all of history, as some would say, then who was Paul?

After Paul died, Christianity kept spreading, slowly first, then very fast. In the next three hundred years, it soaked up other faiths (absorbing their beliefs), and at last Roman emperors decreed it was the state religion. It grew a large bureaucracy, run by wealthy men in velvet robes, built universities and churches, supported learning, preached of peace and sometimes stirred up wars, and advocated moral standards that were widely honored and ignored. It instructed many, many million people that God had promised them that after this world's sorrows they would live again in heaven. Today a third of all the folk on earth describe themselves as Christians.

ISLAM, YOUNGEST OF the world religions, was born in Arabia, not far from where the Hebrew fathers led their flocks, and Moses spoke with God, and Jesus met his fate. Arabia is not so much a peninsula as an island, girdled on three sides by water and on the fourth by desert. Almost all of it is hot and dry and bare. Until recently most of the Arabs were nomads who wandered ceaselessly in search of grass for their camels, horses, sheep, and goats. They lived mainly on milk and on the dates they grew in the rare oases.

Before the time of Muhammad, founder of Islam, the Arabs weren't a single people. Most of them belonged to tribes that often raided other tribes for camels, slaves, and horses, and to let their warriors test their courage. They shared a religion of shrines and spirits, but it appears that even before Muhammad's birth some Arabs were beginning to believe in a single god.

Muhammad was born in about A.D. 570, not among the nomads on the desert sands but in the holy town of Mecca. The great distinction of his birthplace was a well-known shrine that Arabs called the Kaaba (meaning "cube"). It housed many idols, even images of Mary and Jesus. Mecca was also an oasis where camel caravans, bearing frankincense and spices through the desert, halted by the waters of the sacred Zamzam well.

His father had died on such a journey before Muhammad's birth, and his mother died when he was about six. A grandfather took him in charge, but then he too died, so an uncle raised the boy. The family was poor, and Muhammad seems to have worked for a while as a shepherd and then a leader of caravans.

At the age of twenty-five, however, he married the woman he worked for, a well-off widow fifteen years older than Muhammad. Now he had some leisure, and opportunities to look at life. Probably he now observed the social gap between the wealthy, settled Arabs, like his wife, and those who still were nomads. He learned a bit about other religions, perhaps from Christian and Jewish Arabs, of which Arabia had many.

This is what Islam teaches about its founding: Muhammad often meditated in a mountain cave in the arid desert outside Mecca. While he was there, one day in about A.D. 610, the archangel Gabriel appeared to him. Gabriel told him that he, Muhammad, was the Messenger of God, and he told him to "recite." Muhammad answered that he couldn't read, but Gabriel insisted, saying:

> *Recite, in the name of the Lord, who created,*
> *Created man from a clot of blood.*
> *Recite, for the Lord is the most bounteous,*
> *Who teacheth by the pen,*
> *Teacheth man what he did not know.*

In a state of terror, Muhammad hurried home. Feeling cold, he lay down and asked his wife to cover him. Then he heard that voice again, this time saying:

> *O thou wrapped in thy mantle,*
> *Arise and warn!*

Gabriel now began to tell Muhammad sacred truths in poetic prose. For the next two decades Muhammad would continue to hear and then repeat these messages. After his death in 632, Muslims wrote them down and put them in a sacred book, the Qur'an. A worldwide faith arose from revelations by an angel to an Arab in a cave.

That is the accepted story. Some modern scholars think the faith took shape more slowly. They point out that the earliest evidence of the Qur'an's existence dates from 691, the year when Muslims built the Dome of the Rock in Jerusalem. (This famous mosque is on the site from which Muhammad, in a vision, rose to heaven and met with God.) A few Qur'an inscriptions ornament the Dome, and they differ somewhat from the standard version of the Qur'an. These discrepancies suggest that decades after the prophet's death the Qur'an was still evolving.

Muhammad began instructing any Arabs who would listen in the truths that he was learning from the angel. These were: that there is no god but the one God, Allah. That he, Muhammad, was God's messenger and prophet. That a day of judgment was approaching. And that after dying those who followed God's commands would go to paradise, while punishments were waiting for the others. That simple formulation was the heart of the religion that Muhammad was founding. The religion was Islam, or "submission" (that is, to the will of God). Its believers were (and are) the Muslims, "those who submit."

Success as a religious leader came only slowly to the Prophet. His wife and family, another man who would later take his place, and several hundred others soon accepted Islam. Most of them were young men under thirty when they joined. Other Meccans, especially the prominent, were hostile to his teachings. Even those who led his clan opposed him. Perhaps he scared them when he taught that the poor had a rightful claim to the wealth of the rich.

At a trade fair Muhammad talked with men from Yathrib, an oasis north of Mecca. They, and later others from that town, became his followers. They invited him to come to Yathrib, hoping he could settle quarrels between the place's rival factions. Since many of his fellow Meccans were hostile, even threatening to him, Muhammad welcomed the invitation. He sent his Meccan followers to Yathrib. Then Muhammad and a friend slipped out of Mecca, followed desert trails that were seldom used, and safely got to Yathrib in September 622. Muhammad renamed Yathrib Medina, or City of the Prophet.

In Medina, the Prophet and his fellow Muslims preyed on Meccan caravans that crossed the desert. Though few in number they had great success. The Meccans twice sent armies out to crush Medina, but Muhammad's little army held them off. The Muslims grew more sure than ever that God was with them. Muhammad meanwhile fought the Jews in Medina, who wouldn't recognize him as a prophet. He drove away two Jewish clans, and later he attacked another clan, slew the men, and sold the rest as slaves.

In the year 630, the Prophet marched to Mecca and peacefully conquered it. He spared the lives of his enemies. While he did not insist on this, many of the Meccans now converted to Islam. Many tribes near Mecca and Medina allied with him and converted to Islam. When hostile Arabs sent an army to get rid of him, Muhammad routed it. Most of the Arabs allied with him and converted to the new religion.

He brought the warring tribes a faith that bonded and inspired them. "Muhammad is the Messenger of God," says the Qur'an, "and those who are with him are hard against the unbelievers, merciful one to another. You see them bowing, prostrating, seeking bounty from God and good pleasure. . . . God has promised forgiveness and a mighty wage to those of them who believe and do deeds of righteousness."*

It wasn't only piety that fused them but the Prophet's force of character. He was a pitiless fighter but a courteous and humble man. He and his wives—about a dozen of them—lived in simple homes of clay. He liked honey and milk, and could often be seen mending his own clothes. He owed much of his success to the model that he set for his disciples.

When he died in 632, the Arabs badly needed a leader to replace him. He had only lightly glued them together, and after he died many tribes rejected orders from above. To make things worse, the Prophet had not clearly chosen a successor. How does one replace God's Prophet?

At first, at least, his followers chose their leaders well. They picked as "caliph" (which means successor) one of Muhammad's advisers and the father of one of his wives. In just two years this leader crushed the rebels and joined the quarreling tribes in one Islamic state. When he died, the Arabs chose as caliph a man who continued to live as simply as a desert sheikh. He owned a cloak and one patched shirt, and slept on palm leaves.

**Qur'an,* trans. Arthur J. Arberry (1955), Victory 48:35, p. 229.

The Arabs now began a dazzling set of conquests. The world has never seen the like. Riding swiftly on their camels, first they struck their two big neighbors: the Eastern Roman Empire to the northwest and Persia to the northeast. In fifteen years (632–49) they conquered Roman Egypt and most of the Persian Empire. (Later, they would take the rest of it.) The Persian king of kings was murdered by a Persian, and his son escaped to China. Meanwhile these desert nomads transformed themselves into sailors, as the Romans had done so long before when they fought the Carthaginians. They built a fleet of ships and seized some islands that gave them nearly total control of the Mediterranean.

After halting half a century they thrust to the east again and conquered some of India's western coast, and oases on the barren plains of Central Asia. They fought and vanquished Chinese armies, but they didn't push inside the Middle Kingdom. But under an able general, Musa ibn-Nusayr, they battled west across North Africa, all the way to the Straits of Gibraltar.

How did these provincial Arabs win so much so fast? Perhaps the simple answer is, for once, the right one: Muhammad had brought them union and a goal. Muslims were required by God to bring the world his revelation. But the Arabs weren't fanatics bent on winning souls; in fact, they rarely forced their faith on those they ruled. Probably they found it easier to tax infidels than converts, and easier to rule a people split by several religions than a people united by one.

Even so, many millions of the people whom the Arabs conquered did convert. They may have reasoned that the God of Muslims must be strong indeed if his people won so many wars. And perhaps they were drawn to a religion that set out so explicitly the duties of the faithful. Islam also promised them an afterlife in which their faces would shine with joy, and they would praise God "in gardens of bliss," where "immortal youths will serve them with goblets, pitchers, and cups . . . and they may choose fruit of any kind and whatever fowl they desire and chaste companions with eyes of a beauty like pearls hidden in shells."

Now that the Muslims held so much of Africa and Asia, the time had come to deal with Europe. Musa, who had conquered all North Africa, chose a general named Tariq to conquer Spain. In 711 he shipped a force of only 7,000 Muslim converts across the narrow strait that separates North Africa from Europe. They won the overwhelming rock that guards the strait and christened it Jabal Tariq (Mount Tariq), a name that later warped into Gibraltar. Then the Muslims routed a defending army, and Tariq and Musa overran Iberia, which now is Spain and Portugal. (Later, the caliph summoned Musa and Tariq to Syria and accused them of stealing funds. The conquerors of North Africa and Iberia died in obscurity.)

In 732, however, the Muslims rode their horses northward from Iberia, across the Pyrenees, and into "France." At this time France was little but a name, but its ruler somehow raised some fighters dressed, it's said, in wolf skins. In a daylong battle fought in western France, the Muslim horsemen tried repeatedly to rout the French but failed each time. After night had fallen they withdrew from France. This Muslim failure looks, in hindsight, like a turning point for Europe and Islam. In later generations, Europe grew too strong to overcome, and many centuries later Christians drove the Muslims from Iberia.

While Muslims raided France, an Arab navy, at the other end of Europe, tried to conquer strong-walled Constantinople. Perhaps the Muslims had a plan to conquer all of Europe using pincer movements at the same time on the west (Iberia and France) and in the east. But that appears unlikely since by this time there were several Muslim rulers, and they didn't always work together. The Arab ships blockaded Constantinople for five years with no success, paused two generations, then attacked again. But the ancient city didn't fall. Once again it proved to be the bulwark Constantine had wanted when he chose it as his capital four hundred years before. Islam's armies, blocked in Europe in both west and east, never conquered Europe as they had the Middle East, North Africa, and a giant chunk of Asia.

The day of rapid Muslim victories had passed, but at last Islam had joined Buddhism and Christianity as one of the world's far-flung religions. Today more than a billion people follow the religion that Muhammad long ago revealed.

HOW OFTEN IT is said; how true it is. The founders of the major faiths were humble people in forgotten places. But it was they who asked the most important questions. They answered them so vividly that they reached the minds of every later generation.

CHAPTER 8

Europe prepares for its big role.

BACK WHEN ROME was at its peak, most of Europe was merely the northwest corner of the far-flung Roman Empire. And it was not the jewel in the emperor's crown. Europe held little but forests, marshes, Roman forts, villages, and tribes of semicivilized hunters and cattle herders. No one would have guessed that these backwoodsmen would one day lead the world.

That leading role was still far off in A.D. 500, when it was clear that the western half of the Roman Empire had fallen apart. The next millennium, from 500 to 1500, is the period that Europeans of much later times would name the Middle Ages. The name implied that those thousand years were just a dreary slog between Rome's golden age and theirs.

The Middle Ages were more than just a long and dismal trek. However, this is true: life in the *first* half of the millennium, that is, from about 500 down to about 1000, was often violent and wretched. Historians sometimes call that half millennium the Dark Ages. It's

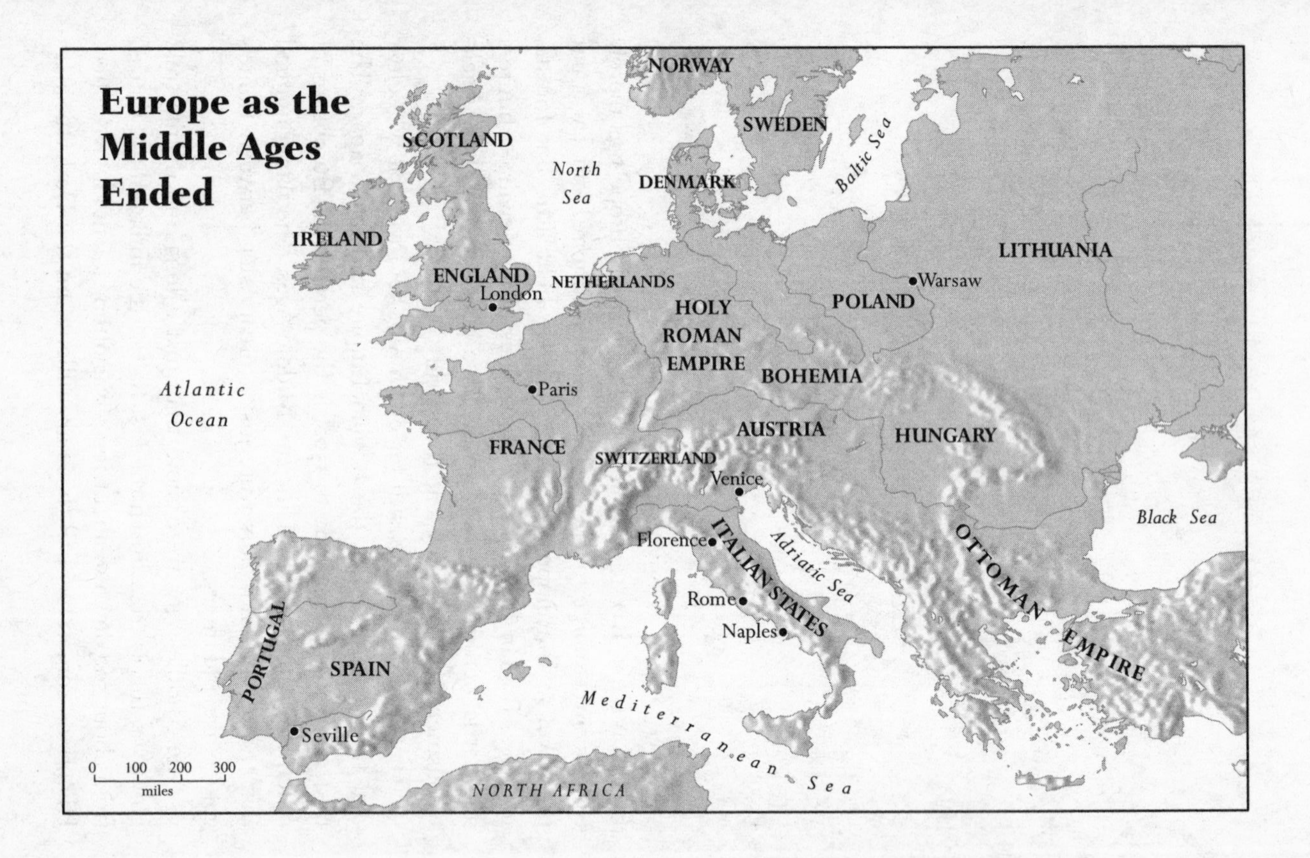
Europe as the Middle Ages Ended
NORWAY
SWEDEN
SCOTLAND
North Sea
DENMARK
Baltic Sea
IRELAND
LITHUANIA
ENGLAND
London
NETHERLANDS
Warsaw
POLAND
HOLY ROMAN EMPIRE
BOHEMIA
Atlantic Ocean
Paris
AUSTRIA
HUNGARY
FRANCE
SWITZERLAND
Venice
Black Sea
Florence
ITALIAN STATES
Adriatic Sea
OTTOMAN EMPIRE
Rome
Naples
PORTUGAL
SPAIN
Mediterranean Sea
Seville
0 100 200 300
miles
NORTH AFRICA

true that even then some kings and emperors tried to maintain order. The greatest of these men was Charlemagne (or Charles the Great), a portly, high-minded warrior-king who conquered much of western and central Europe just before and after 800. He liked to think that he had reassembled what Rome had earlier built, and men of later centuries called his lands the Holy Roman Empire. With many ups and downs his work would last a thousand years in central Europe.

By and large, however, in the Dark Ages any man who owned a fortress and commanded a troop of horsemen ruled the land around him. We know of one such man who owned a two-floor, two-room wooden fort. In the first-floor room he stored his food and weapons, and on the second floor he, his family, his servants, and probably his horsemen ate and slept in just one room. Such a "lord" *might* shield the nearby folk from robbers and invaders; more likely he would grab their land.

Churchmen at a meeting in France in 909 groaned that "every man does what seems good in his eyes, despising laws human and divine. . . . Men devour one another like the fishes in the sea." Many Europeans lost their freedom and became serfs, living miserably in hamlets, forbidden by their lords to leave. A genealogy of a family of French serfs that was prepared for a trial suggests the violence of these times. It ends with "Nive, who had his throat cut by Vial, his lord."

Order was not the only thing that vanished in the Dark Ages. In earlier times, the western branch of the Christian church, based in Rome, had brought ideals and learning to the Roman Empire. Now, however, many of the bishops, and the abbots who ran monasteries, cared more for gold and grandeur than for souls or schools. Bishops were so involved in politics that they often fought in battles. When they did they carried maces instead of swords, because, as clerics, they were not allowed to spill blood but could freely shatter skulls.

Meanwhile, trade contracted, and the towns declined and shrank. In Rome itself, sheep grazed the grass in ancient forums where emperors and senators once had run an empire. Adding to these many woes, Vikings from northern Europe, Huns and Magyars from Asia, and Muslims from North Africa raided and plundered Europe and conquered chunks of it.

AFTER ABOUT 1000, however, life in Europe markedly improved. Kings, great landowners, and some of the towns began to drive the troublemakers out. A good example is a king of France, Louis VI, who was widely known as Louis the Fat, although he may have preferred his other nickname, Wide Awake. Louis spent a good part of his reign patrolling France, which was then a little princedom surrounding Paris, as if he were a policeman on his beat. His biggest enemy was a robber baron, Thomas of Marle, who made a career of his hatred for the Church, and towns, and kings, and used to hang up captives by their testicles. It took King Louis many years to level Thomas's castles and finally to catch him when he was mortally wounded.

In England, the story was different. At the end of the Dark Ages, English kings were sometimes even weaker than their counterparts in France. They could barely influence the wealthy lords with private armies, or make the ordinary Englishman obey the law. But all that changed in 1066, when William, duke of Normandy (in northwest France), took a gamble. With a small army, William "the Bastard" sailed across the English Channel, won a bloody battle on the southern coast of England, and proceeded to master all the country.

Now he was no longer William the Bastard, thank you, but William the Conqueror. His victory was so total that he could dominate the English ruling class and shape the country as he wished. English kings from that time onward wielded far more power than rulers in the rest of Europe. They changed the legal system and collected most of the taxes they wanted. They took pains to have an honest coinage. A chronicle tells us that William's son, Henry I, "bade that all the mint-men [counterfeiters] . . . should lose each of them the right hand, and their testicles beneath."

While the monarchies grew stronger, the Catholic Church revived.

Its leaders once again promoted learning, raised up soaring churches, and pled with Europeans not to slaughter one another. Popes demanded that even kings obey them, haughtily declaring that "the Roman church has never erred; nor will it err to all eternity." They also dreamed of sending Christian armies to recover the Holy Land from the Muslim Turks.

Peddlers once again trudged Europe's rough dirt roads—people called them "dusty feet"—and traders opened shops in little market towns. Among the peddlers was a Scot named Godric. Probably a younger son of peasants, who had left the farm because there wasn't any land for him, at first Godric made a wretched living searching on the North Sea shore for wreckage thrown up by the waves. In about the year 1100 he found a windfall, goods that had washed up from a shipwreck. He sold them and became a peddler, toting pans and cloth and needles on his back. Later he joined a band of merchants, hauling goods from port to port in a rented ship until he made a fortune. Still later Godric heard the call of God and became a hermit.

As prosperity increased (by inches), merchants in the southern cities such as Genoa, Marseilles, and Venice (see map, page 128) dealt in Asian delicacies that wealthy Europeans paid a lot for. These included pepper, cloves, and sugar; frankincense and myrrh; and sun-dried figs and raisins. They also brought in shop-made goods, among them paper and perfume and costly kinds of cloth that European weavers later on would imitate. These included "damask from Damascus, . . . muslins from Mosul and gauzes from Gaza."*

We must not forget the legs that Europe stood on. In the Middle Ages nine Europeans out of ten worked the land and made life possible for the one who didn't. And yet most medieval writers rarely noticed peasants except to ridicule them. "The devil," they would joke, "didn't want the clods in Hell because they smelled too bad." When English peasants rebelled in 1381 against their landlords and the royal tax collectors, they complained, "We are made men in the likeness of Christ, but you treat us like savage beasts." However,

*Henri Pirenne, *Economic and Social History of Medieval Europe* (n.d.), p. 145.

after roughly 1250 serfdom slowly disappeared. Even then the peasants' lives were hard, but at least they now were free.

THE EUROPE SKETCHED above, still poor and primitive, was the Europe that waged the Crusades, a string of wars against the Muslims in the eastern Mediterranean.

In 1095 Pope Urban II journeyed to a Church meeting in France and urged that Europeans wage a holy war against the Seljuk Turks. They were Muslims and they now held Palestine, which was a Holy Land to Muslims as well as Jews and Christians. It looked as if the Turks might also overcome the Christians in the remnant of the eastern half of the Roman Empire that lay between Europe and Palestine.

"Oh, how shameful," cried Urban, "if a people so despised, degenerate, and enslaved by demons [he meant the Turks] should thus overcome a people [Christians] favored with the trust of almighty God, and shining in the name of Christ! Oh, how many evils will be imputed to you by the Lord Himself if you do not help those who, like you, profess Christianity! . . . Now, let those who until recently lived as plunderers be soldiers in Christ; now, let those who once fought their brothers and relations rightly fight barbarians; now, let those who recently were hired for a few pieces of silver win their eternal reward!"

"It is the will of God!" his hearers roared.

Almost immediately, a fiery Frenchman known as Peter the Hermit, barefoot and unkempt but a hypnotizing speaker, raised a mob of fervent French and German peasants. This wild and holy army swarmed from western Germany through central Europe like a plague of locusts, stealing food and burning houses. A remnant of them got to Asia Minor, where the Turks destroyed them.

After Peter's tragicomedy, Europeans waged more orderly crusades. (But of their eight crusades we shall sketch only the first four.) On the first one, wealthy noblemen led corps of horsemen, wearing armor made of ringlets sewn on leather coats. In Palestine they conquered several little kingdoms, and in 1099 they stormed Jerusalem. "And if you desire to know what was done with the enemy who were found there," they wrote the pope, "know that in Solomon's Porch and in his Temple our men rode in the blood of the Saracens [Turks] up to the knees of their horses."

Later, when it looked as if the Turks and other Muslims might retake the Holy Land, a king of France and a Holy Roman (that is, central European) Emperor led a second crusade. Many of their men were killed en route, however, and the monarchs squabbled with each other. Their siege of Damascus, north of Jerusalem, was a failure, and a Turkish ruler soon retook the Holy Land. Almost fifty years later, therefore, the kings of France and England and the Holy Roman Emperor all brought armies on a third crusade. The results were trivial.

The fourth crusade produced a stunning outcome. Venetian merchants who transported the crusaders on their ships (for a price) managed to detour this holy war. In 1204 the crusaders shocked all Europe when they conquered not Jerusalem but wealthy, Christian Constantinople. (Daring troops had stormed across the up-till-then impregnable wall.) They ruled the city and the land around it for a generation, till a local army won them back.

The other four crusades were small and even less important. All in all, crusades were just a foolish, costly failure. They make the point that Europeans weren't yet capable of dominating others, especially a fighting people like the Turks. But they also show a culture that was greedy, bold, and sure that God was on its side. One day such a people might attempt to dominate the world.

THREE ITALIANS HAD a great adventure that would help to introduce the Europeans to the lands that lay beyond their corner of the earth. It began when brothers Niccolò and Maffeo Polo, merchants of Venice, rode on horseback into southern Russia. They planned to sell some jewels to the region's Mongol ruler. (This was in the heyday of the Mongols.) When they learned that a local war was blocking their return route to the Black Sea, they journeyed 1,500 miles southeast to the oasis town of Bukhara, in what is now Uzbekistan. They were now well inside Asia.

In Bukhara they met a Mongol who invited them to join his caravan, which was headed to China. So they rode 3,000 miles to the east, around or over the lofty Pamir Mountains and along the edges of the largest deserts in the world, until they reached Peking. Few Europeans had ever been there. Kublai, Great Khan of China and the grandson of Genghis Khan, received them well. He suggested that they return to Europe and come back again to China with a hundred learned men.

The Polos did go home, and then, in 1271, set out again for China. Instead of the one hundred men that Kublai had asked for they brought along a mere two priests, and even these took fright and soon went home. They also brought along young Marco Polo, Niccolò's twenty-year-old son. After they arrived in China, the Polo brothers no doubt did some trading. But Marco caught the khan's attention, and he actually entered the service of the Mongols. He spent sixteen or seventeen years traveling back and forth across the country gathering information for the khan. But finally the Polos left for home, this time traveling by sea. En route they stopped in Persia to deliver a seventeen-year-old Mongol princess to the prince she was to marry. After they had entered what today is Turkey robbers seized nearly all their goods.

After his return to Venice, the Italian city of Genoa captured Marco in a war. To pass the time in prison Marco told another prisoner about his travels, and this man helped him to retell his story in a book. *The Travels of Marco Polo* had a great success. He wrote it long before the age of printing in the West, but Europeans copied and recopied Marco's book thousands of times.

The import of the *Travels* lies in what he told the Europeans. China, Marco said, had the greatest ruler in the world, messengers who swiftly carried royal orders everywhere, money that (amazingly) was printed, and vast amounts of silk and cloth of gold (most alluring to the European merchants). Japan, which he had only heard about, was very wealthy, "so that no one could count its riches." India, where the Polos stopped on the homeward voyage, was so hot that even kings wore nothing but loincloths. Its people were devout, and India was rich in diamonds, pearls, and spices (also tempting to a merchant). Everyone obeyed the laws, and traveling merchants safely slept outdoors with sacks of jewels beneath their heads. All in all, Marco Polo gave his fellow Europeans appealing glimpses of a wider world.

BEGINNING IN THE later 1400s Europe looked as if it had a destiny within that wider world. For one thing, Europeans now became the leading makers and sellers in the world. Until this time, Europeans had imported costly goods, like silk and paper, that were made in Asia and North Africa. But as the Middle Ages ended, they discovered how to make their own fine goods, the ones that only wealthy folk could buy. These included paper, perfumed soap, handsome woolens, goblets, mirrors, clocks, and (something new) eyeglasses. They made them for themselves and also for the export trade to Asia, North Africa, and the Middle East.

In the case of paper, Europeans now used river-water power to run machines that shredded rags, thus making lint, which used to be the sole ingredient in paper. (In other places workers shredded cloth by hand.) Because they saved on labor costs, Europeans' paper cost less than their foreign competition's. Other examples of Europe's rising economic role: Christian Venice made lamps of glass for use in mosques in Muslim lands, adorning them with phrases from the Qur'an. And European foundries made the cannons that the (Asian) Turks employed in 1453 to smash the walls of Constantinople.

European merchants thrived and some got very rich, far richer than that Scottish merchant Godric, who had sailed his ship from port to port four centuries before. The biggest of the biggest were the Fuggers, in the south of what today is Germany. In the late 1470s the brothers who controlled the Fugger business found they needed help. The two convinced their youngest brother, Jacob, who had planned to be a monk, to join them. Jacob proved to be a business genius; soon he ran the firm. The Fuggers traded everywhere in velvets, spices, jewels, and guns, and mined for copper, silver, mercury, and gold. Jacob's motto was "I want to profit while I can," and he was known as "Jacob the Rich." He wasn't sure that God would overlook his greed, so in his final years he tried to make up for it by building houses for the poor.

In these same years when businessmen were on the rise, Europeans saw another major change, the making of the printed book. Printed books were something like computers in our times: an invention that transformed the way we humans store and spread our knowledge. Before the 1400s, copyists made books by hand at great expense. A person who desired a copy of a book had first to borrow someone else's copy, and then find a scribe to do the work. The scribe would copy each and every word, writing with a goose quill pen, in lampblack ink, on pages cut from sheepskin. To copy a Bible took about fifteen months.

In the middle 1400s German craftsmen found a better way to make a book. (They probably borrowed the idea partly from the Chinese, who for a thousand years had been printing books with wooden blocks.) A printer first poured molten metal into molds. The result was many little metal sticks, each of which had the shape of a letter on one end. A compositor then set a book in type by picking out the sticks (the letters) that he needed to form words, and grouping them in lines. After he had several dozen lines of type he locked them and set them on a printing press. Another man rubbed ink across the type, laid a piece of paper on it, and then pulled a lever. A mechanism pushed the paper down against the ink-rubbed type, thus printing all those lines of words. Then the workman printed other pages of the book, and a binder sewed them all together.

In the early 1450s Europe may have boasted just one printer, a German named Johannes Gutenberg. He produced what may have been the earliest printed book, a handsome Bible. But only several decades later printers were at work in all the major cities. Since books were so much cheaper now, many Europeans learned to read, and printers rushed to offer them the books they wanted. In the 1470s an Italian printer apologized for his careless work. Other men were printing the same book, he said, so he had to rush his book through the press "faster than you can cook asparagus." Another printer had a placard on his door that said, "Talk of nothing else but business, and dispatch that business fast."

Printers turned out countless sacred books, a lot of easy reading, and a bit of everything else. Some books would shape the future since they made their readers think. These books included tracts of Protestant reformers, sly attacks on kings, medical textbooks, travel diaries, business handbooks, and theories about the earth and sun. Long ago a historian observed, "He who first shortened the labour of copyists by device of *Movable Type* was disbanding hired armies, and cashiering most Kings and Senates, and creating a whole new democratic world."

THESE BUSY DECADES also were the age of what we sometimes call "new monarchs." Kings were trying, more perhaps than rulers ever had before, to dominate their lands. Usually their tactics were to raise their revenues from taxes, use these funds to raise an army, and employ the army to suppress the wealthy nobles. (The wealthy nobles sometimes maintained armies of their own and fought against their kings.) Rulers wanted all power in their hands.

In France in 1422, Charles VII, a boy of seventeen, inherited the throne and the problems that his father, who was often mad, had never solved. A major problem was the noblemen, the castle dwellers who would take no orders from the king. The other problem was the English, who had raided France since early in the 1300s in what was now the "Hundred Years' War." The English ruled great chunks of northern and southwestern France, including the capital, Paris. Charles lacked the money for an army strong enough to drive them out. Most Frenchmen didn't care if he did or didn't.

Charles looked like the last man in France who could solve these problems. He was puny, timid, and gloomy. Because the English held his capital, he lived south of it in the provincial town of Bourges. He was too indolent to drag himself to Reims, where kings of France were always crowned, so the mocking French had named him "king of Bourges."

Then, however, France encountered something new, the force of national pride. In a village in northeastern France, a teenage girl named Joan brooded over "the pity that was in the Kingdom of France." She heard the voices of saints telling her that God had chosen her to help her king win back his land. She went to the court of Charles, and she persuaded this pathetic man to let her join his little army. Amazingly, with utter confidence that God was on their side, Joan inspired the troops and got them fighting.

While Joan was fighting north of Paris, though, the English captured her. Despite what she had done for him, Charles made no attempt to save her, and a Church court tried her as a religious heretic. The judges were Frenchmen who backed the English, and they had no trouble deciding that Joan was the tool of the Devil. The English and their French collaborators burned her at the stake. A secretary to the English king cried, "We are lost! We burned a saint!"

Surprisingly, Charles finally pulled himself together. He called meetings of the French "estates" (churchmen, nobles, and well-off commoners) and got them to permit him to collect a special wartime tax. (This tax was later called the *taille,* or "cut.") He spent most of the resulting income building up his army, and he used the army to repel the English. By 1453 the French had won their country back, and ended the long war.

Charles and the kings who followed him always needed money, but they knew that the estates would not allow a tax in peacetime. So they ceased requesting their permission to collect the taille; they simply did it. Because they had the troops that they had hired with the money they had earlier collected, the kings could stifle any protests. Their income from the taille became a major source of kingly power and one reason why France, for centuries, was Europe's greatest power.

After Charles had died, his able son Louis XI spent his revenues on soldiers, bribes, and smart advisers, and he used them to domesticate the nobles. He could be very cruel. For years he kept some enemies shut up in iron cages that a bishop had designed for him. (Louis later held that bishop in one of his cages for fourteen years.) The king's immense ambition was to overcome his mortal foe, the wealthy duke of Burgundy. France became a single country when, in 1477, soldiers in the pay of Louis crushed the army of the duke. Two days later, searchers found the body of the duke, stripped by looters. His head was split from scalp to chin.

While Charles and Louis won control of France, England seemed to fall apart. Mighty nobles fought each another for the throne, or at least to rule the man upon the throne. Near the finish of these civil wars, Richard, duke of Gloucester, seized the throne from his twelve-year-old nephew (and may have had him murdered). But then a wealthy noble, Henry Tudor, who had a shaky claim to the throne, raised an army and in 1485 defeated Richard (who died in battle, with the crown of England lying on a nearby bush). Then Parliament, the English legislature, made Henry king.

One by one, the king got rid of rivals. He exiled or beheaded some survivors of the wars, obliged the others to disband their private armies, and seized his enemies' estates. The sister of the late boy king had a better title to the throne than Henry, but Henry married her. That took care of that, but hostile claimants to the kingship were another matter. Among them was a boy named Lambert Simnel, who pretended to be an earl with a title to the throne. Some nobles rallied to his banner, but King Henry crushed the rebels, and he made the boy a scullery worker in his kitchen. Another boy pretended to be the younger brother of the late (perhaps murdered) boy king. Backed by several European kings he thrice invaded England, but Henry caught and hanged him.

Henry made finance his special project. He closely checked the royal ledgers, initialing every page with "H" as he approved it. With his own great fortune, and the lands he took from enemies, and the taxes he persuaded Parliament to approve, he made himself ten times as rich as England's richest lord.

In Spain the story of the rise of royal power differs for this simple reason: until the end of the Middle Ages no country known as

"Spain" existed. Three small kingdoms nearly filled the Iberian Peninsula: Portugal on the west, Castile in the center, and Aragon on the east. (A little Moorish, Muslim, kingdom held the south.) But in 1469, Isabella, who was going to inherit Castile from her half-brother King Henry the Impotent, married her cousin Ferdinand, who was going to inherit Aragon. When she later inherited Castile and he Aragon, they in effect united the two kingdoms to form one country, known by the ancient name of the region, Spain.

That was the easy part. The bigger problem was the usual one: making everybody stay in line. So Ferdinand and Isabella did the things that other kings were doing: collecting taxes, choosing loyal governors, and smashing castles of unruly nobles with cannons. They also ordered towns to hire squads of archers and shoot down bandits in the hills.

With "unbelievers" the king and queen were harsh. In 1492 they conquered the remaining Muslims in the south of Spain, and drove them out. In the same triumphant year the two expelled the Spanish Jews, robbing them of most of what they owned. They turned their temples into churches and their graveyards into pastures. The "Catholic monarchs" (as everybody called them) were certain that by ousting Jews and Muslims they were pleasing God. They may have also felt, if they didn't put it into words, that by ridding Spain of other faiths they made their subjects one united people.

BY THE FINAL decade of the 1400s western Europe's monarchs had their homes in order. Now they could afford to add a room or two. We shall tell this story (briefly), because it explains so much about the Europeans' fighting skills, and the role they were about to play in world affairs. In 1494 King Charles VIII of France (the son of Louis XI) resolved to seize the realm of Naples in the south of Italy. He had inherited a doubtful claim to Naples, and he claimed to want it as a base from which to wage a war against the Muslim Turks. What he really craved was land and glory. But for France to conquer Naples, even if it could, would be a foolish move, since the two were far apart. With little trouble Charles's army marched through Italy to Naples and defeated it. But other countries started to oppose him and, just then, his troops began to melt away, ravaged by disease. So the king and army marched back home, bearing loot and syphilis. They slithered past an allied army that almost trapped them in a mountain pass.

Charles's failure should have taught his fellow kings to ponder well before they started foreign escapades. But of course it didn't. They felt the lure not just of Naples but of all Italian cities. Italy was not an infant nation, as theirs were, but a group of city-states (like ancient Greece) that were temptingly rich and weak. So the wars that Charles began continued, with European countries fighting one another over Naples, Florence, Venice, and Milan. The fighting over Italy ended in 1529, with Spain in possession of Naples and Milan. But these Italian wars were followed by a general European one. European warring would continue, though with pauses, until 1945.

War was both a chronic sickness and a force in shaping Europe's domineering ways. Because they fought so much, the Europeans got quite good at killing. Already by the time of the Italian wars, the days of men in armor clanking down the battlefield on horses were long gone. Men, meaning gentlemen, still fought on horses, but the infantry were more important. The soldiers most admired were the hardy Spanish swordsmen and the dreaded Swiss, who fought in clusters holding long, sharp pikes, like porcupines.

It was guns, however, that began to star on battlefields. As we said in chapter 5, the Chinese long before had taught the world what happens when you light a mix of charcoal, sulfur, and saltpeter. When Europeans learned about gunpowder, they used it first to fire big balls of stone or iron from heavy cannons. These guns were hard to move, and they might explode and kill their crews, but they battered down the walls of castles.

By the early 1500s, gunsmiths turned out lighter cannons that were easier to haul from place to place. Gunners fired them now not just at castles but also at opposing armies. Captains fastened them to ships and used them to blast their way into enemies' harbors. At about the same time, gunsmiths also started making iron tubes, carried on the shoulder, that used gunpowder to shoot bullets made of lead. With one of these little portable cannons (or early muskets) a soldier could kill an enemy from two hundred yards away.

Armies of the time were pitiless. Here (condensed) is how a well-informed Italian pictured Charles VIII's attack against a fortress: "This was a strong position, well-supplied and manned. But the French bombarded it for a few hours and then assaulted it with such ferocity that they took it by storm that very same day. Then, because of their natural fury and in order to discourage other places from resisting by this example, they slaughtered great numbers. After having committed every other possible kind of atrocity, they set the buildings on fire.

"This manner of warfare filled the whole kingdom [of Naples] with terror; because in victories [before this], however achieved, the cruelty of the victors usually did not go beyond disarming the defeated soldiers and setting them free, sacking towns taken by force and taking prisoner the inhabitants until they paid a ransom, but always sparing the lives of those who had not been killed in the heat of battle."*

TAXES, WEAPONS, LIES, and slaughter—in the early 1500s these unpleasant subjects filled the thoughts of Niccolò Machiavelli. This famous student of politics grew up in Florence (see map, page 128) while the able Medici family ruled the city (and while two other gifted Florentines, Leonardo da Vinci and Michelangelo Buonarroti, were learning to paint and sculpt). After an unusual Medici (a bungler) was driven from the city, Machiavelli served an anti-Medici government for fourteen years as Florence's "secretary." He had a part in almost all that Florence did. He helped run a war against the nearby town of Pisa, and he went on missions to a king of France and to a

*Francesco Guicciardini, *History of Italy,* ed. John R. Hale, trans. Cecil Grayson (1964), pp. 185–86.

Holy Roman Emperor. He learned a lot while representing Florence at the camp of Cesare Borgia, the ruthless son of the pope, while Cesare waged a war in central Italy. He witnessed Cesare's pleasure on the day he caught and caged some men who had betrayed him, and strangled several others.

In 1512 the Medici returned to power and Machiavelli lost his job. They suspected him of plotting, and had him tortured on the rack, but when they found no evidence against him they released him. Just to play it safe he lived outside the city for a while. There he quickly wrote two books, a long one about ancient Rome, which no one reads today, and a short, electrifying one called *The Prince*.

At heart he liked republics, governments like that of ancient Rome before the emperors. In other words, he favored rule by men whom the well-off and well educated chose. But, as he once explained, he had witnessed Italy overrun by one French king, plundered by another, torn apart by Spain, and humiliated by Swiss pikemen. A republican government, he concluded, was not the answer for Italy. Only a hard-hitting despot could unite and save it, and Machiavelli described this man. He would be as cruel as Cesare Borgia, and would lie like Cesare's father, Pope Alexander VI. (No one, Machiavelli wrote, "pledged his word more solemnly, or kept it less" than the pope.) And like King Ferdinand of Spain, Machiavelli's prince would talk "of nothing but peace and honor," but do the things he had to.

Machiavelli wasn't just a cynic. He believed that a good prince, once safely in power, could bring about a good society. But idealism is not what endures of Machiavelli. What he taught his fellow Europeans is that an able ruler focuses not on good ends but on harsh means. When Machiavelli died in 1527, the Italian wars were near the end, and it was clear that Italy was not to have his kind of prince. Half of Italy was soon to fall to Spain, and the rest would stay divided and count for little.

What a shame that Machiavelli, the hard-boiled Italian, never had the opportunity to talk with Thomas More, the high-minded Englishman! Both of them saw very well the strengths and flaws of Europe's infant nations. More's career was far more brilliant than

Machiavelli's. While More was still a boy, a chancellor of England rightly predicted that he would "prove a notable and rare man." By his early twenties he was already a thriving lawyer, and at twenty-six he had a seat in Parliament. Here he argued so effectively against a bill proposed by Henry VII that the ruler fined and jailed More's father, to teach the son to watch his words.

After Henry died, young Henry VIII appointed More a councillor and ambassador to France, and later he made him his lord chancellor. The chancellor and king were friends, but More was wise regarding royal friendships. Once he told his son-in-law: "I believe he [Henry] doth as singularly favor me as any subject within this realm. Howbeit, . . . if my head would win him a castle in France it should not fail to go."

Three years after Machiavelli wrote *The Prince,* More produced a little book he named *Utopia.* Apparently he wrote it mainly to amuse himself and certain friends. As the book begins, More tells about a conversation he supposedly had had with a weather-beaten traveler. This man, a Portuguese, tells him that he once had visited an island named Utopia. (Utopia is a Greek word meaning "nowhere.") The extraordinary people of this place had done away with all the ills that troubled Europe.

One such problem was the foolishness of kings, who dreamed of military glory and whose advisers were all yes-men. The traveler asks More to imagine a king of France who is determined to seize Italy and also expand his northern borders. Suppose, he says, someone told the king that he would only wreck his kingdom with these wars, and he should "concentrate on the kingdom that his ancestors handed down to him, and make it as beautiful and prosperous as he can . . . love his own subjects and deserve their love . . . live among them and govern them kindly, and . . . give up all ideas of territorial expansion, because he has more than enough to deal with already."* Of course the king would *not* appreciate advice like this; it might be wiser not to offer it.

*Thomas More, *Utopia,* trans. Paul Turner (London: 1988), p. 59.

Utopia, the traveler tells More, had no kings. Instead, an elected mayor and elected councillors made the big decisions. And the Utopians rarely went to war. They hated war and considered it subhuman, "although human beings are more addicted to it than any of the lower animals." Utopians would fight only to defend themselves or to help the victims of a tyrant. If they absolutely had to wage a war, they paid their neighbors, the Venalians (that is, venal), to do their fighting for them. (More had in mind the Swiss, whose soldiers often fought for pay for other countries.) Utopian councillors spent their time on the essentials: water, sanitation, health care, and education.

The Utopians practiced communism. Everything was owned in common; everybody wore the same kind of clothes; and everyone exchanged houses every ten years. No landlords and moneylenders lived off the labor of others. Everybody shared the chores, but prisoners (the few there were) did the hard and dirty work. No Utopian was so silly as to wear jewelry, and they might use gold to make their chamber pots. They had no lawyers, since (wrote More, himself a lawyer), "they consider them a sort of people whose profession it is to disguise matters." Men and women who wished to marry were shown to each other naked, so that both would know what they were getting.

What gives *Utopia* its bite is not the answers More supplies but doubts he raises. Unlike Machiavelli, he wasn't dazzled by the new monarchs. He knew the Henries, father and son, too well for that. He knew the dangers that arose from having too much power in one man's hands, and he knew that countries ought to focus not on war but on matters that matter.

More himself became a victim of a king. In 1535 Henry VIII, who was then taking over the Catholic Church in England, ordered that his subjects take an oath recognizing Henry's second marriage. More refused to take the oath. Doing so, he said, would mean agreeing not only that Henry (not the pope) was head of the Church, but also that the state was the be-all and end-all of our lives. When he would not do as Henry had commanded, he was tried for treason, convicted, and beheaded. His head was impaled on a spike on London Bridge.

Machiavelli had described the often brutal nations that were taking shape, and he hoped that Italy would be one of them. In contrast, More had warned of dangers that the crystallizing nations raised. Quite predictably, Europeans continued on the path *The Prince* had charted, and few of them ever dreamed about a high road to *Utopia*. It was the tough and grasping Machiavellians—the merchants, the monarchs, and the military—who were about to explore and exploit the world and seize as much of it as they could.

CHAPTER 9

We find each other.

UNTIL FIVE HUNDRED YEARS ago human beings weren't spread throughout the world, like jam on bread. Not at all. We lived in scattered clusters, like the hives of bees a farmer sets along the border of a field. These human clusters mostly lay in one broad band that ran from Europe and North Africa through the Middle East, and on to India and Indonesia, and then China and Japan. Smaller clusters lay along the American highlands, from Mexico to Chile. The rest of the earth held only scatterings of gatherers and part-time farmers.

Europeans, North Africans, and western Asians knew of one another, and to some extent had traded with and sometimes fought each other. Thanks to travelers like Marco Polo, Europeans dimly knew about the Chinese, at the other end of Eurasia. But in general contacts between the human clusters were few or none at all. Often peoples of one continent had never heard, never even dreamed of, other continents and the people living on them.

Now, all of this would change. We humans would discover one another. Europeans, for the most part, would do the finding, and do it fairly quickly. They started at about the time when other Europeans

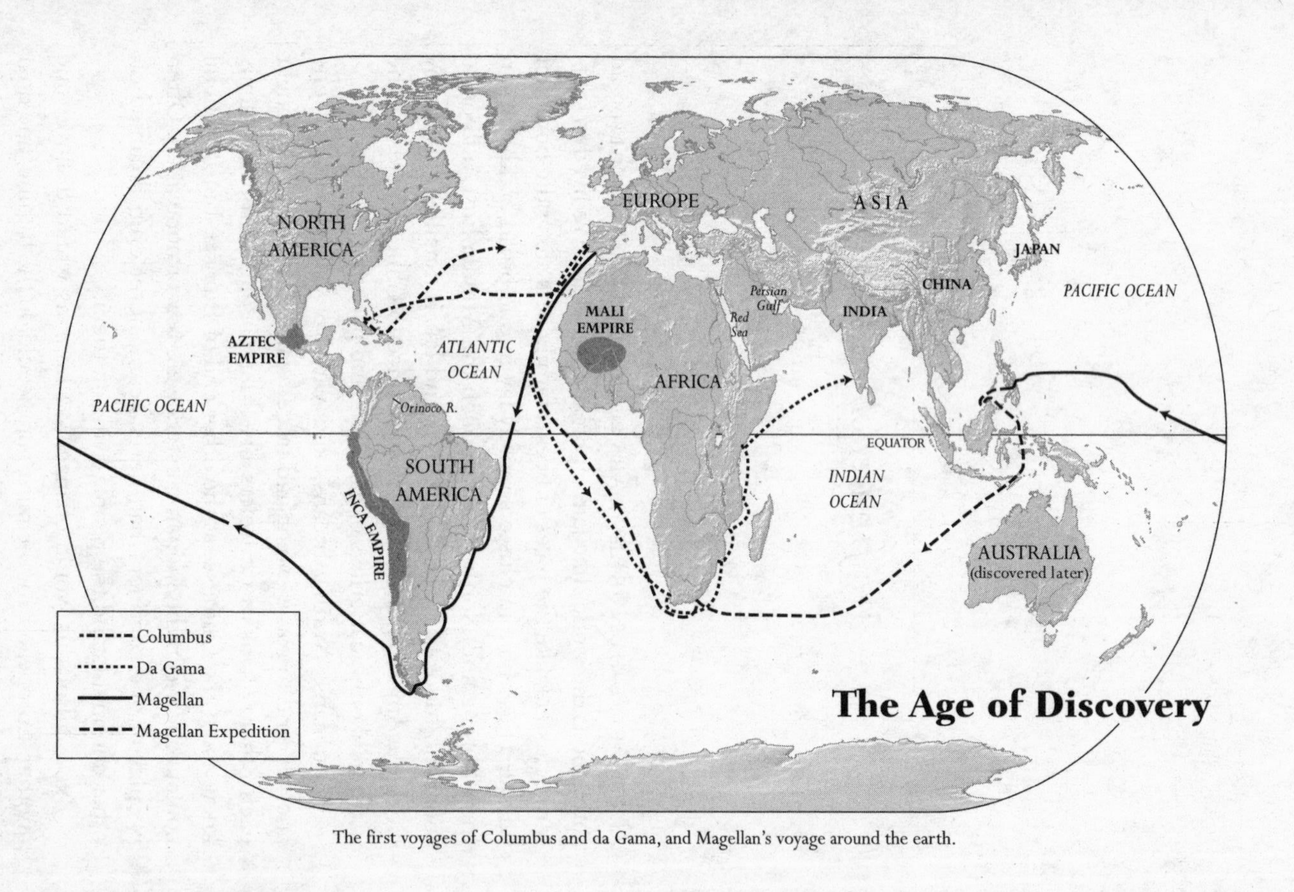

The first voyages of Columbus and da Gama, and Magellan's voyage around the earth.

started printing books and making war with cannons, and they finished about three centuries later.

IT WAS SMALL and backward Portugal that led the way. The man who launched this country's effort was the vigorous Prince Henry, a younger son of King John I. At age nineteen, Henry played a part in Portugal's conquest of Ceuta, a Muslim town in Africa, not far from the Strait of Gibraltar. The Portuguese found quantities of gold there, and this was most alluring. For a long time Europeans had heard about the wealth in gold of kings in northwest Africa. Now Henry learned that caravans of North Africans often trekked far south across the Sahara Desert to the Senegal River, and bartered for gold with the men who dug and panned it.

Intrigued, the prince decided to explore the northwest coast of Africa, where few if any Europeans had set foot. He had his reasons. Desire for gold was one, of course, and perhaps he would discover new exotic goods for Portuguese to trade in. Possibly he'd even find the realm of Prester John, a legendary Christian king. For ages Europeans had fantasized about this rich and fiercely anti-Muslim king, who was thought to govern somewhere in Africa or Asia. If Henry found King John, or John's descendants, he or they might join the European struggle with the Muslims.

In 1419 Henry left his father's court at Lisbon and moved to Cape St. Vincent, the rocky southwest tip of Portugal and, for that matter, of Europe. Here he gathered men to help him carry out the search he had in mind. They were men of both theory and practice, and included geographers, astronomers, sea captains, navigators, compass makers, and shipbuilders. They came from all the European coastal countries, and were later joined by tribesmen from Africa's west coast.

Then the prince sent out explorers. Year after year they sailed due south to Africa's northwest coast, then inched their way south along the shoreline, each explorer venturing farther than the last. For a while Cape Bojador, which juts from south Morocco, blocked them because tradition held that just below it lurked great danger—the seas were said to boil, the air to be too hot for life. Fifteen of Henry's expeditions set out to round the cape but lost their nerve. Finally, a captain doubled it, and found clear skies and placid water.

Henry's captains went on making small advances. They sailed along the coastline of the desert, seeing only nomads, and below the desert they followed the coasts of bush lands. Later, other captains sailed past steamy jungles, glimpsing now the villages of blacks. When they anchored off the shore, they bartered with the Africans for gold. They also asked for news of Prester John but, not surprisingly, they didn't find him. For Henry's pleasure, captains brought him back not only gold and maps but ostrich eggs, dried and salted elephant (which Henry managed to chew and swallow), and a dozen Africans to be stared at by the Portuguese, and touched, and then returned to their homes. Soon the Portuguese began to trade in human beings; they started by bringing two hundred Africans to Europe to be sold as slaves.

Henry died in 1460 after sending expeditions south for more than forty years. By the time he died his ships had reached the point just above the equator where Africa's coastline bends and runs due east a thousand miles.

The king of Portugal granted a businessman, Fernaõ Gomes, a monopoly of African trading on condition that he also send his ships exploring past the point that Henry's captains had already reached. Gomes's captains sailed eastward along the "Grain" (or Pepper) Coast, the "Ivory" Coast, the "Gold" Coast, and the "Slave" Coast. When they reached the corner where the coastline bends and once again runs south they turned to starboard and explored southward almost to the mouth of the Congo River. They brought home slaves, ivory, pepper, ebony, and gold.

In the 1480s, Portugal's King John II took charge of the explorations, but he changed their goal. Africa was very well, said John, but what he really wanted was to find a water route to "the Indies." Wealthy Europeans loved the precious goods that came from south and eastern Asia, the far-off region Europeans knew of vaguely as "the Indies." The wealthy paid a lot for rubies, emeralds, silks, and perfumes, and for spices such as pepper, cloves, and cinnamon. These costly goods had always traveled west on land and water routes through Asia, the Middle East, and the Mediterranean.

As John knew well, this business in imported goods made money for Italians and others but not for Portuguese. The shape of much of Africa was still unknown, but John believed that ships could sail around it, and in this way reach the Indies. If they found this ocean route, then, given its location at Europe's southwest corner, Portugal would be ideally placed to use it. So John's explorers stopped their cautious, step-by-step advances and started moving in giant strides down the southwest coast of Africa.

In 1487, Bartolomeu Dias was told to pass the farthest point yet reached and, if it could be done, to sail around the tip of Africa. When his ships had sailed well down the coast of Africa, a tempest blew them even farther south than the continent. After he had sailed back up to Africa, Dias discovered that the coastline now was bending north and east! Clearly he had sailed around the bottom of the continent. He wanted to continue, but his men had had enough, so he turned around and started home. This time he saw the cape that he had earlier sailed around without seeing it, and he named it Cape of Storms. Later, King John renamed it Cape of Good Hope "for the promise it gave of finding India, so desired and for so many years sought after."

And now, on to India. To lead an expedition, John's successor, Manuel the Lucky, chose a Portuguese aristocrat. Vasco da Gama was an expert seaman and tough enough for any venture. (On one occasion he would chop off the hands, ears, and noses of Muslim traders and fishermen and send them to the hostile king of Calicut, advising him to cook them as a curry.) He sailed from Portugal in 1497 with four ships that had been designed expressly for sailing in the South Atlantic's stormy waters. He sailed around the cape and up the eastern coast. (See the map on page 148.) He was now in the Indian Ocean, where European ships had probably never been. When he reached the coast of what today is Kenya, the sultan of a seaside town permitted him to hire a pilot, and this man guided da Gama eastward to the Indian port of Calicut. (Arabs had been trading on this route for a thousand years.)

Local merchants greeted the exotic strangers with the words "May the Devil take you! What brought you here?" To which the Europeans answered: "We came in search of Christians [Prester John] and spices." The king of Calicut, reclining underneath a gilded canopy and spitting betel juice into a gold spittoon, also asked why they had come. He was insulted that da Gama hadn't brought him costly gifts, rather than some strings of coral, lumps of sugar, two barrels of rancid butter, and the common cloth the Portuguese were used to trading with in Africa. The local merchants scorned da Gama's trading goods and didn't want to barter for them. Only with much trouble did da Gama buy some cinnamon and ginger, and then he left for home.

The two-year expedition almost seemed a failure. Not only were the cargoes he came home with small, but da Gama had lost two of his four ships. His men had wrecked one on the outward voyage, and they burned the other on the homeward one. (Scurvy, a disease resulting from a lack of fruits and vegetables, had killed so many sailors that da Gama couldn't man the ship.) Of 170 men who had sailed to India, 55 returned. But so valuable were spices that his cargoes (the cinnamon and ginger) sold for sixty times the expedition's cost.

Knowing how to sail to India was not enough. The next task for the Portuguese was to conquer land in Africa and Asia. They didn't want to conquer kingdoms, which would have been too big a task. What they needed were some far-flung bits of land along the route. Some of these must be in Africa, so ships could stop and load supplies, and the others had to be in Asia, in the "Indies," and would serve as ports to trade in and naval bases for preventing trade by other countries.

Planting bases on the sparsely peopled coast of Africa proved fairly easy, but doing it in Asia was another matter. Here they had to fight. A grizzled Portuguese soldier, Afonso de Albuquerque, sailed there with a fleet. His ships were armed with cannons, and they blew the Arabs from the seas.

Albuquerque conquered Goa, off the Indian west coast, and made this island Portugal's headquarters. To the west of India he captured Muscat and Hormuz, a town and island that together controlled the entrance to the Persian Gulf. By holding them, the Portuguese could block one major sea-land-sea trade route to Europe. Albuquerque failed, however, when he tried to conquer Aden, at the Red Sea's southern entrance. As a result, the Portuguese would never fully block other countries' merchants from sailing up the Red Sea toward the Mediterranean. On the other side of India, to the east, Albuquerque took Malacca, on the Malay Peninsula. From here the Portuguese controlled a strait through which ships had to sail to reach the spice islands in eastern Indonesia, or to go on to China and Japan.

Despite its mighty efforts, Portugal reaped rewards for only half a century. Then its fortunes faded. The little country lacked resources, and its people numbered fewer than a million. They simply couldn't build and man sufficient ships to do the two essential tasks: to journey back and forth to Asia, and to fight off ships of other lands. Later, other countries muscled in and used the route to Asia that the Portuguese had found. The little country's moment in the sun was past.

But we must say this about the Portuguese: they played a major role in charting the earth and linking its human clusters. Their poet Luís Vaz de Camões would write admiringly about his countrymen, "If there had been more world, they would have found it."

IN 1476 A redheaded young Italian was a crewman in a convoy sailing up the coast of Portugal. He was the son of a weaver and had been a pirate several years before. His name was Cristoforo Colombo, and he later called himself Cristóbal Colón, but the world remembers him by his name in Latin, Columbus. When the convoy was just off Portugal, his ship took fire. Columbus leaped into the ocean and seized an oar floating near him. Using it to hold himself up, he kicked his way to shore close to the place from which Prince Henry had run his explorations. Local people dried and fed him, and helped him join his younger brother, Bartolomeo, who was then in Lisbon.

At this time the Portuguese were still exploring Africa, and the brothers made their livings making charts that showed the latest findings. Christopher also sailed on voyages, including one that ventured northward to the Arctic Circle. It wasn't long before he caught the exploration fever.

Soon he reached a remarkable conclusion: the shortest water route to Asia was not the around-Africa path the Portuguese were then exploring. (See the map, page 148) Instead, he thought, the shortest route was west, across the Atlantic Ocean. (In fact, he thought, this might turn out to be the only water route to the Indies. At this point, the Portuguese hadn't proved that one could sail around Africa.) Columbus was assuming, by the way, what every educated European knew: that the earth is round.

He was wrong, of course, in thinking that the shortest ocean route to Asia lay westward, across the Atlantic. He didn't know that two big continents sprawled right across that route. Also, he believed that Asia is bigger than it is and earth is smaller. He therefore pictured a bigger Asia stretching way around a smaller earth, leaving room for only a narrow, easily crossed Atlantic Ocean. That slender ocean, so he thought, was all that lay between Europe and Africa on one side of it and Asia on the other. He reckoned that the distance from Africa to Japan was roughly 2,500 nautical miles. In fact, it's more than four times that long.

Columbus asked the king of Portugal to fund an exploration voyage across the Atlantic, but the king turned him down. Columbus then turned to Ferdinand and Isabella, king and queen of Spain. A year went by before they saw him, but after they did they named a committee of "learned men and mariners" to advise them. The committee tried for four years to get clear information from Columbus, and then they advised against the project.

The queen, however, leaned in favor of it. She liked Columbus, so it seems, and realized that an easy route to Asia might be profitable for Spain. It also might provide a chance to save the souls of millions of benighted pagans. That opportunity appealed to Isabella and Columbus, who were both intensely pious. The queen decided to back the voyage.

Columbus's expedition left from Spain in three small ships in August 1492. (On the same tide other ships were hauling off from Spain the last of several thousand Jews whom Ferdinand and Isabella were driving out. Expulsion was their way of thanking God for victory over the remaining Muslims in the south of Spain.)

Columbus and his crewmen journeyed down the coast of Africa to the Canary Islands. They stayed there for a month and then headed west, across the Atlantic. The voyage was easy, since a trade wind pushed them steadily along, and the weather, Columbus wrote, was "like April in Andalusia. . . . All that wanted was to hear the song of the nightingale." True, his sailors grumbled some. Some feared to go so far in unknown waters and worried that they'd never find the west wind that they'd need to blow them back to Spain. Columbus lied a little, telling them that they were sailing shorter distances each day than he knew they were.

Early in October they were cheered by seeing land birds flying southwest. They peered at the horizon, every sailor hoping he would win the cash reward promised to the first who spotted land. On October 12, before the dawn, a lookout saw what seemed to be a cliff gleaming in the moonlight. He yelled, "*Tierra! Tierra!*" After sailing little more than a month, Columbus and his men had crossed the ocean.

What they came to was a coral island in the Caribbean Sea, east of the isthmus linking North and South America. As the ships approached the land, the sailors saw a sparkling shoreline, lofty trees, and nearly naked people. The islanders feared the monsters nearing shore, and they vanished in the forest. Columbus and his captains went ashore, and there they kneeled and thanked God for their safe arrival, and Columbus formally named the isle San Salvador (Holy Savior).

The islanders returned, and timidly they offered gifts. Columbus, thinking he was in the "Indies," called them Indians, and he wrote that they were "poor in everything." However, some were wearing little golden pendants hanging from their noses. The Spaniards focused keenly on these pendants, and were sorry to learn that the gold had come from some other island. The Indians prepared them food that Europeans hadn't ever tasted, such as maize (or corn) and cassava bread. (The roots of cassava plants are a staple food in many tropical countries.) Columbus noted "how easy it would be to convert these people—and to make them work for us." The natives, meanwhile, must have viewed the dirty, woman-hungry, white intruders with distaste and fear.

After he had explored San Salvador for a couple of days, Columbus headed south. He came upon a larger island, which was really Cuba, but he took it to be part of China, the kingdom of the Gran Khan (Great Khan), whom he had read about in Marco Polo's *Travels.* He misunderstood an Indian word (*Cubanacan*) to mean "El Gran Khan," and tried in vain to find the ruler. The Spaniards also met some Indians who were carrying, Columbus wrote, "a firebrand in the hand, and herbs to drink the smoke thereof." The herbs were huge cigars, which they passed around so each could take a drag. The sailors, meanwhile, found many of the women friendly and obliging.

From Cuba Columbus journeyed east, and came upon the northern coast of Hispaniola (now Haiti and the Dominican Republic). From its position, east of what he thought was China, he decided this must be Japan. Marco Polo had recorded that Japan was rich in gold, and, sure enough, the Spaniards found a little gold in the riverbeds. This matter clearly needed following up, so when one of his ships was wrecked, Columbus founded a colony with the surplus men. Leaving them to hunt for gold, he sailed back to Spain.

When he reached the Spanish court, Columbus was an instant wonder. He claimed, wrongly but sincerely, to have reached "the Indies" (specifically China and Japan) and found them rich in spices, cotton, timber, gold, and slaves. Those whom he convinced overlooked a crucial fact. His only proof of New World wealth was golden baubles and some shrubs and roots he wrongly thought were spices. He had also brought some Indians, dressed in plumes and holding parrots, and they intrigued the Spaniards. No one guessed that what Columbus had explored was not the wealthy Indies but some islands on the fringe of two as yet undreamed-of continents. Two impressive civilizations flourished on those continents, but Columbus wouldn't live to see them.

He would make three other voyages to the New World, but all were failures since he never found an Asian sea route, though he thought he had. A host of things went wrong. One time sea worms tunneled so many holes in his ships that for a year Columbus and his men were stranded on an island. Another time a Spanish enemy seized him and sent him back to Spain in chains. He found no gold that was worth the effort of collecting. The Indians on Hispaniola, goaded by the Spaniards' cruelty, murdered nearly all the men he left there. When he shipped some Indians back to Spain, they proved to be no use as laborers. He suffered badly from arthritis and at one point from malaria. Worst of all, the Spanish court began to doubt that he'd found anything of value. The queen stood by him, but her husband found Columbus and his fruitless trips a bore.

On all his voyages, however, Columbus kept on adding to the Old World's knowledge of the New (and the New World's tragic knowledge of the Old). Voyage number three produced a startling find. Columbus headed farther south than he had done before, and he came upon an island and a fair-sized bay beyond it. From the land beyond the bay, a giant river (the Orinoco) flowed so strongly that the water in the bay was wholly fresh.

For such a mighty river Columbus could only imagine one explanation: it issued from a mighty land. "I believe this is a very great continent, unknown until today," Columbus wrote in his journal. "[And] if this be a continent, it is a marvelous thing, and will be so among all the wise, since so great a river flows that it makes a fresh-water sea of forty-eight leagues." (A league equals three miles.) He was right. He had found the continent that would later be named South America.

AFTER COLUMBUS FOUND the New World, others started to explore it, and other finds resulted. Only five years after Columbus's first voyage, John Cabot headed west from England in a ship so small that eighteen men could sail it. Although Italian (his real name was Caboto) he was sailing for the king of England, Henry VII. The king had authorized a voyage to "all parts, countries and seas . . . unknown to Christians." Cabot probably was looking for a northern route to the Indies, around the lands Columbus had discovered. He made a landing on the eastern coast of Canada, or on Newfoundland Island, and noticed signs that humans lived there. No doubt the local Indians had seen the pale-skinned strangers and the monster that they lived in, and stayed away.

Back in England, Cabot told King Henry he had found a part of Asia, and the thrifty king rewarded him with ten pounds. Cabot made another journey west, this time with four ships, but the expedition vanished and was never seen again. In time the experts realized that on his first voyage Cabot had found another continent, North America.

Not long after Cabot's second voyage, Amerigo Vespucci, an Italian ship captain and geographer, voyaged down the coast of the continent that Columbus had discovered. Later, someone published forged accounts of Amerigo's travels, based partly on his letters, and these were widely read in Europe. As a result, a German scholar honored Amerigo (and slighted Columbus) by naming the newfound southern continent for him: America. Later it was renamed South America, and the other continent, the one that Cabot found, became North America.

Now the question rose, what lay beyond these continents? A Spaniard, Vasco Nuñez de Balboa, found the answer. Spain had planted a colony in what we know today as Central America, the corridor of land that links the two Americas, and Balboa was its governor. An Indian told him that this land, which the Spanish hadn't yet explored, was just a slender isthmus, and beyond it lay an ocean. Guided by some Indians, Balboa led an expedition through the swamps and jungles and across the mountains. Twice they had to fight their way past hostile Indians. Balboa climbed the final hill alone, and when he reached the top he saw, still far away, a vast expanse of water. He called his men, and they knelt, gave thanks, and built an altar out of stones. They trekked down from the mountains to the shore of the Pacific, where Balboa ceremonially claimed for Spain "all that sea and the countries bordering on it." Like many a discoverer, however, Balboa met a dismal end. Six years after he had crossed the isthmus, the governor who succeeded him tried Balboa on a phony charge of treason and beheaded him.

IT NOW WAS clear that the discoverers (including some we haven't mentioned) had found two continents that were linked by an isthmus and were not the Indies. Beyond these continents and isthmus was more water but perhaps, the Europeans thought, not too much of it. Beyond that water, so they hoped, lay the longed-for Indies. To reach them, people wondered, could one sail far north, and pass around the northern continent? Considering the fog and ice and bitter cold, that seemed unlikely. Somewhere up there Cabot's ships had vanished. It might be wiser, so they thought, to journey south and hope to get around the southern continent, or better yet, find a strait that led right through it.

A Portuguese named Ferdinand Magellan (Fernão Magalhães) was certain that the southwest route to the Indies was feasible. A battle-seasoned officer, who limped from an old wound, Magellan had fought for Portugal in its African-Asian empire. So he was familiar with the far-off "Indies" that he wanted to sail to by a western, not an eastern, route.

The expedition needed royal backing, so Magellan went to see the teenaged king of Spain, Ferdinand and Isabella's grandson. He told King Charles that he could find a route to the Indies around the southern continent and across the ocean that Balboa had already claimed for Spain. Earlier, at the request of Portugal and Spain, the pope had drawn an imaginary line around the earth, so that each country could claim whatever lands it found within its half. They had stuck to this agreement, although often they were hostile to each other.

Magellan told the king that the Spice Islands in Indonesia—or some of them at least—lay within the Spanish half of earth, not the Portuguese one. A short, safe route to the Spice Islands (shorter and safer, so he claimed, than the Portuguese route around Africa) would clearly be a boon to Spain. He won the young king's backing for an expedition, just as Columbus, a quarter century before, had won the backing of Charles's grandmother, Isabella. The Portuguese tried to stop the expedition but they failed.

Magellan and a fleet of five ships sailed from Spain in 1519. (The route is indicated on the map, page 148.) They sailed to Africa and part way down its coast, and then they crossed the South Atlantic Ocean and began a journey down the coast of South America. The trip was full of wonders. In Brazil they marveled at the ways of Indians, so different from anything they knew. Either they went naked and painted their bodies or they dressed fantastically in parrot feathers. They ate pineapples and sweet potatoes, slept in hammocks, and hacked canoes from logs, using axes made of stone. When the Europeans anchored off their shores, Indians swarmed aboard the ships and pilfered what they wanted. The startled Spaniards saw a woman coolly seize a nail a little larger than a finger and stash it in her vagina. She jumped the rail and swam to shore.

When the five ships reached what now is Argentina, far below the Equator, the air was cooler. They traveled up the River Plata, hoping it would prove to be a strait that reached the sea Balboa had discovered. No such luck. They journeyed farther south and reached the cold and treeless coast of Patagonia. Fearing that it might be even colder farther south, they spent the winter here.

Troubles rose. Magellan's officers were mostly Spanish, and some of them now mutinied against Magellan, in part because he was a Portuguese. Although they took him by surprise, Magellan quickly won control. He pardoned all the mutineers but three, beheaded one of these, and marooned the others in this dismal place. But mutiny was not his only problem. While reconnoitering, the vessel *Santiago* hit a reef and broke apart.

When the winter ended, the four remaining ships sailed on. They reached a waterway that seemed to be a strait, and Magellan decided to try it. The ships began to thread their way among a baffling maze of islands, and up and down the many channels that often proved to be dead ends. The crew of *San Antonio* mutinied and returned to Spain.

The three remaining vessels inched their way along the strait, short of food and water, battling the tides and cold and fog. At night they saw the lights of Indian campfires, but the Indians kept out of sight.

After nearly forty days of wandering they reached the western end. Joyfully they fired their cannons. Even iron-willed Magellan wept as they sailed into the open water of the ocean they had searched for. They headed north, and up the hilly coast of what today is Chile. Then Magellan left the coast and headed west across the ocean. Now, he must have thought, he couldn't be so far from the Spice Islands. He didn't know that he was sailing on a sea so big it swathes a third of the earth. From where he was, the distance to the islands was 12,600 miles.

Even when they started sailing west their food supplies were low, and now they wouldn't set foot on land for ninety-nine days. The only land they came upon was barren islands. They ate what little rat-fouled biscuit they still had, and the maggots in it. Now starving, they scraped up crumbs in biscuit barrels, ate the rats, chewed on sawdust, boiled and gnawed the leather of the yardarms. Their drinking water turned yellow and stank. Many of them died of hunger, thirst, and scurvy.

At last they reached Guam Island. The islanders came out in boats, climbed aboard, and dashed around the ships, stealing what they could. Weakened though they were, the Spaniards pushed them off. They followed them to a village, where they stole their fruit and other food.

Then they reached the islands that were later named the Philippines (for Spain's King Philip II). Magellan had brought with him an interpreter from the Malay-language region of southeast Asia. When this man spoke to the islanders, it was clear they understood his words. For Magellan, this conversation must have been a thrilling moment, for now he knew he had succeeded. By sailing west from Europe he had nearly reached the waters of southeast Asia that the Portuguese had reached a decade earlier by sailing east. Since he had himself been one of those Portuguese, he had now been almost around the earth.

Then he made a bad mistake. When an island chieftain, Lapu-Lapu, converted to Christianity, Magellan joined him in a war against the chieftain of another island. (One of his Spanish captains later explained the reason. With forty men, he said sarcastically, Magellan "went to fight and burn the houses of the town of Mactan to make the King of Mactan kiss the hands of the King of Cebu, because he did not send him as tribute a bushel of rice and a goat.") Magellan trusted in his European fighting skills, but the Mactan islanders drove the Europeans back. As his men retreated to their boats, Magellan, with his sword, fought to shield them. Pierced by a poisoned arrow, slashed by scimitars, and speared in the face, he fell upon the sand. A dozen warriors leaped upon him and killed him.

Lapu-Lapu now had second thoughts about his new religion and his European allies. He invited twenty-seven of them to a feast and slaughtered them. The others quickly sailed away. But they were short of men, so they burned the ship *Concepción*.

Two ships out of five were left. In these they sailed for many weeks among the islands, pirating. But at last they journeyed to the Spice Islands, which had been their destination all along. Here they bought and loaded cargos of cloves.

Victoria set out for home. No one on her wanted to return the way they came and face that fearsome South American strait again. So her captain, Juan Sebastián de Elcano, set his course to westward, hoping to get through the Asian seas that hostile Portugal controlled, around the tip of Africa, and sail from there to Spain. The other ship, the *Trinidad,* required repairs and left the Spice Islands later. Unlike Elcano, her captain had decided to sail east, back to South America. But after many troubles *Trinidad* returned to the Spice Islands and was captured by the Portuguese. Then a storm destroyed the ship.

Of the fleet of five that had sailed from Spain only *Victoria* now survived. She was badly rotted after years at sea, and was sailing in south Asian waters patrolled by hostile Portuguese. Elcano managed to elude the Portuguese, and he got to Africa, sailed around the Cape, and started up the western coast. Suffering from hunger and disease, his men and he at last arrived at the Cape Verde Islands, off what is now Senegal. The islands were Portuguese, but at the greatest risk they stopped and traded some of their cloves for rice. But then they had to flee, and they left behind them thirteen men, who were captured by the Portuguese.

In September 1522, *Victoria* arrived in Spain, three years after she had left it. Eighteen weary, hungry Europeans, all that were left of nearly 250, and four Malays they had hired on, staggered onto land. They were "weaker than men have ever been before."

And here's the sequel to this dreadful, splendid voyage around the earth: when the cargo of *Victoria* was sold, it paid the voyage's total cost. (Cloves were costly; men were cheap.) Magellan's name was given to the strait in which the expedition suffered for so long. The thirteen men abandoned on Cape Verde were released, and four sailors from the *Trinidad,* captured in the Spice Islands, also got back to Spain. In the years that followed, many men who had survived the voyage no doubt went to sea again. Elcano, captain of *Victoria,* was honored by King Charles, who added to his coat of arms a globe, and the words "You were the first to encircle me." He later died on a Pacific crossing. *Victoria,* which had sailed around the earth, should have been preserved and honored. Instead, her owners patched her up and sent her to the New World twice. Returning from the second voyage, *Victoria* and all aboard her sank.

Magellan's famous voyage did not result in Spanish rule of the Spice Islands. (He was wrong in any case; they lay within the half of the world that the treaty assigned to Portugal.) Just the same, his journey taught important lessons. Earth is bigger than the scholars had believed, and a colossal ocean stretched from the Americas to Asia. Columbus had been right in thinking one could get from Europe to Asia by sailing west, but the journey was too long. The voyage's most important consequence was this: humans now knew more about the people of the other clusters, and how to reach them.

LONG AFTER THE Magellan voyage, the peoples of Eurasia, Africa, and the Americas still were unaware that earth held two more continents. Australia and Antarctica lay so far south of customary routes of trade that ships had never happened on them, and fog and ice packs shielded Antarctica from view.

In the 1600s, Australia's isolation neared an end. Ships of other nations had replaced the Portuguese on the Africa-Asia trade route, and some Dutch sea captains found a southern route to Indonesia. From the southern tip of Africa they sailed straight east, and at the proper moment they veered north, caught a breeze, and raced right up to Indonesia. However, if they turned too late they might smash against a giant mass of land. Having found this hostile place, several captains partially explored it, but they didn't grasp that they had come upon the west side of a continent. Since it looked unprofitable, they left it pretty much alone.

A Dutch explorer, Abel Tasman, sailed around most of Australia in the 1640s without ever seeing it. This sounds absurd, but Tasman was only following his sailing orders, and he did prove a negative. Europeans of his time believed that a gigantic "Great South Land" existed below the equator to somehow balance the huge land masses north of it. Tasman's voyage established that Australia, however big it was, wasn't big enough to be a part of this mythical place.

More than a century elapsed before Europeans took another look, and then they did it on Australia's other side. In 1768 the

British navy and Britain's Royal Society sent a scientific expedition to the south Pacific. In command was "Captain Cook"—James Cook—a farmhand's son who already had a brilliant Navy record. One of Cook's assignments was to find that still-alluring Southern Continent. After first exploring the islands of New Zealand, Cook sailed west and reached Australia's scarcely known eastern coast. Sailing northward he explored along the coast for two thousand miles, surveying as he went. He had to thread his way between the shoreline and the Great Barrier Reef, and once a prong of coral pierced his sturdy ship and nearly sank it.

As they sailed along the coast, the Europeans sometimes glimpsed the native folk, the Aborigines. As the ship sailed by, the Aborigines would gaze out at it and then look away, dismissing it as too monstrous to be understood. When the sailors came ashore, the Aborigines would often slip away, but sometimes they were shyly friendly. Presents held no interest for them; when given clothes, they accepted, then abandoned them.

Cook and Joseph Banks, the voyage's leading scientist, both made notes about the Aborigines. Banks observed how few they were, a sprinkle of humans on an endless land. One tribe they came upon numbered only twenty-one: twelve men, seven women, and a boy and girl. The Aborigines were small and thin, completely nude, and dirty. (Banks wet a finger, rubbed it on a man, and discovered that his skin was chocolate brown.) They slept in huts made out of branches, big enough for four or five. A fire might smolder in the hut, but only to prevent mosquitoes, for they rarely cooked their food. Women did the work, even toting heavy burdens on the march, and the men often beat them.

Cook found the Aborigines "far from disagreeable . . . a timorous and inoffensive race, no ways inclinable to cruelty." He almost envied them. "They may appear to some to be the most wretched people on Earth, but in reality they are far happier than we Europeans. . . . They live in a tranquility which is not disturbed by the inequality of condition: the earth and the sea of their own accord furnish them with all things necessary for life."

Before he sailed away from Australia's northernmost tip, Cook anchored by a little island and was rowed ashore. Standing where he saw the mainland, he raised the British flag and claimed Australia's eastern seaboard for King George III.

AFTER COOK SUBMITTED his report about the voyage, the people of six continents knew of one another. Now the only undiscovered continent was Antarctica, which of course held no humans whatsoever. After he explored Australia, Cook received another mission: to learn what lay at the southern end of earth. Could the Great South Land be there? In the early 1770s he sailed with two ships far, far south of Australia, into frigid waters and "pinching cold." Icebergs, white and blue, as lofty as the dome of London's St. Paul's Cathedral, crunched and toppled all around the ships. The expedition searched and searched for land. At one point, though they didn't know it, they were less than seventy-five miles from Antarctica, but pack ice blocked them, threatening to crush their wooden ships. They had to sail away.

Half a century later, several nations, caught up by rivalry and science, raced to learn what lay beyond the ice. In 1820 Russians almost surely were the first to see Antarctica; they sailed around it. Not much later, a team of British scientists set foot upon the continent. In the early 1900s other groups began to venture inland, using sledges, hardy Asian ponies, a balloon, and even an early automobile. In 1911 a Norwegian expedition, using skis and dog teams, reached the South Pole.

EUROPEANS NOW HAD "found" four continents and a sea route to two others. Is that important? After all, the people they discovered had already known exactly where they were. But the linking of the human clusters would, in the future, help to shape the world.

CHAPTER 10

The New World falls to the Old one.

FOR THE TWO great Indian empires of the Americas, that day was the beginning of the end, that day in 1492 when frightened Indians on a Caribbean island noticed winged monsters nearing, hid behind the trees, and watched as pale-skinned strangers landed on their shore. Of course, those island Indians lived far from the two big Indian civilizations. Just the same, the white-sailed vessels were a dreadful omen for the Aztecs and the Incas. Now, for just a moment, they would enter the main current of world history and then vanish from it.

IN CHAPTER 1 we guessed at how the New World Indians' tale began. A handful of Siberians, we said, bearing spears and berry baskets, may have wandered eastward on the tundra plain that then, like a bridge, linked Asia and North America. At the end of the land bridge they found themselves in a continent they had never known of. In the centuries that followed, those Indians' descendants scattered through that land and the continent below it. In the meantime, glaciers melted, oceans rose, and stormy water cut off North America again. As a result, the Indians were isolated from everybody else on earth.

Oceans separated the Indians from the Old World for at least ten thousand years. During those millennia, as we have seen, Europeans, Asians, and Africans had learned to raise crops and animals, live in towns and cities, build, write, use metals, make calendars. Meanwhile, just like them, the Indians of the Americas also learned to farm, settled down, and began to live in cities. What is puzzling is that, having no contact with the Old World, they made these changes only some three thousand years later than the Old World people did. Seen against the one to two hundred thousand years that modern humans have been on Earth, three thousand aren't a lot.

We simply don't know why farming began in the New World so soon after it began in the old one. Did the climate of the world suddenly get harsher everywhere at the same time, forcing everyone to quickly find out how to grow food instead of gathering it? Prehistorians have found no proof of this. Did an Old World farmer sail to the New World and teach the Indians how to raise crops and tend cows? That's unlikely. Perhaps the discoveries of farming, and then of civilization, were just an astounding coincidence.

If we don't know *why* farming began in the Americas when it did, we do know *where.* It happened in a concentrated region. Think of the two Americas together as an hourglass, broad at the top, broad at the bottom, but narrow in the middle. The area of rapid change was in or just above or just below that narrow middle where the two great masses joined, in what today are Mexico, central America, and a portion of Peru.

The Indians' biggest find was how to grow a crop of corn, or maize, which they transformed from a humble weed into a benefactor of humankind. (Prehistorians, digging in ancient Indian shelters, have found remains of husks and kernels that illustrate how corn developed.) Before long the Indians were also raising other noble plants that the rest of the world had never tasted: potatoes, sweet potatoes, peanuts, cocoa, avocados, tomatoes, chili peppers, squash, and string and lima beans. They also raised two not so noble ones: tobacco and the shrubs that yield cocaine.

After they began to farm, the peasants in the middle region, where the continents joined, had more to eat, and their number rose. They settled down in villages, cleared the land around them, drained the swamps. Some of the villages grew into towns and cities, and they organized as city-states with all the things we cannot do without: politicians, soldiers, priests, and tax collectors.

Striking proof of how these cities thrived are the remnants of Teotihuacan, which stood not far from where Mexico City sprawls today. While the Roman Empire flourished—but many of the world's great cities weren't yet born—150,000 people lived here. They built two temple pyramids so huge that later migrants to this place supposed that gods or giants made them.

OF ALL THE ancient peoples, the most accomplished were the Mayas. And yet they flourished in an uninviting place. The northern half of the narrow isthmus that connects the two Americas was and is a tangled jungle, hot and damp, and full of things that bite. But here it was that the Mayas, when they began to farm, settled down. At first they lived in villages, but in time their numbers rose, and many a dusty village became a handsome city. Each of these was built around a square, flanked by palaces and temples. On one side of the square the Mayas usually had a court where they played their popular ball game, and grandstands for the spectators. On another side, a step pyramid would rise, crowned by temples that towered above the nearby jungle.

Many Mayas lived quite well. We know about the living standard of one Mayan village because of a recent scientific dig on a site below a volcano in what is now El Salvador. This was a village of three hundred Mayas (or near-Mayas, differing a bit in speech). One day 1,500 years ago disaster struck. This happened during summer, as we know because the fruit on nearby trees was ripe.

It was late afternoon, and the farmers had returned from working in the fields. Some had just begun to eat, dipping fingers in their bowls of cornmeal mush and cherries, and drinking cocoa. Others had already eaten, but as yet they hadn't washed their bowls or taken out their sleeping mats. Then they heard a roar like thunder and they knew that the volcano was exploding. They fled and saved their lives, but they lost their village, which was buried under sixteen feet of ash.

As they dug away the ash, archaeologists discovered that these Mayan farmers had not only thatch-roofed houses of their own but a chapel, a village hall and storehouse, a shop for making knives and tools, and a sauna. They wove the cotton they raised into cloth and made their clothes, and lived on vegetables and fruit, shellfish, turkeys, deer, and dogs. The food they didn't need they doubtless carried to Copán, a Mayan town some sixty miles away, and bartered for volcanic glass (to make their knives) and finely painted pots. In one of the humbler houses in the village archaeologists found over seventy of these handsome pots. These ordinary peasants fared better than most peasants of their time in Europe and Asia, and better than the peasants of El Salvador today.

The Mayas flourished from about A.D. 250, when Rome ruled one end of Eurasia and China ruled the other, till near the end of Europe's Dark Ages. Until recently, all we knew of Mayan history came from buildings, paintings, pots, and carvings found in vine-clad ruins in the jungles. But experts now can read the Mayas' glyphs, the picture-words they carved on buildings.

This much is clear: The Mayas didn't have just one united kingdom. On the contrary, they lived in about forty towns or cities. As in ancient Greece, a Mayan kinglet often ruled a single town and the villages around it. Many rulers had engaging names, such as Curl-Nose, Animal Skull, Lady Six Sky, and Smoking Squirrel, and they often fought each other.

In Bonampak, a Mayan city-state in what today is Mexico, dramatic murals tell what happened when the ruling family had a child.

In one, the ruler holds the baby up before the nobles. Trumpets blare, dancers listen for the beat of drums, and revelers prepare to feast on crayfish, carp, and corn. In another, women of the royal family prick their tongues and let their blood drip down on bits of paper, which will then be burned and offered to the gods. In a battle scene the king, with a bleeding enemy's head suspended from his neck, wages war against his neighbors. The hard-eyed ruler and his warriors seize their captives by the hair and slaughter them to please the gods and thank them for the heir.

A thousand years ago the Mayan towns and cities fell apart. Long periods of drought may have ruined them. The stuff of civilization—cities, law courts, rituals, business, ball games, paintings—vanished. In many places squatters occupied the buildings for a while, but then the jungle slithered in and hid the palaces and temples. Everything was gone except the Mayan folk themselves, who still were farming in the villages. Only recently have archaeologists pushed away the vines and found the ruins, and tourists come to view them. A tourist who beheld the Mayan ruins in Chiapas, Mexico, asked her guide, a local man, "Where did all the people go?" "She was talking to a Maya," the guide later told another visitor. "We're still here. We never left."

When Europeans reached the New World (as they called it) after 1492, the Mayan cities and some other ancient cultures had expired, but others (as we'll see) had risen in their places. No one knows how many Indians there were. Different experts have guessed the total number, in the two Americas, at anywhere from 40 to 100 million. (Europe's population then was roughly 80 million.)

Among the newly civilized Indians were the able, brutal Aztecs (see Aztec Empire in the map on page 148). They had earlier left their homes in northern Mexico, perhaps because they grew too numerous to survive there. They turned into nomads, dressed in shirts and loincloths made of palm leaf fiber and sandals made of plaited straw, and they wandered southward to the Valley of Mexico. Despite its name, the "valley" is a high plateau encircled by volcanos. Until five hundred years ago the land was mostly forests, lakes, and swampy plains.

The Aztecs couldn't find a place to live, because several city-states already held the fertile land around the lakes. So they squatted in several places near the lakes until the rightful owners drove them out. Not only did the owners want to keep their land, they also were repelled by the Aztecs' brutality, about which many tales survive. According to one story, the ruler of the Culhuá people made the big mistake of granting the Aztecs his daughter as a wife for their chief. When the ruler came to the wedding he discovered that the Aztecs had honored him by sacrificing his daughter to their gods. An Aztec priest was wearing her skin. Understandably, the Culhuá drove away the Aztecs.

They resumed their wandering till they reached two vacant, swampy islands off the southwest shore of Lake Texcoco. In this seemingly hopeless site these able people settled down and built a city. They named it Tenochtitlán, "beside the prickly pear cactus." They built giant causeways to link the islands to the mainland, and they dug canals across the islands that would serve as streets for their canoes. From the bottom of the lake they scooped the soil and shoveled it on rafts and man-made islands where they grew their crops. They built houses, palaces, and temples, and an aqueduct to bring them drinkable water from the hills. Before long, Tenochtitlán was one of the largest cities in the world.

When the Aztecs settled on their islands, nearby kings were always making war on one another. This belligerence proved helpful to the hungry new arrivals. At first they fought as mercenaries for their neighbors, but later on they turned against their bosses, one by one. They fell upon them suddenly, burned their temples, and divided up the prisoners, the booty, and the women.

Sometimes they used milder methods. Once they marched to a nearby state and "invited" the ruler to bring his lords to their capital and adore their god, Hummingbird on the Left. The ruler swore he'd never even let his whores dance before the god, but when he saw he had no choice, he groveled in the dirt and pleaded for his life.

It wasn't long before the Aztecs held a rich and ample empire. At its peak it stretched across what now is Mexico from the Atlantic Ocean to the Pacific, and it held perhaps five million people. In size and population it was roughly equal to a Western European country of its time. Every year the Aztecs forced their subject towns to send them porters bearing gold; and also corn, tomatoes, beans, and squash, turkey, deer, and dogs; and humans they could kill to please their gods.

Yes, human sacrifices. Like many other Indian peoples then, the Aztecs held that they must give their gods the food they wanted: human hearts and blood. If they failed to carry out this sacred task, their victories would cease, the sun would die, and life would end. It was partly from this need to feed the gods that Aztecs always sought new lands to conquer, since these would bring new pools of victims for their knives. They claimed it was an honor to be slaughtered for a god.

Every year they prodded about 15,000 naked prisoners up the steps of temples. As each captive reached the top the clerics forced them down on slabs of stone, split their chests, ripped out their beating hearts, and fed their blood to idols of their gods. Later on the priests and other dignitaries sometimes made a dinner of a victim.

Bloodstained though they were, these former nomads turned into a civilized people. That is, they learned the arts and skills the Maya once had known and which many peoples of the "Old World" also knew. True, they never learned the art of smelting metals (they edged their swords with volcanic glass), and they didn't know the use of wheels. But they had a government that worked, handsome buildings, painters, poets, astronomers—even historians. By farming with intelligence and squeezing tribute from the folk they ruled, the Aztecs lived quite well. Or at least their noble families did, richly housed, brightly clad in cotton cloth and feathers, fed on dogs and chocolate, and certain of an afterlife as gemstones, clouds, or many-colored birds.

And yet the Aztecs tore the hearts from living humans. Can we argue that in the Aztecs' case human sacrifices were themselves a proof of their attainments? Something other Indians also did, the new-rich Aztecs did much more, as if to prove how well they'd learned what was expected of a people who were civilized.

IT WAS WHILE the Aztecs were at their peak, conquering and sacrificing others, that Columbus reached the New World's eastern edge. After he had sailed home and reported his findings, others got on ships and journeyed to the new-found lands. Many of these men were Spaniards who had fought in Europe's wars; they were restless seekers of adventure, ravenous for treasure, land, and humans they could rule.

Among them was Hernán Cortés, whose most daring deed until he sailed from Spain had been his nearly fatal leap from a balcony when his lover's husband came home early. In the New World young Cortés took part in the Spanish conquest of Cuba. (His reward was land and Indian slaves.) He was clever, brave, often decent, sometimes cruel and greedy.

In 1519 Cuba's Spanish governor chose Cortés to lead an expedition to the central American mainland. He gathered ships and troops and sailed to the eastern coast of Mexico, where the Aztecs' lands began. His forces were 508 soldiers, about 100 sailors, 16 horses, several little cannons, and 13 muskets. When they reached the mainland, Cortés burned the little ships. Without the ships there could be no return to Cuba, so his soldiers hadn't any choice but to conquer or be killed.

From the steamy jungles near the shore Cortés led his army up and through the rugged hills, toward Tenochtitlán. Many Indians on their way abhorred the Aztecs, so they joined the Spaniards, fed them, hauled their guns, and fought beside them. Other Indians tried to drive them back, but the Spaniards had an edge: the skills and weapons they'd acquired in European wars.

Whereas the Indians charged in yelling mobs, Spaniards fought in ordered ranks. Whereas Indians fought with lances, swords of hard volcanic glass, and bows and arrows, Spaniards fought with horses, crossbows, pikes, and swords of steel. They had weapons that not only killed but terrified the Indians: sticks that barked and killed, logs that roared and killed by scores, and, worst of all, terrifying beasts that had two heads and many legs. What's more, the Spaniards had the will to win. They never doubted that they fought for God against Satanists, sadists, and sodomites (Aztec priests), or that the Aztec ruler's treasures might as well be theirs as his.

When they reached Tenochtitlán, Cortés so charmed the Aztec ruler, Montezuma, that he invited the Spaniards to stay there as his guests. Then they took him prisoner and divvied up his gold and silver, pearls and jewels. They invited themselves to a festival, and when the drums were beating and the chanting reached its peak and six hundred Aztec nobles were dancing, the Spaniards struck. They lopped the drummers' hands and heads off, and they killed the noblemen and stripped their bodies of their gold.

However, when the Spaniards, pious Christians, started to destroy their temples, the outraged Aztecs rose against them. A bloody battle followed, in the course of which the Aztecs wounded their submissive king with rocks and arrows. Montezuma died soon after, hastened on his way perhaps by Spanish knives. To escape, the Spaniards fought their way along the causeway from the islands to the mainland, losing a third of their number.

But nothing would deter the Spaniards, who, remember, had no ships in which to sail home. Two weeks after they had fled Tenochtitlán, the Spaniards crushed an Aztec army. Then Cortés collected reinforcements, built some boats (for island fighting), and began a battle for Tenochtitlán.

By now the Spaniards had some silent allies fighting by their side, more destructive than their guns. In their bodies they had brought from Europe germs of smallpox, measles, and dysentery. Even for the Europeans these diseases could be deadly. How much more lethal were they for the Indians, so isolated from the Old World and its epidemics. Unlike the Spaniards, they had no immunity at all to Old World microbes, and their bodies served as splendid hosts.

Smallpox, whose virus one could spread by simply breathing, was the worst. At the start of 1519 it had swept through Hispaniola Island, killing most of its Indians. It had reached the mainland shortly after Cortés, by no coincidence of course, and raged across the Aztec Empire, often killing from half to all the people in a town. And now, while the Spaniards struck Tenochtitlán, smallpox also hit the city. It largely spared the Spaniards, who were probably immune, but it wiped out nearly half the city's defenders.

As the Spaniards and their Indian allies took the city street by street, they razed it house by house and dumped the rubble in canals. They raped the women, killed the Indians by the hundreds of thousands, and probed their bodies searching for their hidden gold. When they stopped, the earth was drenched with gore and strewn with maggot-covered limbs and bodies.

The stench was so atrocious that the gagging victors moved outside the ruined city to a place along the shore. At just this moment wine from Spain arrived, and pigs from Cuba, and the Spaniards held a victory party. Next day they climbed a hill that overlooked the lake. A priest conducted mass, and they gazed down on the wound they'd made when they ripped out the Aztecs' heart.

BACK WHEN THE Aztecs had won their empire in the north, the Incas to the south had built their equally ill fated one. The Incas' heartland perched high up in the Andes Mountains, on the western side of South America.

The Incas left no written records, so we know their story chiefly from the tales their "memorizers" or tradition-tellers had passed on. Before they started on their conquests the Incas had been much like other Andes peoples. But under two extraordinary kings they overwhelmed their nearest neighbors. Then they conquered farther tribes and stationed troops among them. They added conquered soldiers to their army.

Soon the Incas held almost all of the Andes Mountains and the western coastal desert. Their empire stretched 2,500 miles from the

Equator down to what is now south-central Chile. They ruled perhaps as few as six or as many as 12 million people. These included llama herders high up in the mountains, where the air was thin and visitors could only gasp for breath; potato farmers in the valleys; naked hunters in the jungles, whom the Incas rarely glimpsed; and fishermen in towns along the coastal desert.

The Incas brooked no protests and they wanted no revolts. They often broke up tribes they conquered, forcing many of these tribesmen to resettle elsewhere, far from home. Then they brought in loyal Indians to take the places of the tribes they'd moved.

But what a contrast with the Aztecs! The Incas could be harsh but they were also decent, or at least prudent, and they wanted all their subjects clothed and fed. They required the "have" areas in their empire, which produced ample potatoes and llama wool, to send their surpluses, even herds of llamas, to "have-not" areas. They resettled many Indians from the highlands in the valleys so that they could grow sufficient corn (or maize) and coca (a beloved drug) for the people of the lands they came from. In ways like these the Incas built a state, turned their former enemies into loyal subjects, and made them prosper.

The Incas knew what the Persians and Romans had known before them: that an empire must have first-rate communications. So they built no fewer than 19,000 miles of roads linking the provinces to Cuzco, their capital. Incan roads ran through swamps and deserts, up the sheerest cliffs, through tunnels in the mountains, and over bridges made of interwoven vines that the Incas slung across the deepest gorges. Over these roads and bridges relays of messengers ran 150 miles a day, bringing messages they memorized.

AMONG THE FORTUNE-SEEKERS arriving in the Americas was Francisco Pizarro, a Spanish soldier's bastard son. Unlike Cortés, who in his teens had studied law, Pizarro as a boy had tended pigs. He had probably fought for Spain in Italy, and he may have been a bandit for a time. After he reached the New World, Pizarro went to Panama, and he was with Balboa when the discoverer crossed the isthmus and saw the Pacific. For several years Pizarro was the mayor of Panama, and he made a modest fortune.

When he heard that somewhere south of Panama was a wealthy empire, Pizarro scented treasure. In the 1520s he led two expeditions down the coast. They learned that they were indeed on the edge of a great empire, rich in gold and silver, which they named Peru (see Inca Empire on map, page 148). *Peru* is probably a corruption of Virú, the name of a river.

Before he struck the Incas, Pizarro returned to Spain and convinced King Charles to authorize his plan. Charles, who now was Holy Roman Emperor as well as king of Spain, made Pizarro governor of the yet-to-be-conquered lands. Pizarro returned to Panama with four of his brothers, and in 1531 he sailed to Peru with 180 men and 37 horses. Two hundred more soldiers led by Pizarro's partner, Diego de Almagro, joined him later. Even then Pizarro's forces were smaller than those Cortés had had in Mexico.

Without knowing it, however, he had picked the perfect moment to attack. Five years earlier, smallpox had spread from Central America down to Peru. Just as it had helped Cortés in Mexico, it would help Pizarro in Peru by cutting down the Incas. What's more, Peru's late emperor, before he died, had divided the empire between the rightful heir and his favorite son, Atahualpa. The two half-brothers then fought each other and Atahualpa had won, but events would show that he had weakened the regime.

When the Spaniards reached Peru, they headed inland. They climbed the pleasant foothills and the dark forests of evergreens at the base of the Andes. Then they ascended the steep slopes of the mighty mountains, all the way to Atahualpa's headquarters. Pizarro sent a message to the emperor, proposing that they meet on the town's big central square. He hid his soldiers and his cannons near the square, where Atahualpa couldn't see them. Soon the emperor arrived in splendor, with an escort of several thousand men, lightly armed. Not far away he had another 25,000.

"Where are the strangers?" Atahualpa asked. A Spanish priest advanced alone, a Bible in his hands. Using an interpreter he explained the Christian creed and the power of King Charles. He urged the Incan to agree to Jesus as his god and Charles his master, but Atahualpa flatly told him he would not. "Your god, you say, was put to death by the very people he created. But mine," he said, and pointed to the sun, "my god still lives in the heavens and looks down on his children." When the priest offered him the Bible, he threw it to the ground.

At once Pizarro waved a scarf, the signal that the Spaniards had agreed on. Cannons roared and gunsmoke billowed on the square, and the handful of Spaniards sprinted out and slaughtered most of the bewildered Indians. Pizarro and some horsemen seized the emperor.

The Spaniards held him hostage, just as Cortés had done with Montezuma. The Incas couldn't help him. Like Cortés the Spaniards wanted treasure, so Atahualpa promised he would fill the room where he was held with gold as high as he could reach. He did this, but the treasure didn't save his life. With astounding chutzpah the Spaniards told him he was guilty of the murder of his own half-brother and that he had schemed against them. They offered him a choice: remain a heathen and be burned or become a Christian and be strangled. He chose the latter.

When they heard about his death, the Incan army near the city panicked and retreated. Pizarro marched to Cuzco, the Incan capital, and captured it without a fight. The Spaniards seized a huge amount of royal treasure, and they added this to what they'd squeezed from Atahualpa. They melted down the precious metals, sent a fifth to Charles in Spain, and divided up the rest. Later, they allotted every soldier an estate of land, with Indians to work it for him. Every soldier now was rich, though few would live for long to enjoy their wealth.

The beginning was the easy part. Now the Spaniards had to win the rest of the Incan empire, two thousand miles of desert, forests, and the Andes. The Indians outnumbered the Spaniards a hundred to one, and the odds would have been much worse if new diseases hadn't scythed so many down. Just as with Cortés's invasion, the Spaniards' battle skills and swords of steel triumphed over poorly commanded Indians armed with weapons made of stone and copper. Just the same, it took the Spaniards years not only to defeat the Indians but to crush revolts in distant mountain valleys.

The Spaniards' biggest enemy was themselves. Apparently Pizarro cheated Almagro when the Spaniards shared the Incan loot. When Almagro rebelled, a brother of Pizarro caught and killed him. Almagro's son and some friends suspected that Pizarro planned to slaughter them as well, so they moved first. They surprised Pizarro as he ate his dinner and stabbed him to death. King Charles sent an agent to Peru, and this man caught the young Almagro and had him put to death. Charles then sent a governor to Peru, but another brother of Pizarro killed him. Of Pizarro's four brothers, three were killed and the other died in a Spanish prison. Most of the ordinary Spanish soldiers became as rich as they had dreamed they might but suffered violent deaths.

The Spaniards couldn't govern what they greedily had grabbed and plundered. All the things the Incas had constructed—order, sharing, roads and bridges, just to name a few—fell apart.

PIZARRO AND CORTÉS began a long and dismal story. The Spaniards took the finest land, made slaves of many Indians who had somehow stayed alive through wars and epidemics, sweated them to death in silver mines, tried to crush their religions, and made their lives so wretched that for generations many of them chose to have few children or none at all.

Half a century after Spaniards conquered Mexico, an Aztec matron visited a friend not far from flattened Tenochtitlán. She wanted to congratulate the younger woman on the birth of her two sons. A document survives that tells about the visit. The older woman calls the children "precious jewels and emeralds." And precious they were just then, when many Indian women had no children, and babies were miscarried or died in infancy. "Hardly anyone who is born grows up; they all just die," she says. She thinks back to her youth, before the Spaniards and the worship of the new god Jesus, when the Aztecs were as many as the ants. "But now everywhere our Lord is destroying and reducing the land, and we are coming to an end and disappearing. Why? For what reason?"*

In fairness, it wasn't only men from Spain who did bad things to Indians in North and South America. From Portugal, next door to Spain, and later England, colonists arrived in other places in the Americas, and they often treated Indians as the Spaniards did.

In a way, the Indians had their revenge on the Old World. Not all historians agree, but these appear to be the facts: One of the very few diseases from which the Indians suffered before the arrival of the Spanish was syphilis, apparently in a mild form. When Columbus reached the New World in 1492, his sailors probably caught the disease from Indian women. In any case, syphilis apparently arrived in Spain and went from there to Naples, which had close ties to Spain. It now had changed its nature, as diseases sometimes do, and was often deadly.

When a French army conquered Naples in 1494 (as described in chapter 8), the soldiers and their women caught the new disease. Later, when they withdrew from Naples, they scattered it through Italy and France. From there it raced through Europe, Africa, and Asia. Each country named it for the place from which it had had the honor of contracting it. The French called it "the Naples disease," and the Italians "the French disease." The English called it "the French pox" or "the Spanish disease." As it spread through eastern Europe, Poles named it "the German disease," and Russians in turn called it "the Polish disease." In the Middle East it was "the disease of the Franks [Europeans]." The Chinese called it "the disease of Canton" (one of their port cities), and then the Japanese called it "the Chinese disease." All these countries paid a price for Europe's conquest of the New World.

*Alfred W. Crosby, *The Columbian Voyages, the Columbian Exchange, and Their Historians* (1987), p. 25.

CHAPTER 11

We suffer famine, war, and plague.

ANY SPECIES' MISSION is survival, but we humans have done more than just survive. We did what God, according to the Bible, ordered us to do: "Be fruitful and multiply, and fill the earth." However, let me make this clear: *this* chapter deals with reasons why we humans formerly increased about as fast as sloths climb trees. In chapter 16, we will look at humans' later rapid rise.

Until quite recently, our numbers grew extremely slowly. Ten thousand years ago, when our ancestors were hunters and gatherers, they probably numbered between only five and ten million. Ten millennia later, in the 1600s, at the start of the global population explosion, the people of the earth numbered roughly half a billion. Yes, this was a big increase; our number had doubled six or seven or eight times. But the *rate* of increase had been slow. Each doubling had required about fifteen hundred years.

In a way, it's strange that formerly our numbers increased so slowly. Until modern times married couples who lived long enough would often have from five to seven children. Given this ability for making babies, you might think that long ago we would have swamped the Earth.

The reason why our numbers used to increase so slowly isn't hard to find: about as fast as some of us were born, others died. Through most of our history, births have, on average, only slightly outnumbered deaths. That doesn't mean that our increase, though slow, was steady. Not at all. At least till recent times, humans always went through waves of increase and loss. We slowly increased until we were too many and troubles started, and then we decreased. If we showed this story as a line upon a graph, it would look like a series of self-canceling rises, crests, and falls. Only if we followed it for centuries, maybe millennia, would we notice that the wavy line was very slowly rising.

Take, for example, China, and the changes in its population. For most of early Chinese history we know nothing of the rises and the falls. But in the 1200s and 1300s there must have been a giant fall. The Chinese suffered first the bloody Mongol conquest, and (later) civil war, famine, and epidemics. Even if we don't have the numbers, we can be certain that these disasters killed the Chinese by the tens of millions.

Then their number rose. In the 1400s and 1500s, under the Ming dynasty, the Chinese probably increased quite fast. At the end of those two centuries a historian wrote disgustedly about the government's official census figures. These showed southeast China's population having risen only 20 to 30 percent. "During a long period of peace," he wrote, "the population must have grown and multiplied. . . . The empire has enjoyed, for some two hundred years, an unbroken peace unparalleled in history. During this period of recuperation and economic development the population should have multiplied several times." Modern experts basically agree with him. The Chinese, and not only those in the southeast, must have increased rapidly under the Ming dynasty.

The long rise ended in the 1600s, when there were endless bandit wars, peasant risings, and invasions by the Manchus. The number of Chinese surely fell again, but we don't know how many died in battle or of hunger or disease, and the data are so poor we'll never know.

The census figures for the early 1600s are especially unbelievable. To illustrate: not long before the fall of Peking in 1644 a government official, not expecting the collapse, drew up arbitrary census figures for a number of years *after* 1644!

So no one really knows the number of Chinese before these disasters or after them. The losses clearly were terrific, but demographers can only guesstimate them. One expert thinks China lost a quarter of its people, and another says it may have lost a third.

This terrible decline was followed by another rise. With so many Chinese dead, farms were plentiful for those who had survived, so they lived well and married and had children. By the early 1700s, China's population seems to have climbed back to where it had been before the horrors of the 1600s began. It continued rising throughout the 1700s until there were again too many people, and the stage was set for yet another fall.

The Chinese story fits the classic pattern, common everywhere till recent times. Human beings (in any given area) would increase for decades, even centuries. Finally they would reach the point of surfeit, when many didn't have enough to eat. At this point everything that could go wrong would do so: famine (or at least malnutrition), civil wars (which fed on misery and hunger), and disease. The fall was sometimes slow, sometimes quick and dreadful.

Our gradual increase over many thousand years was linked to those cycles of gain and loss happening all over earth. Again and again, a rising flood of humans would be followed by a *nearly* equal ebb. The gain made in one century would be *nearly* wiped out in the next. Nearly wiped out, but *not quite.* For at the end of the cycle of slow gain and rapid loss a little plus remained, a small addition to our numbers. It was these tiny bonuses at the end of each cycle that made our total number rise, from the time when farming began down to roughly 1750. We increased with the speed of a glacier.

EUROPE'S POPULATION HISTORY, like China's, is full of rises and declines. But for Europe we have fuller information on the causes of death, as well as better population figures. We can better see how losses nearly canceled out the gains, and just what caused those losses.

Europe's Dark Age, full of turmoil, ended in about the year 1000. In the next three centuries, down to about 1300, the average person's life improved, and, as a result, the population rose. The clearest evidence of this increase is the way the towns were growing. For security, people often built walls around their towns. Later, and this is the important point, they built new walls to enclose the houses that had been built just outside the town *since* the earlier walls were made. These new walls are evidence of population growth. They tell a tale of rapid population rise in the centuries after the Dark Age.

Europeans also were increasing farmland, or using it better, to feed their growing numbers. Near Milan, in northern Italy, they built a network of canals to deal out river water so that farmers could raise several crops of hay. In the Netherlands, people garnered many thousand acres from the sea by building dikes to hold the water back. In Germany they drained their marshes, and in France they leveled forests.

By about 1300, however, Europeans had increased too much. Whole regions suffered famines since they had too many mouths to feed. And famines were only the most shocking signs of an agriculture that was overstrained. We have no census figures, but no doubt many died each year of malnutrition and diseases that thrive on it.

Another killer now appeared, ghastlier than famine. In the fall of 1347, European merchants who were trading in the Black Sea region (like the Polo brothers, fifty years before) took fright when plague broke out among two warring armies. (One side catapulted bodies of their dead inside their enemies' camp, to share the dread disease with them.) So the European merchants climbed aboard their ships and sailed back home. They probably carried with them rats infested with plague-infected fleas. Plague soon raged in several ports on Europe's southern coast. Then it hurried inland, and in 1348, 1349, and 1350 plague sometimes crawled and sometimes leaped through much of Europe. It soon acquired a name: the Black Death.

In southern Ireland a churchman kept a chronicle that tells what it was like to live and die in a time of plague. "That pestilence," John Clyn wrote, "deprived villages and cities, and castles and towns of human inhabitants, so that scarcely a man was found to dwell therein; the pestilence was so contagious that whosoever touched the sick or dead was immediately infected and died . . . many died of boils and abscesses, and pustules on their legs and under their armpits; others frantic with pain in their head, and others spitting blood. . . . And lest things worthy of remembrance should perish with time, and fall away from the memory of those who are to come after us . . . so have I reduced these things to writing; and lest the writing should perish with the writer . . . I leave parchment for continuing the work, if haply any man survive, and any of the race of Adam escape this pestilence and continue the work which I have commenced."

Clyn lived long enough to write another entry while plague still raged around him. Then these words appear in his chronicle, in another hand: "Here it seems the author died."*

The Black Death probably spared villages that were isolated from the outside world by miles of dusty roads. In the towns, however, it killed anywhere from an eighth to two-thirds of the people. At least a quarter of all Europeans died from plague. In one Italian town a chronicler recorded, "there was not left [even] a dog pissing on a wall." For about a hundred years, plague returned again and yet again, now here, now there. In a small Italian town not far from Rome, a chronicler observed: "The first widespread pestilence took place in 1348 and was the worst." But then he added: "Second pestilence, 1363. Third pestilence, 1374. Fourth pestilence, 1383. Fifth pestilence, 1389." Another hand added the words, "Sixth pestilence, 1410." In most of Europe, plague kept coming back until about 1450.

And then the European population cycle turned around again; the population rose. The reasons for the rise seem fairly clear. Plague

*Charles Creighton, *A History of Epidemics in Britain from A.D. 664 to the Extinction of Plague* (1891), p. 115.

was now less common (no one knows the reason), and France and England had ended their Hundred Years' War. Most important, so many people had earlier died of plague that there was land and food enough for those still living. And a well-fed population was more resistant to disease.

Europeans noticed that their numbers were increasing. A chronicler in Germany wrote that "hardly a nook, even in the bleakest woods and on the highest mountains, is left uncleared and uninhabited." Another noted that "all the villages are so full of people that no one is admitted. The whole of Germany is teeming with children." An ambassador to France wrote in 1561 that the country "is heavily populated . . . ; every place has as many inhabitants as it can well have."

We have even better evidence for this rise: beginning at this time in European history we have solid population numbers. In some countries a few records have survived of the numbers of "fireplaces" or family units paying taxes, and these totals were rising. Italian towns were even counting individuals, not "fireplaces," and these censuses also show big rises. Judging by this data, it looks as if Europeans in the year 1500 probably numbered between 80 and 85 million. A century later they had reached 100 to 110 million. This is an increase of about a quarter.

However, just like China at the same time, Europe was headed for trouble. This had to happen, given what we know about population cycles, and finally it did. By about 1600 Europe's population rise was at an end. In some places the human increase only slowed and almost halted; in others populations actually declined.

AN ANCIENT CHRISTIAN prayer requests: "From famine, war, and pestilence, deliver us, O God." We will see the way these population spoilers worked, mainly in the seventeenth century.

Famine was always a threat. In these times Europeans, like most people in the world, depended on a grain crop as the mainstay of their diet. Whether the grain was wheat or barley, corn or millet, the supply was crucial to survival. Famines and great loss of life took place when either too much rain or too little of it spoiled the all-important harvest. Suppose that in a given region there were so many people that the grain supply was barely adequate even in years when the harvest was good. In such a place a dismal harvest might cause a famine and great loss of life. A war could make a famine lethal. Armies ate the food on hand, burned the growing crops, and blocked the roads that farmers' wagons bearing food had to use.

Shortages of food were common everywhere, especially where too many lived, or the soil was poor, or the climate harsh. For example, "Old" Castile, in central Spain, is semiarid. (According to a legend, Jesus visited the town of Ávila and when he saw the sun-baked land around it, he wept.) Spaniards say the climate is "nine months of winter and three months of hell." In this region poor harvests and high grain prices used to occur roughly once a decade.

The history of Beauvais, a town in northern France, shows how famines killed. In the year 1600 about 12,000 people lived here. Most Beauvaisians were clothworkers, laborers, or servants, and they lived in crowded, tottering houses around a cathedral that had a habit of collapsing.

Like many places, Beauvais had a problem with its food supply. The nearby peasants barely raised enough to feed themselves and the town. What made the food supply so risky was that farmers mostly raised a single crop: a mix of wheat and rye. They grew no other crops to give themselves insurance against disaster, and a frigid winter or a rainy summer could trim a year's supply of food to almost nothing. When the crops were good, the area could feed itself, but when they failed the price of bread would rocket up. Shortages would follow, and many would die of hunger and disease.

Imagine the plight of the ordinary textile worker, probably a weaver. He earned about seven and a half sous a day. Of these, at least one sou went for taxes, rent, and payments to his church and trade association. This left six sous for bread, the staple in his family's diet. At the price for which bread sold when times were good, he could buy enough for himself, his wife, and one or two children.

Suppose, however, that the price of bread should double, or triple, or even quadruple. Between the early 1670s and the early 1700s that often happened. Poor harvests caused the price of grain, and therefore the price of bread, to double in twelve of those years, triple in four of them, and quadruple in three. When prices rose this way, the family could not afford to eat. Records show that in the same years when the harvest failed the number of deaths in Beauvais would jump to three or four times the normal number.

This is what happened to one family, the Cocus, all five of whom had work as spinners and weavers. In the summer of 1693 a poor harvest made the price of bread rise steeply. At the same time, the failure of the harvest caused a crisis in the textile business. So work was scarce and family incomes fell just when the price of bread went up. It is easy to guess what happened to the Cocus. At first, says the historian Pierre Goubert, they probably used the coins saved up for rainy days, and then they pawned their few belongings. Then "they began to eat unwholesome food, bran bread, cooked nettles, moldy cereals, entrails of animals picked up outside the slaughterhouses." Soon they were starving, weak, and listless, and suffered from "pernicious and mortifying fevers." As winter began, the Office of the Poor had the Cocus on its list. Three months later the youngest daughter died, and two months after that the eldest daughter and the father. "All that remained of a particularly fortunate family, fortunate because everyone in it worked, was a widow and an orphan. Because of the price of bread."*

The second of the spoilers, the life-takers, was war. In Europe, with so many quarreling nations, war was almost constant. Only in three years between 1540 and 1640 was there no warfare anywhere in Europe or on the Mediterranean. As we saw in an earlier chapter, warfare at this time became more deadly. Armies still used pikes and swords but now they also carried muskets, and battles often ended with the winners slaughtering the losers.

*Pierre Goubert, *Beauvais et le Beauvaisis de 1600 à 1730* (1960), pp. 76–77, as translated in Peter Laslett, *The World We Have Lost* (1965), pp. 112–13.

Europe's Thirty Years' War is the classic illustration. Picture armies fighting here, marching there, fighting, wintering, marching, and fighting once again. Since countries paid their soldiers poorly and rarely gave them clothing, food, or shelter, the troops were forced to loot and pillage to maintain themselves. They robbed and murdered travelers, and they fried peasants on their stoves until they told where they had hidden food and horses. Since the wretched peasants often gave up farming, famine soon was everywhere. Meanwhile, armies ruined towns and cities. They besieged Leipzig five times and Magdeburg ten, burning it to the ground in 1631. Worst of all, the armies spread bubonic plague and typhus, which took a heavy toll of both the soldiers and civilians.

Parts of what today is Germany were shockingly depopulated. In a region stretching from Berlin to northeast France, many rural areas lost four out of five of their people to wounds, hunger, or disease, or because of migration to safer areas. In Germany as a whole, a third of all the townsfolk and two-fifths of the peasants died.

EVEN DEADLIER THAN war and famine in the 1600s was pestilence (by which we mean not only epidemics but also common illnesses). Earlier, in the Middle Ages, the worst of the epidemic diseases had been plague. Now a mafia of diseases fed on crowds of people: not only plague, but smallpox, dysentery, and particularly typhus.

We know today that body lice can spread typhus. Typically a victim is infected when he or she scratches skin where an infected louse has left its feces. Typhus microbes in the feces penetrate the body, and as their toxin multiplies the victim has a sudden headache and a fever. Often she or he is then enfeebled, covered by a rash, then in shock, coughing, breathless, then delirious, comatose, and dead.

Typhus broke out anywhere that lice could thrive: in prisons, ships, and battle zones—any place where people lived in crowds, and where they rarely changed their clothes, and huddled under filthy straw or blankets. In 1586 typhus struck a law court in the

English town of Exeter. Local people later traced the path the disease had taken. First, soldiers had thrown captured foreign sailors, sick with typhus, in the castle jail. The sailors infected other prisoners, Englishmen awaiting trials, and some of them died. Then the remaining English prisoners, even though they were so sick they couldn't stand, were put on trial. Bailiffs helped some of these prisoners to walk to the courtroom, wheeled some of them in barrows, and carried others in their arms. Naturally, the bailiffs then fell ill. During and after the trial typhus felled the judge and jurors, and from the courtroom it spread to the population at large.

Europeans knew very well that diseases always threatened them. For centuries children have danced to the English nursery rhyme that goes

Ring-a-ring o'roses
A pocket full of posies,
Ashes! Ashes!
We all fall down.

The subject of this poem is plague. A red ring on the body was a symptom of it; people carried posies (in this case meaning herbs) to ward it off; and they burned trash (hence ashes) to counteract its evil vapors. Just the same, "all fall down."

People died by thousands during major epidemics but would also die in ordinary times from humble curses such as dysentery and obscure "sweats" and "fevers." A good source of information on causes of death is the London Bills of Mortality, which were weekly reports of burials in London churches. During the 1600s the compilers of these bills began to specify the cause of death. They were not expert diagnosticians (no one was), and probably they often confused one disease with another. But the bills provide a rough idea of the causes of death in one large city. During the quarter-century from 1661 to 1686 the bills blamed nearly two-fifths of all deaths on epidemic diseases such as plague, typhus, smallpox, measles, and "gripping of the guts," meaning dysentery. But they blamed a larger amount, three-fifths of all deaths, on ailments that were always there: childhood illnesses, tuberculosis, and diseases of old age.

Poverty, dirt, and ignorance account for much of the loss of life. Towns and cities grew like mushrooms and turned into toadstools, and streets were paved with garbage, mud, and dung. Families crowded into one or two rooms and got their water from dirty wells and rivers. An English coal miner told an investigator that the men's "legs and bodies are as black as your hat." And he said, "I do not think it usual for the lasses [in the mines] to wash their bodies; my sisters never wash themselves." As fast as cities killed their residents, country people poured in to take their places as scullery maids, stable boys, apprentices, laborers, and prostitutes. Many of these would also die of the diseases of the cities.

Dirt and trash were everywhere. An Italian artist once explained in a handbook how to ready a wooden panel for painting. Rub it with a fine powder made from chicken bones, he said, "and the older they are the better." And where would one find such bones? Take them, he wrote, "just as you find them under the dining table."

The writer Tobias Smollett once gave this picture of the milk that was sold on the streets of Edinburgh, Scotland. It was, he said, "carried through the streets in open pails, exposed to foul rinsings discharged from doors and windows, spittle, snot and tobacco quids, from foot passengers; overflowings from muck carts, spatterings from coach wheels, dirt and trash chucked into it by roguish boys for the joke's sake, the spewings of infants, who have slobbered in the tin cup measure, which is thrown back in that condition among the milk, for the benefit of the next customer; and, finally, the vermin [fleas and lice] that drops from the rags of the nasty drab that vends this precious mixture, under the respectable denomination of milkmaid."

Dirt and squalor weren't found only in the cities. A well-known scholar, Desiderius Erasmus, described the typical country inn in Germany. "When you have taken care of your horse you come into the Stove Room, boots, baggage, mud, and all, for that is a common room for all comers. . . . In the Stove Room you pull off your boots, put on your shoes, and, if you will, change your shirt. . . . One combs his head [for lice?], another . . . belches garlic. . . . In my opinion nothing is more dangerous than for so many to draw in the same vapor . . . not to mention the farting, the stinking breaths . . . and without doubt many have the Spanish or, as it is called, the French pox [syphilis], though it is common to all nations."

Doctors weren't much help. Like everyone, they knew plenty about rats, fleas, and lice but not how they spread disease, and they knew nothing about viruses and bacteria. They had only a few drugs that could help, such as digitalis (for the heart) and mercury (heaps of which they prescribed for syphilis, constipation, and too many other things). The Italian writer Giovanni Casanova maintained that "more people die at the hands of doctors than are cured by them." When Charles II of England had a stroke in 1685, officials called in all the well-known London doctors. They bled a pint of blood, applied hot iron to his head, and made him drink a potion made from skulls. He died.

With famine, dirt, disease, and ignorance against them it's no wonder that so many died so young. John Colet, an English scholar, was the oldest child of a wealthy businessman. He had twenty-one younger brothers and sisters, and all of them died before he reached the age of thirty-two. His contemporary, the German artist Albrecht Dürer, was one of eighteen children, and only three of them, it seems, reached adult years. A wealthy English couple, Edward and Judith Gibbon, produced seven babies, six boys and a girl. They named all the boys Edward in the hope that at least one would survive and carry on this name, which was a tradition in the family. The five younger boys died in infancy. The oldest barely survived a number of childhood illnesses, grew up, and wrote a famous history of *The Decline and Fall of the Roman Empire*.

The story of one particular peasant family illustrates the dire effects of "ordinary" diseases. The family, whose name was Zuzek, lived on a rocky plateau in what is now northeastern Italy, and in the early 1800s they were poor. They lived in a small stone house surrounded by thirteen other houses on the side of a hill, and they worked thirteen little plots of rocky land. A tax collector reported that the people of their village were "poorly nourished, but strong, and able to bear up under the toil of the farm."

In 1800 Tomaz Zuzek married Marina Gabrovic, and in the next seventeen years they had eleven children. Eight of the eleven died within a year of birth. In his funeral register, the local priest recorded the causes of these deaths, or what the family told him were the causes. Three died of "weakness," two of "ordinary" causes, one of "swollen throat," and two of what may have been tuberculosis or some other wasting disease. Marina, their mother, died at thirty-six, soon after giving birth to her last child. She died during a famine, which may have contributed to her death.

Two of the daughters grew up and left the village, and one son, Matija, reached adult years and inherited the house and land. At the age of eighteen he married another eighteen-year-old, Marijana, and this couple proceeded to have twelve children. Eight of these died within a matter of days or at most a year, and one lived only to twelve. The priest recorded these causes of death: "premature birth," "weakness," "natural," "weakness," "pneumonia," "scarlet fever," "ordinary," "dysentery," and "consumption." Four of the twelve children reached adult years, but one of them died at twenty-one of tuberculosis. Matija died at forty-two of typhoid fever, leaving Marijana, also forty-two, pregnant with their last child.

Altogether, these two couples had a total of twenty-three children, and of these only six lived past the age of twenty-one. Many births, many early deaths. From the standpoint of the survival of our species, their story is actually positive. Four parents left behind them six children who might themselves have children.

And that brings this story to the early 1800s. By now our species had lived through countless rounds of rise and fall. We had increased in spite of famine, war, and pestilence, and now we humans numbered about a billion. But we had only begun the global surge of population that has lasted till today.

CHAPTER 12

We discover who we are and where we live.

NOT SO LONG AGO, our beliefs about our place in the grand scale of things were lofty and satisfying. Or at least, Europeans' notions were. And that is the important thing, because it was Europeans who were going to find out who we are and where we really live.

Europeans had inherited their understanding of the universe from ancient astronomers, and especially from Ptolemy, an Egyptian who did his work while the Roman Empire flourished. Ptolemy summed up what scientists of his time believed about the universe in a book that later became known as the *Almagest,* "the greatest compilation."

As Ptolemy explained it, the universe was made up of earth, at the very center, and the sun, the moon, five planets, and the stars. Only earth was made of solid, heavy matter; the others out of something weightless. Around the earth, which didn't move, the others endlessly circled, each of them fastened to an invisible, forever-spinning globe. These globes were concentric—one inside the other, like Chinese boxes. The moon was on the smallest globe, nearest to the earth. Mercury's globe enclosed the moon's, Venus's globe enclosed Mercury's, and then came the sun, Mars, Jupiter, and Saturn, each one on its globe. All the stars were fastened to a single globe, the eighth.

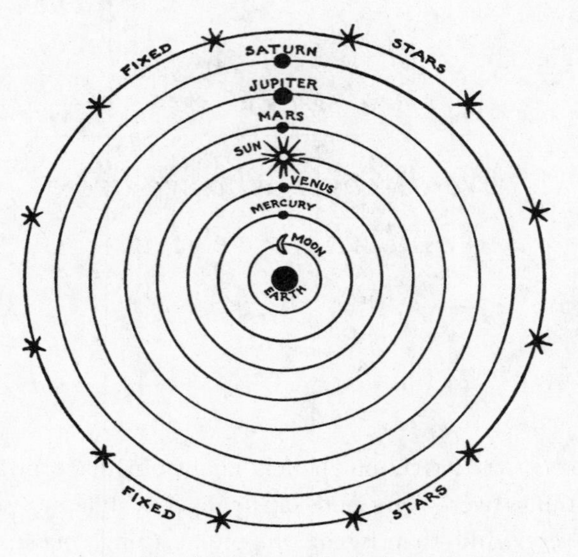

The older view of the universe

As the centuries passed after Ptolemy, Christians decided that beyond the globe of stars was the realm of God. It was his angels who kept the globes revolving and forever bearing the heavenly bodies in perfect circles around the earth.

Ptolemy's picture of the universe was satisfying because it fitted with what everyone could see, or thought they did. The earth indeed appeared to stay inert, and the heavenly bodies did appear to rise and set each day or night, as if they were circling around us. To most people this was obvious, just common sense, and the earth-centered explanation of the universe became the widely accepted one.

Not only was this view commonsensical, it was also pleasant to believe that we were at the center of all God's creation, and that the heavens circled around us, and that the sun, moon, and stars shone only to give us light. The whole conception made us feel important, as if we mattered.

The idea that the heavens circled earth fitted nicely with another pleasing European, or Christian, belief: namely, that human beings are a special creation of God. Our understanding of this matter came mostly from the two creation stories at the beginning of the Bible. In one of these stories, God first creates the fish, the birds, and the "cattle and creeping things and beasts of the earth." Then, separately, he creates men and women "in his own image, in the image of God," and he gives them "dominion" over the other living things. In the second creation story he first creates Adam, or man, in a special way, breathing "into his nostrils the breath of life." And then he creates Eve, again in a special way, shaping her out of Adam's rib. But he simply "forms" the animals and birds, and brings them to Adam so that he may name them.

Both these stories taught that animals and humans were created separately. Animals were one thing; humans another. What's more, humans were superior, and they were the deity's chief concern. So taught the Bible, the word of God. And, just as with Ptolemy's picture of the earth-centered universe, this notion of a separate, special creation was one that humans willingly believed. No one needed to convince the humblest farmer that he was better than his ox.

PTOLEMY'S TEACHING about an earth-centered universe reigned for nearly 1,400 years. Popes, professors, painters, and poets learned it when they went to school and later taught it to the following generations. Even the plowman, who knew that he was better than his ox, probably also believed, if he thought about it, that the land he plowed stood still and the sun was circling around it.

The earth-centered universe was not so elegant and simple as it sounds. Rather, it was quite complex, and this is why. The ancient scientists, Ptolemy and others, had noticed that when they observed the planets against the background of the "fixed" or fastened stars, the planets were not always where they should be if they moved around the earth in simple circles. Mars, especially, often seemed to go in the wrong direction, so much so that the ancient Egyptians called it "who travels backward." What's more, the planets sometimes moved quite fast and sometimes slowly, and they were sometimes bright and sometimes dim.

To account for such anomalies, the old astronomers had come up with far-fetched explanations. Here is one example: they decided that each planet was fastened not directly to the globe that carried it around the earth, but rather to a smaller globe, called an epicycle, that was fastened to the bigger globe. (Remember: all these globes, large and small, were transparent.) This epicycle, carrying the planet, rotated independently, and sometimes in an opposite direction from the main globe. That, the astronomers explained, is why the planets sometimes seemed to go in the wrong direction.

Then it turned out that one epicycle per planet was not enough. To account for all the wrong places in which planets were to be found, the astronomers had to imagine that some of the planets were fastened to epicycles *fastened to other epicycles* that in turn were fastened to the globes. When someone explained this system to a medieval king in central Spain, he said, "If the Lord Almighty had consulted me before embarking upon Creation, I should have recommended something simpler."

In the early 1500s, while Machiavelli was exploring politics and Magellan was sailing around the world, a Pole named Nicholas Copernicus began to rethink the earth-centered universe. As a young man, he had studied Church law and medicine in Italy, and had also heard some lectures on astronomy. Back home again, he became an assistant to a bishop, his uncle. He also served the bishop as a personal physician, thought up schemes to reform the local currency, and painted his self-portrait. For a while he had a mistress, which (since he served a bishop) got him into trouble.

Meanwhile, purely as a hobby, Copernicus studied the motions of the stars and planets. For half a century, from his early years in Italy until he lay upon his deathbed, he pondered the ancient view of the universe. He was a conservative man, and he accepted much of Ptolemy's view, even with its troublesome epicycles.

Copernicus decided, however, that we should make one major change in our view of the universe, a change that was literally earth-shaking. Our earth, Copernicus decided, was really just another

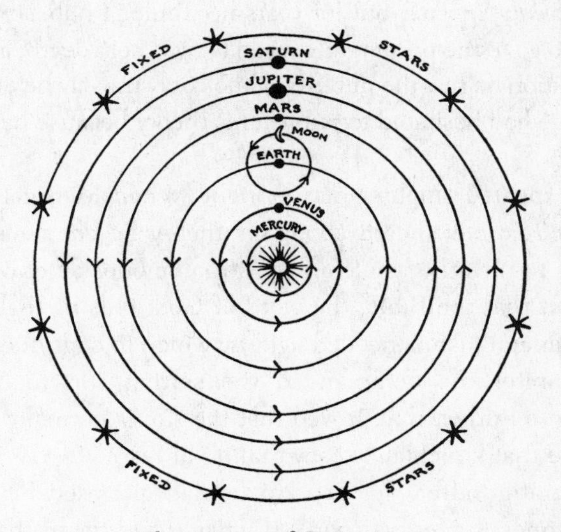

The universe of Copernicus

planet circling the sun. It was not the earth but the sun that stood still at the center of the universe, while earth was out on sphere number three. And why did Copernicus reach this daring conclusion? Did he have some startling data about the movements of the planets? No, in fact he hardly ever did any stargazing at all. He had only made a brilliant or a lucky guess.

Copernicus was only suggesting what he thought was a better explanation of the same data that Ptolemy and others had used. At first he seems to have believed that putting the sun at the center, instead of the earth, had the advantage of reducing the number of epicycles needed to explain what he called "the entire ballet of the planets." He liked the simpler picture because he had learned from ancient Plato that truth is always simple. In fact, however, it turned out that Copernicus needed *more* epicycles to explain the movements of the planets around the sun, not fewer; about fifty, instead of forty. Nevertheless, he thought the sun-centered universe was simpler and more elegant than the earth-centered one.

Copernicus wrote a book about his ideas called *On the Revolutions of the Heavenly Spheres,* but for years he wouldn't publish it. He permitted it to go to the printer only when he was near death, and according to tradition he saw the published book only the day he died in 1543. Probably he had hesitated to publish his theory because he was afraid of ridicule.

If he expected laughter he was right. When Martin Luther, the Protestant reformer, heard about his theory he complained, "That fool wants to turn the whole art of astronomy upside down." Luther pointed out that the Bible, the word of God, tells us that when the Israelite general Joshua needed light by which to continue slaughtering Canaanites, he commanded the setting sun to stand still. According to Luther, that proved that the sun is normally in motion. How could that knucklehead say that it's at rest?

A generation after old Copernicus had glimpsed his book and died, Johannes Kepler was born. If anyone had heard that this particular baby would one day chart the courses of the planets he would have thought it unlikely. Kepler's father was a German ne'er-do-well and wife-beater, who one time "ran the risk of hanging" for some unrecorded crime. He would later leave his family and die in exile. Much later, Johannes's mother was indicted as a witch, and narrowly escaped burning at the stake. In small ways, though, this couple introduced Johannes to the heavens. When he was six years old his mother took him up a hill to see a comet, and when he was nine his parents once called him outdoors to see the moon eclipsed.

Johannes seemed like poor material for greatness. He nearly died from smallpox and was always sickly. Worst of all for a future astronomer, he suffered from multiple vision. Nothing was easy for him. At the age of twenty-one, he writes, he "was offered union with a virgin; on New Year's Eve. I achieved this with the greatest possible difficulty, experiencing the most acute pains of the bladder."

Everything was against him but this: he was brilliant. As a result, he had, at government expense, a good high school and university education. He tells us that he wrote plays and learned long poems by heart, read the works of Aristotle in Greek, and "argued with men of every profession for the profit of [my] mind." He "explored various fields of mathematics as if [I] were the first man to do so [and discovered things] which [I] later found had already been discovered."

He also learned about Copernicus's "hypotheses," which put the sun at the center of the universe. Decades later Newton would show that neither Copernicus nor the old astronomy had it right. The sun does not go around the earth, or the earth around the sun; they go around each other. But Copernicus's explanation enraptured Kepler. He was deeply pious, and believed Copernicus had glimpsed a beauty in the heavens that was worthy of God, who had made them.

Kepler's big opportunity came in 1600 when a Danish astronomer invited Kepler to join his research staff. Tycho Brahe was widely known not only as a man who gazed at stars but also because, having lost the bridge of his nose in a duel, he had replaced it with another made of gold and silver. For twenty years he had made many thousands of observations of the movements of the planets. That was his great achievement: those many, many fixings of the planets as they journeyed through the skies. His measurements were not only precise but continuous; they showed where the planets went, and how long it took them to get there. It was as if Tycho had replaced still photos with a movie. Before his time astronomers had had only a limited number of observations to work with, but now, providing Tycho made his measurements available, they would have long series of them.

When Kepler joined him, Tycho's working years were over; he had little time to live. He probably sensed that what he needed was a greater scientist than he. He needed someone who could take that raw material he had gathered, that wealth of observations, and discover in it the structure of the universe. Probably he guessed that Kepler was that person. But he hated to give his life's work to another person, so he only occasionally threw Kepler scraps of information. On his deathbed, though, he bequeathed him all his data. Just before his death Brahe kept repeating, "Let me not seem to have lived in vain."

Kepler focused on Tycho's data, especially what the older man had learned about the movements of Mars, the planet that was never where it should be. He started by assuming that the planets traveled around the sun, as Copernicus had said, but without those complicated epicycles.

In 1609 Kepler published his discovery that Mars (like other planets, he assumed) does not orbit the sun in a perfect circle. It moves in an ellipse, the shape you get if you slice a cone not straight but at an angle. If the planets orbit in ellipses, not in circles, that means that they are not attached to spheres. (A sphere is the result of rotating a circle, not an ellipse, around one of its diameters.) Circular motion and invisible spheres had been key features of the older view of the universe, but out they had to go. Kepler was unhappy with this finding because he had grown up with the old idea of spheres and perfect circles. The ellipse, he felt, had nothing to recommend it except truth. He compared it to a load of dung that he had to bring into the heavens as the price for ridding them of a vaster amount of dung.

Later Kepler also found that a planet travels faster in its elliptical orbit when it's closer to the sun than when it's far away. And he found that the shorter a planet's mean distance from the sun the faster it completes its orbit. All these findings he expressed precisely, using numbers.

Kepler knew how much he had accomplished. He had described the structure of the universe, and he was ecstatic about it. "The die is cast," he wrote, "and I am writing the book [about his discoveries]—to be read either now or by posterity, it matters not." His book, he said, could wait a century for a reader, just as God had waited ever since he created the universe for a Kepler to see the wonder of it.

While Kepler was finding laws about the motion of the planets, Galileo Galilei was studying the heavens in a different way. Galileo was a mathematics professor at the University of Padua, in Italy, and a maker of compasses and surveying instruments. In 1609 he heard that a Dutchman who made eyeglasses had found that if he held two lenses in a certain way and looked through them at the same time, they greatly magnified distant objects. The Dutchman had, in short, invented the telescope. Galileo at once began to make one for himself.

Then he did what today seems obvious: he pointed a telescope up and looked at the night sky. He was not the first to do this, but the first top-notch observer. What he saw amazed him. Scientists had thought the moon, our nearest neighbor, was absolutely smooth. Galileo found it "full of hollows and protuberances, just like the surface of the earth itself, which is varied everywhere by lofty mountains and deep valleys." He saw four moons circling Jupiter, and many new stars—"more than ten times as many" as anyone had ever seen with the unaided eye. He discovered that the Milky Way was not just a vague white smear across the sky but "a mass of innumerable stars."

Galileo reported on his sightings in 1610 in a pamphlet called *The Messenger from the Stars.* It immediately made him famous. In the same year a poet in far-off England recorded that Galileo had "summoned the other worlds, the stars to come neerer to him, & give him an account of themselves." His university offered him a permanent job and a huge raise, but the Grand Duke of Tuscany (in Florence) offered to make him his court philosopher and mathematician. Of course he took the job in Florence, since it freed him from annoying colleagues and dim-bulb students.

Galileo's findings partly proved that Copernicus and Kepler were right: the earth was not the center of the universe. But *The Messenger*'s main effect was to excite people about astronomy and stir up arguments. Many astronomers sided with him, but some did not. They had spent their lives learning the old view of the universe and didn't want to see it challenged.

Galileo loved a fight, and he took to calling his opponents "mental pygmies" and "hardly deserving to be called human beings." Two professors at his university hadn't even deigned to peer through his telescope. When one of them died a little later, Galileo wrote that he "did not choose to see my celestial trifles while he was on earth; perhaps he will do so now that he has gone to heaven."

The Catholic Church began to take an interest. Theologians had never taken a stand about the structure of the universe because the matter hadn't seemed important. But Galileo championed a view of the universe, Copernicus's, that seemed not only new but shocking. Many churchmen who had never even heard of Copernicus now learned that he had fathered these disturbing ideas. An Italian bishop wanted Copernicus thrown in jail and was surprised to learn that he had been dead for seventy years.

The Church's leading theologian talked with Galileo in Rome and gently warned him that it was all right to discuss the sun-centered universe "hypothetically." But to say that the sun "in very truth" was at the center would be "a very dangerous attitude." It would arouse philosophers and theologians and tend to "injure our holy faith by contradicting the Scriptures." The official had in mind such passages in the Bible as the one Luther had cited against Copernicus, in which Joshua tells the sun to "stand thou still."

For almost two decades after he was warned, Galileo kept his fingers off his troublemaking pen. In 1632, however, he published a book that was bound to stir up trouble. Intended not for experts but for laymen, it was called *A Dialogue on the Two Great Systems of the World*. In the *Dialogue* three men discuss the old and new views of the universe. One of them is a persuasive scientist, clearly Galileo, who explains the new, sun-centered view. The other two are an intelligent layman and a pious, stupid one, who sticks to the old earth-centered universe, and makes himself look foolish. Galileo calls him Simplicio.

Galileo was summoned to Rome and questioned by the Inquisition, the Catholic body that combated "false" opinions. After several months, the Church came down against him. The *Dialogue* was not to be read by anyone, it said, and it censured Galileo for non-compliance (to the theologian's warning) in teaching the Copernican view. Told to admit his errors and confess his disobedience, he knelt and did so.

He went home to Florence and spent the rest of his life there under house arrest. During his last four years he was blind, perhaps from having peered at the sun through his telescope. Not long before his death he wrote a friend, "this earth, this universe, which I, by marvelous discoveries and clear demonstrations, have enlarged a hundred thousand times beyond the belief of the wise men of bygone ages, henceforward for me is shrunk into such small space as is filled by my own bodily sensations."

Despite the Church's stand, opinions changed. By the time of Galileo's death, every educated person had heard, and many believed, that the earth was not the center of the universe, and that the heavens seemed to stretch forever into space. Astronomers also knew that Kepler had described how planets moved.

No one knew as yet what force held the planets and the stars together and made them move the way they did. Some scientists had offered explanations. For example, William Gilbert, an Englishman who had studied magnetism, speculated that the same mysterious force that made a compass needle point to north also held the planets in their courses.

In 1642, the year Galileo died, the man was born who would bring this question to a kind of resolution. This scientist was Isaac Newton, the son of an English farmer who couldn't sign his name and who had died before his son was born. Newton grew up in a country house, went to the village school, and spent much of his boyhood drawing and making water clocks and little windmills, powered by a mouse. Then he went to Cambridge University, where he proved to be astoundingly brilliant. While other students were chasing girls and foxes, Newton invented the well-known binomial theorem, which states that, for any positive integer *n,* the *n*th power of the sum of two numbers . . . well, never mind.

In 1666, when he was twenty-three, plague broke out in Cambridge.

To escape it, Newton went home and stayed there a year and a half. And how did he fill his time? Well, for one thing he invented calculus, a branch of mathematics that deals with the effect of changes in one of several variables. (He would need this tool to analyze the pull of planets on each other.) *And* he developed his ideas about how forces here on earth act on moving bodies. *And* he discovered the composition of white light. *And,* as if all that were not enough, he began to calculate that force that Kepler had described, that glue that holds in place the stars and planets, that drawing power we now call gravitation.

When someone asked him how he made all these discoveries, Newton answered simply, "By thinking upon them." He was generous enough to admit that he had built upon the work of other scientists, such as Kepler and Galileo. "If I have seen further," he said, "it is by standing upon the shoulders of Giants."

Decades later, Newton told someone that he began to think about gravitation one day in 1666 or 1667 when he was lying in his orchard, and saw an apple fall. He asked himself why it fell straight toward the center of the earth. A "drawing power" pulls it, he surmised, and that power must be in proportion to the sizes of the earth and the apple. Newton had no idea *what* the drawing power was. It might be magnetism, as Gilbert had suggested; or whirlwinds of invisible matter, as the French scientist René Descartes had guessed; or the omnipotent hand of God. Not knowing *what* the force was, how hard it must have been to figure *how* it worked!

Newton concentrated on the moon. By rights, the moon should wander off. But earth, which is bigger, attracts it just enough to keep it orbiting our planet. Newton imagined the moon in its orbit as if it were a stone that a farm boy whirls around his head in a sling and then lets fly against a rabbit. Then Newton worked out *how* to calculate the force needed to keep an object here on earth moving in a circle, like the stone in the sling.

Using this formula, Newton calculated the force that the earth exerts on the moon to keep it in its orbit, that is, to keep the moon from flying away from earth. And then he showed that earth's pulling-in force on the moon precisely equaled the moon's flying-away force. This led him to formulate a law describing the gravitational pull between all heavenly bodies, such as the planets and the sun. This force, he said, is inversely proportional to the square of the distance between them. All of the universe depends on this measurable force that moves masses through space and time. (However, physicists now know of major exceptions to this law.)

Perhaps that does not sound like such a big achievement. After all, Newton had not identified the force that holds the universe together. He did not pretend to know what that binding-together force is, and neither do we, to this day. But he had, as he said, demonstrated "the frame of the System of the World." He had tied together with mathematics the findings of Kepler, Galileo, and other scientists.

What Europeans had learned in the century and a half from Copernicus to Newton was unsettling. We humans are not at the center of the universe. Rather, we live, as the astronomer Carl Sagan once wrote, on an "insignificant planet of a humdrum star lost in a galaxy tucked away in some forgotten corner of a universe in which there are far more galaxies than people."

Some found it alarming to imagine a universe stretching far beyond what one could see with his eyes, or even with a telescope. That space apparently was endless and nearly empty, and the sun, the planets, and the stars rolled on silently, indifferent to us. In such a universe it was hard to imagine a fatherly God concerned above all with the doings of humans he had created in his image. Blaise Pascal, a French scientist and devout Christian, famously described his anguish at the thought. He felt "engulfed in the infinite immensity of spaces whereof I know nothing, and which know nothing of me. I am terrified by the eternal silence of those infinite spaces."

Most of those who learned of this new view of the universe, however, felt no sorrow at losing their favored place in the universe. On the contrary, poets acclaimed what had been learned. Alexander Pope wrote:

Nature and Nature's laws lay hid in night:
God said, "Let Newton be!" and all was light.

An Italian writer charmingly explained the findings of the astronomers in a book called *Newtonianism for the Ladies.*

The report of the finding that earth revolved around the sun spread around the world. In the 1600s, while Kepler, Galileo, and Newton were at work, Catholic priests from Europe were serving the emperor of China as astrologers, absurdly predicting the future for him by studying the positions of the stars and planets. But these men were also missionary teachers, and we know that during most of the 1600s they also taught the Chinese the new, sun-centered view of the universe. They also introduced the telescope, and they printed astronomy books in Chinese.

When their teachings reached Japan, the Japanese quickly accepted the new sun-centered idea. Japanese scholars explained that the sun was really an ancient god of the Japanese, "the god who rules the center of the heavens." So the brand-new science was really their ancient faith.

Many hailed Newton as the greatest scientist of any age, and Newton, who was anything but modest, probably agreed. But a little before he died he told a friend, "I do not know what I may appear to the world, but to myself I seem to have been only a boy playing on the sea-shore and diverting myself in now and then finding a smoother pebble or a prettier shell than ordinary, while the great ocean of truth lay all undiscovered before me."

IT WAS SHOCKING enough to learn that a human being is merely a passenger on a small planet who during his lifetime takes a few dozen trips around a local star.* Bigger news was on the way.

The leading bearer of these tidings was Charles Darwin, who was

*I paraphrase from Carl Sagan and Ann Druyan, *Shadows of Forgotten Ancestors: A Search for Who We Are* (1992), p. 30.

born in western England less than a century after Newton's death. Unlike Newton, Darwin came from a brainy family. Someone once described a great-grandfather of Darwin as a "Person of Curiosity" who had found a "a human Sceleton [*sic*] impressed in Stone." A great-uncle wrote *A Concise and Easy Introduction to the Sexual Botany of Linnaeus.* His paternal grandfather, Erasmus Darwin, when he wasn't chasing women or stuffing himself at the dinner table (from which a crescent had been cut to fit his awesome belly), had written a great deal about medicine and botany. Another grandfather was Josiah Wedgwood, a famous pottery manufacturer. And though he hated the sight of blood, Darwin's father made a good living as a doctor.

As a boy Darwin was an average student, but he watched birds, collected rocks and beetles, and did so many chemistry experiments that his friends had named him "Gas." He also loved to hunt, and his disgusted father told him, "You care for nothing but shooting, dogs, and rat-catching, and you will be a disgrace to yourself and all your family." He studied medicine in Scotland for a while but didn't like it, so his father decided to transfer him to the Church, which was the last refuge of wellborn dullards.

Darwin went to Cambridge, as Newton had done, and he studied theology. On the side, however, he took science courses, and these turned him into a keen naturalist. A botany professor took him on field trips and taught him how to observe living plants. At one point Darwin read with great excitement Alexander von Humboldt's *Personal Narrative of Travels to the Equatorial Regions of the New Continent.* (The continent was South America.) Many years later Darwin wrote, "My whole course of life is due to having read and re-read as a youth [Humboldt's] 'Personal Narrative.' "

It looked as if Darwin would spend his life as a country parson, hunting beetles six days a week and preaching on Sundays. But then, out of nowhere, came a splendid opportunity to follow Humboldt's path. On his botany professor's recommendation, he was invited to serve as naturalist on a scientific expedition. A ship called the *Beagle* was to sail down the east coast of South America and up the west coast, doing navigation research, and then to sail around the earth. The voyage would last about two years. Charles's father first opposed this detour from his journey to a parsonage, then grudgingly agreed.

The voyage lasted not two years but five. Whenever the *Beagle* touched shore, Darwin walked the beach and climbed the hills, gathering specimens and scanning everything in sight. He asked himself a host of questions. Why did coral island atolls form in circles? Why are fossil seashells to be found two miles high in the Andes? Why do mockingbirds and finches on one of Ecuador's Galapagos Islands have sharper beaks than the same species on the next island? (Had God taken all that trouble when he created them?) And (in the back of his mind) were humans merely another kind of animal?

When he was back in England, Darwin thought no more of serving God. (He married a wealthy Wedgwood cousin and lived comfortably on her pottery fortune and what his father left him.) For twenty years he led a double life. In fact he was writing a great book. But as far as most friends knew he was writing lesser pieces and pursuing his hobby, observing worms.

He discussed with almost no one what was really on his mind, "the species problem." Why, he asked himself, had there been in the past, and why were there now, so many, many kinds of animals and plants? The Bible said that God created every form of life in just a week, but was that true? It seemed unlikely. For instance, the ancient armadillos, whose fossils Darwin had observed, had disappeared and been replaced by modern, slightly different ones. So God's creation hadn't been the end of their story. Species seemed to come and go, and this, said Darwin, was the "mystery of mysteries."

Darwin was not the only person who was wondering how species, including humans, had evolved. The question was in the air, and scientists in several countries were asking it. In 1840 an English novel called attention to the topic. A fashionable lady tells a friend what she has read. "You know," she says, "all is development. The principle is perpetually going on. First there was nothing, then there was something; then—I forget the next—I think there were shells, then fishes; then we came—let me see—did we come next? Never mind that; we came at last. And at the next change there will be something very superior to us—something with wings. Ah! That's it; we were fishes, and I believe we shall be crows."

Darwin found a clue to the answer he was looking for in a book by Thomas Malthus, another Englishman who had escaped a career as parson. In a well-known *Essay on Population,* Malthus had written that human beings increase faster than they can increase the production of their food. The number of humans therefore rises until they run out of food, at which point famine, war, and sickness violently stop the rise. When that happens, many die but not all. Some survive.

Reading Malthus, Darwin wondered if that struggle to survive didn't influence whole species. Imagine a litter of piglets, suckling from their mother. If most of them are rough and tough but one is meek and weak, which will fail to nurse and therefore starve to death? The runt, of course. The others (till the farmer baconizes them) will live to parent more piglets with the useful traits of size and greed. But the flaws of the runt—its gentleness and weakness—will vanish, and a good thing that will be. Darwin called this process "survival of the fittest." If it happens often, evolution will occur. Pigs in general, as a species, will grow greedier and bigger.

Darwin could explain the fact of evolution, but he didn't know exactly how it happened. Slowly he accepted the opinion of a French biologist, the long-dead, long-named Jean-Baptiste Pierre Antoine de Monet, chevalier de Lamarck. Lamarck had held that animals developed traits when they needed them, and passed these acquired traits on to their offspring. If a giraffe, for instance, stretches his neck to reach the highest leaves, his neck grows slightly longer. He passes that extra bit of length on to his offspring, and after many generations all giraffes have necks like . . . giraffes.

But Lamarck's conjecture about acquired traits raises problems. For example, Jews and Muslims have been circumcising their sons for thousands of years, but Muslim and Jewish boys still are born with foreskins. As we shall see in chapter 24, a monk in Austria-

Hungary was working out how plants and animals inherit traits. Darwin didn't know about his work.

Just as Darwin was completing his book about the origin of species, he received a shock. Another Briton, Alfred Wallace, mailed him an essay on his own theory of the origin of species, and it perfectly summarized the theory Darwin had been working on for twenty years. For Darwin this looked at first like a disaster, although he had developed his theory in much more detail. The two men civilly agreed to have reports on their work read in 1858 at a meeting of a scientific society. Although they reported an immense finding, Darwin and Wallace got scant attention. As 1858 concluded, the president of the society reported to the members that the year had produced no "striking discoveries."

The next year Darwin sent a publisher his book, which he called *On the Origin of Species by Means of Natural Selection, or, The Preservation of Favoured Races in the Struggle for Life.* The publisher was not enthusiastic. When he asked the editor of a journal to give him an opinion the man advised him not to publish. The subject was controversial, he declared, and Darwin ought to write a book on pigeons, on which he was known to have some clever thoughts. He added, "Everyone is interested in pigeons."

But the publisher went ahead with the *Origin.* It immediately sold out and the publisher reprinted it. It was widely reviewed and translated into many languages—even Japanese, Darwin happily reported.

What Darwin had to say did not please everybody, and it shocked the clergy of the Church of England. After all, the works of God were their concern. Darwin was diminishing the role of God, who, the Bible said, created all the species in three days. Darwin claimed that nature followed never-changing laws, just as Newton had explained that stars and planets do. That raised a troubling question: if God wasn't needed to run the heavens *or* our earth, then what *did* he do with his time?

The churchmen and some others also were upset to read that God had not created humans "in his own image," as the Bible said.

Darwin hinted, only hinted, in the *Origin,* that humans had evolved from other—most said "lower"—beasts. A hint was quite enough; they knew that Darwin was implying apes. Some were just amused. *Punch* magazine ran jokes such as: "I could a tale [readers were meant to think also of *tail*] unfold." "Could you? Then lose not a moment, but go instantly to Mr. Darwin. He will be delighted to see you." But churchmen didn't think descent from apes was funny.

Six months after *Origin* appeared, the British Association for the Advancement of Science assembled at Oxford University to discuss the book. Darwin hated controversy so he stayed away. The leading speaker was his enemy, Samuel Wilberforce, Bishop of Oxford. Widely known as "Soapy Sam," Wilberforce was clever on his feet, never troubled by a lack of facts.

In a large lecture hall, packed with scientists, ladies, and Oxford students, the bishop smoothly ridiculed the theory of evolution. He appealed to the audience's gallantry, asking whether women, as well as men, were descended from beasts. Then he turned to T. H. Huxley, a scientist friend of Darwin who was there to defend the *Origin.* Unwisely, Wilberforce asked if it was through his grandfather or his grandmother that Huxley claimed descent from a monkey.

Then came Huxley's turn. First, he soberly defended Darwin's views. He explained that Darwin didn't mean to link apes and humans directly but only to show that both had descended, over many thousand generations, from a common ancestor. And then—a famous moment in the history of ideas—Huxley answered Wilberforce's granddad question. He wouldn't be ashamed, he said, of "having an ape for his grandfather." He *would* be ashamed, he went on, looking at Wilberforce, to have a human ancestor who plunged into scientific matters about which he knew nothing, and who used "aimless rhetoric," "eloquent digressions, and skilled appeals to religious prejudice" to divert attention from the point at issue.

One didn't speak like that to bishops. Students stood up shouting. A woman fainted and was carried out.

Huxley won only a skirmish, not a war, and many pious Christians would be slow to change their minds on evolution. But the clash of

Wilberforce and Huxley (whom people nicknamed "Darwin's bull-dog") dramatized the issue. They alerted many to a new opinion on the origin of species and, it followed, on the beastlike nature of us humans. Many English scientists quickly accepted the theory of evolution, probably saying to themselves, "Of course! Why didn't I think of that?" Within a decade after *Origin* appeared, the natural science examinations at Cambridge University no longer asked for "evidence of design" in nature. Instead they asked for an analysis of the struggle for existence. In other countries it took Darwin's theory about a generation to win many converts.

Meanwhile, Darwin had more to say on evolution. A decade after he produced the *Origin,* he wrote another book, *The Descent of Man.* Now he boldly stated that the long-ago ancestor of humans was "an aquatic animal, . . . with the two sexes united in the same individual." Our more recent ancestor was a "hairy quadruped, furnished with a tail and pointed ears, probably arboreal in its habits." This sounded very like an ape, but *The Descent of Man* caused little stir. The *Origin* had readied everyone to hear the worst.

Let Darwin tell us what he thought about the human story. At the end of *The Descent of Man* he wrote, "Man may be excused for feeling some pride at having risen, though not through his own exertions, to the very summit of the organic scale; and the fact of his having risen, instead of being placed there aboriginally, may give him hope for a still higher destiny in the distant future."

CHAPTER 13

Here and there, the people rule.

NOT SO LONG AGO, democracy on planet earth was rare. Far and wide, the few decided for the many, and usually these few were a ruler and an upper crust of wealthy, landed men. It was they who gathered taxes, wrote the laws, branded thieves, and went to war. True, the kings and gentry didn't always get along, and much of "history" is the story of their wars with one another. But on one thing they agreed: the only people with the right to rule were those who owned the land that others worked. Almost no one dreamed of giving power to the vile and dirty mob: the artisans and weavers, the odds and sods, the country clods. Your ancestors and mine.

In the later decades of the 1700s, though, the long, long reign of kings and gentry neared the end. The world began an age of democratic revolutions, which aimed at putting power in the hands of nearly everybody, rich and poor, male and female, red and white and black.

. . .

SURPRISINGLY, THE FIRST of these revolts took place in that far-from-everybody continent, North America. We must give the background.

Two centuries earlier, this continent appeared on maps of the world as just a blur. Great Britain claimed to own a lot of it, and in the 1600s handfuls of Britons, especially Englishmen, left their homes and sailed in little ships across the ocean to North America, where they settled on the eastern coast. Just beyond the sandy shores were virgin forests, the beginnings of three thousand miles of lightly peopled land. Other settlers followed, and the migrant trickle turned into a river. As well as Britons, it included Germans, Frenchmen, Dutch, and Africans.

Except for the Africans, the new arrivals took what land they wanted, or they "bought" it from the woodland Indians. The Indians hadn't any notion what it meant to "own" the land they lived on, but the new arrivals quickly conned them into "selling" it. Indians sold Manhattan Island (heart of modern New York City) to some Dutchmen for cloth and trinkets worth about a pound and a half of silver. The settlers chopped down trees, erected cabins, and planted corn (that is, maize). Indians showed them how to grow this grain, which was new to Europeans, in the interval before the new arrivals killed them off or drove them west.

The British settlers were ordinary folk. Isaac Allerton, for example, had been a tailor in England. He came to Massachusetts with his family and other "Pilgrims" who were seeking land to farm and the right to worship as they pleased. Allerton served his fellow Pilgrims as their business agent until they decided that he "plaid his owne game" and "hoodwinckte" them, and drove him out. Then he made and lost a fortune trading furs, tobacco, slaves, and rum, up and down the coast.

Another immigrant, Gabriel Leggett, came from southern England, where he probably was a younger son who had no chance of being left the family farm. He came to New York colony and married the daughter of a farmer who had "bought" his land in what is now New York's South Bronx from Indians. Someone said of Gabriel that he had a "notories ill behaved & wicked meletious nature." Just the same, he prospered as a farmer and by selling timber to builders in what would later be New York's Harlem section.

Unlike most humans elsewhere, these transplanted Europeans lived with one another on an equal footing, in a rude democracy. Travelers from other countries were surprised to see that workers not only ate their meals with their bosses but expected to be served the best food in the house. When the wealthy American printer, Benjamin Franklin, traveled in France, the gap between the rich and poor appalled him. He wrote, "I thought often of the Happiness of New England where every Man is a Freeholder, has a Vote in publick Affairs, lives in a tidy, warm House."

Not all Americans, however, enjoyed respect and equal rights. At best, the whites had mixed opinions of the Indians. They admired the Indians' forest skills, but whites who lived on lonely farms were prone to think the only good one was a dead one. Sometimes whites attempted to enslave the Indians, but they simply vanished in the forests. The Indian tribes moved west, away from whites. Only the fearsome Iroquois, east of the Great Lakes, held their ground.

In the eyes of the colonists another group of Americans ranked far below the Indians. These were the imported blacks. When the settlers needed farmhands, many of them purchased blacks—slaves from Africa. The story of these slaves was uniformly tragic. In their homeland, gangs of other blacks had stormed their homes, shackled them, and death-marched them to the coast. White slave dealers there had bought them from their captors, loaded them on ships and chained them down between the decks in heat and filth, and brought them to the New World. One such captain of a slave ship was the Englishman John Newton, who later wrote the hymn "Amazing Grace."

The shock of capture, followed by the horror of the journey to the New World, was too dreadful to imagine. A slave ship captain once recorded that "to our great Amazement about an hundred Men

Slaves jump'd over board." His sailors rescued two-thirds of the slaves, but the others "would not endeavour to save themselves, but resolv'd to die, and sunk directly down."

Slavers first imported African blacks to the English colonies in 1619, only a dozen years after the first Englishmen arrived there. A white settler in Virginia wrote about it casually, "About the last of August came in a dutch man of warre that sold us twenty Negars." Slavery soon was common in the south and not unusual in the north. For example, the New York farmer Gabriel Leggett owned a dozen slaves. In Philadelphia, Benjamin Franklin, later a champion of freedom, had four slaves. One was named Othello.

It was whites of course who shaped the laws concerning slaves. Owners of slaves could, in effect, do anything they wanted to them, even bludgeon them to death. The slaves, of course, served not for years and not for decades but for life, as did their children and their children's children ever after.

By the latter half of the 1700s Great Britain owned thirteen colonies in North America, in addition to Canada, which it had recently won from France. The colonies stretched all the way from Maine's frigid lobster shoals to Georgia's Okefenokee Swamp. Great Britain's king and Parliament (the legislature) tried to rule them from three thousand miles away.

Problems rose when Britain found itself in debt. The government decided, sensibly, that its American colonists should pay their share of taxes. It also demanded that Americans should trade only with Britain, thus enriching their mother country and not its rivals, France, the Netherlands, or Spain. And it ordered that the colonists not provoke the Indians by moving into Indian lands beyond the Appalachian Mountains.

Americans rejoined that they could handle their affairs quite well alone. Why, they asked, did Parliament, in which Americans weren't represented, have the right to gather taxes from them and instruct them what to do? To this the British answered that Parliament represented America just as much as it did all of Britain. True, they said, Americans sent no representatives to Parliament, but neither did the thriving English town of Liverpool. Both places, so they said, had "virtual representation" in Parliament. Members of Parliament spoke not only for the places that elected (or appointed) them but also for the whole.

Americans protested and refused to pay the taxes. In the town of Boston, in the colony of Massachusetts, men opposed to Britain's tax on tea disguised themselves as Indians, boarded British vessels after dark, and dumped 342 chests of tea leaves into the water. Britain hesitated to use force against the colonies, but King George III declared, "We must either master them, or totally leave them to themselves." In 1775 British soldiers clashed with American volunteers near Boston. Thus began a minor war that had immense results.

Delegates from the thirteen American colonies gathered in the spring of 1776 at Philadelphia, determined not to yield to Britain. Early in July the delegates declared to all the world that the colonies were no longer under British rule; they now were independent.

They explained the reason in a document they called "The Unanimous Declaration of the Thirteen United States of America." To a world that only knew hard-fisted kings and haughty gentry, the declaration must have been astounding. It said, "We hold these truths to be self-evident, that all men are created equal, that they are endowed by their Creator with certain unalienable Rights, that among these are Life, Liberty and the pursuit of Happiness." To secure these rights, "Governments are instituted among Men, deriving their just powers from the consent of the governed." And "men" (not including slaves) have a right of revolution. "Whenever any Form of Government becomes destructive of these ends [Life, Liberty and the pursuit of Happiness], it is the Right of the People to alter or abolish it."

The principal writer of the declaration was Thomas Jefferson, a young Virginia planter. Jefferson had it all: charm, friends, a hilltop mansion he had designed, research gardens, books, facility in seven languages, and a gift for writing prose that sings. He also had a moral conflict: he loved democracy but owned 150 slaves. (As DNA tests have shown, Jefferson fathered at least one child with a slave. This woman was herself the daughter of a slave-owner, Jefferson's father-in-law.) Jefferson knew very well that owning slaves contradicted his own words about the rights of men. He hated slavery and said so, but he didn't hate it enough to free his slaves.

As commander in chief the rebels chose the capable George Washington, who, like Jefferson, was a wealthy Virginia planter and slaveholder. Years before, Washington had fought beside the British against the French and Indians on the colonies' west frontier. When bullets whistled around him he found "something charming in the sound."

As the war began, it was hard to see how either side could win. There was too much America for Britain to suppress, and too few Americans to keep them from suppressing it. For several years the little war was indecisive. The turning point arrived when France, which earlier had lost a war with Britain, joined the Americans. Men and ships from France made the difference. The fighting stopped in 1781 after Frenchmen and Americans cornered a British army in Virginia and forced it to surrender. According to a legend, as the British troops gave up their arms their bandsmen played a tune called "The World Turned Upside Down."

When the war had started, the thirteen colonies joined to form a government of thirteen states. But the feeble "Continental Congress," as the government was called, had shown that it could barely carry on a war or regulate the trade between the thirteen states. When the war was over, the states decided they must write a better "constitution," or framework of a government.

To reach agreement wouldn't be a snap. In the recent war the thirteen delegations in the Congress hadn't always pulled together. They were different kinds of people. Other regions saw New Englanders as "dam'd Yankees" who wore black stockings and had rasping, whining voices. Washington once said they were "an exceedingly dirty and nasty people." For their part, many Yankees saw the southerners as slave-abusing, bourbon-sipping dandies.

With a government to build, delegates from all the states but one (Rhode Island) gathered once again in Philadelphia during the spring and summer of 1787. Laborers strewed gravel on the street outside their meeting room so that the delegates might concentrate undisturbed by clattering hooves and rattling wagons. Washington, the hero of the just-concluded war, presided, sitting on a high-backed chair. The flies were bad and the weather hot.

It took the delegates four months of wrangling to produce a constitution. Although it filled only four parchment pages, each was packed with crisply worded fundamental laws. The simple fact that it was *written,* which today seems natural, was in fact remarkable. Other governments had no such thing; their kings and gentry followed ancient, unrecorded rules or none at all. In the coming decades, democratic rebels elsewhere in the world would often follow the American example, writing constitutions that embodied social views.

The United States Constitution starts out boldly with the then-surprising words "We the People." Right from the beginning, the shapers of the government made clear that it rested on consent by all. That they stated this was no surprise. Hadn't the Declaration of Independence affirmed a decade earlier that "all men are created equal"? If all were equal, all must share in governing.

In fact, however, the delegates were far from all-out democrats. These "Founding Fathers" didn't trust their children, or at least not all of them, when it came to choosing men to govern. They stipulated that the president was not to be elected directly by all the "People" but rather by "Electors" chosen by the states. They hoped that these electors would be prudent men of means—"the better sort," not sodden sailors or benighted bumpkins.

Senators as well would not be chosen directly by the ordinary voters (whom the delegates considered "less fit judges"). State legislatures would elect them. (But the framers might have made the Senate much less democratic. One delegate, New York's Alexander Hamilton, wanted senators to be chosen only from men of property, and to serve for life.) The delegates decided that only "Representatives," the other "house" of Congress, would be chosen by the ordinary voters—men only, of course.

The delegates gave little thought to letting slaves be citizens, which would have meant the end of slavery. At this time one in five Americans were slaves. This fact embarrassed many delegates, whether from the North or South. No one had to tell them that the Declaration of Independence stated that "all men are created equal."

But the northern delegates knew better than to urge that slaves be citizens. The southerners would never have agreed, or if they had the people of their states would later have refused to vote in favor of the constitution. So they reached a compromise: not to end slavery but to forbid importing slaves after 1808. Some may have dreamed that after that date slavery would fade away.

At last their work was done. As they finished old Ben Franklin gazed across the room at the high-backed chair from which Washington had presided. The sun was painted on the chair back. Franklin remarked that he had "often and often in the course of the session" looked at that sun "without being able to tell whether it was rising or setting; but now at length I have the happiness to know that it is a rising and not a setting sun." When he went outside, a woman asked him, "Well, Doctor, what have we got, a Republic or a Monarch?" "A Republic, madam," he answered, "if you can keep it." After much debating up and down the coast, the states approved the Constitution, and the ship of state was launched.

More democracy was on the way. The first Congress and the states amended the Constitution by adding to it guarantees of certain rights "of the people." These included freedom of speech and "the right of the people peaceably to assemble, and to petition the Government for a redress of grievances." Later on the country dropped those limits that the framers had set on elections of presidents and senators. By doing so, it gave full citizenship to all the "People."

All the People that is, but Indians, slaves, and women. Those were rather big exceptions. In the next half century, slavery, far from withering, would thrive and spread. In the early 1860s slavery would cause the North and the South to fight a bloody Civil War. The war resulted in the end of slavery, but its effects would linger. Only a century later, after bitter struggle, did all blacks truly get the right to vote.

Indians would fare no better. White Americans would push them west and pen them in on barren lands the whites disdained. Many Indians would die of drink, disease, and despair. Only a century and a half after the signing of the Constitution would Congress grant them citizenship. Only later still would it repeal the laws that made it legal to hold Indians as virtual prisoners on "reservations."

At about the same time various states would grant women the right to vote. In 1920 Americans amended the Constitution to enfranchise women throughout the country.

WHEN THE UNITED STATES carried out its revolution in the 1770s and '80s, it was just a forest outpost on the western edge of European civilization. Compare America with France, in 1789, on the eve of *France's* revolution. France was Europe's strongest, richest power and its cultural leader. It was in the language of the French that Austrian and Russian envoys threatened one another with a war. Everyone who counted learned from France what clothes to wear, what food to eat, what books to read, what thoughts to think. If the king of France began to build a palace, other rulers summoned architects. If the king of France had a pretty, witty mistress, other rulers . . . did their best. (The mistresses of Britain's George I were called the Maypole and the Elephant.)

On the eve of France's revolution, in 1789, the king of France was Louis XVI. He had inherited the throne fifteen years earlier, at the age of nineteen, when his grandfather died of smallpox. ("What a burden!" Louis exclaimed when he heard the news. "At my age! And I have been taught nothing!") Louis was a shy and decent man who relished eating, hunting, and tinkering with locks. He and his queen, Marie Antoinette, and their little son lived twenty miles from Paris at Versailles. Their ostentatious palace, more than a third of a mile long, held not only the seat of government but apartments for a thousand courtiers and four thousand servants.

France was deep in debt. The government was spending half its revenues just to pay the interest on the money it already owed. Louis's ministers tried to increase revenues by taxing nobles, clergymen, and government jobholders, all of whom were exempt from taxes. But the noble law courts blocked this sensible reform. So Louis reluctantly agreed to summon the Estates General in the spring of 1789. The Estates were an ancient, almost forgotten body of nobles, churchmen (bishops and priests), and well-off commoners, called the "people." Kings had sometimes summoned them to give consent to taxes, but hardly often; the Estates had last assembled 175 years before.

After the Estates had gathered at Versailles, the third estate, the "people," raised a parliamentary issue. The ancient custom was that each estate (or "order") voted as a block. The practical effect of this was that the nobles and the clergy (bishops and priests) could outvote the "people." That may have been acceptable in 1614, but by 1789 France was a modernizing country with a growing middle class of lawyers, businessmen, and lesser landlords. Their deputies would not agree to have their voices count only a third.

Events moved fast. Boldly, the third estate renamed themselves the "National Assembly" and invited the nobles and the clergy to join their sessions. When Louis therefore barred the third estate, now the National Assembly, from their usual meeting place, they gathered in an indoor tennis court. There they took an oath never to disband until they had given France a constitution. The king considered using force against them, but reliable troops were far away, and he gave in. He told the noblemen and clergy to join the National Assembly and vote "by head," rather than by "order."

A revolution had begun. It would move steadily to the left, against the nobles and the clergy and toward democracy, but also toward violence. One force that drove it on was hunger. Hail and drought had cut the harvests during 1788, and by the spring of '89 the price of grain was high and bread was hard to find. The problem was momentous, since lower-income Frenchmen nearly lived on bread. Hungry workers rioted, and housewives laid siege to bakers. In the countryside, peasants poached the deer and rabbits in their landlords' woods and raided convoys hauling wheat on country roads.

In July, several hundred Paris workmen and shopkeepers stormed a fortress-prison known as the Bastille. They wanted weapons to defend the city from King Louis's troops, who were known to be gathering in the Paris area. They killed the fort's commander and some soldiers, and freed the prisoners—all seven of them. In response, Louis pulled his other troops out of the city. The fall of the Bastille excited millions of the French; to them it proved the power of a determined people.

Meanwhile, rumors spread that wealthy lords were sending squads of thugs to massacre the peasants. This mass delusion, this "Great Fear" as it was called, spread from village to village. In parts of France the peasants grabbed their hoes and pitchforks, anything that they could use as weapons. They broke into the landlords' country homes and barns, looking for the grain they thought the rich were hoarding, and also for the records of the dues and rents that they had had to pay their lords for centuries. They burned some barns and lynched a few aristocrats. Many others fled to cities and later left the country.

The National Assembly knew it must grant something to the peasants, preferably at the expense of the detested noblemen and higher clergy. At a dramatic evening meeting during August the nobles—the liberal nobles who were present—agreed to surrender their right to the dues that some peasants had to pay their lords. The Assembly also agreed that everyone, including nobles and privileged townspeople, would pay taxes, and no one would be made to pay the Church a tax on income.

The National Assembly then approved a Declaration of the Rights of Man. "Men are born and remain free and equal in rights," it said. "These rights are liberty, property, security, and resistance to oppression." Those words are much like those of the American Declaration of Independence, which Americans had written only thirteen years before. Do they show that the American Revolution influenced France's revolutionaries? They do. However, it is also true that for some time now ideas of equal rights had been in the air in western Europe. People known as "democrats" had already been skirmishing with aristocrats.

In autumn, bread was still expensive, though the harvest had been good. The lively new Paris newspapers spread the story that the queen, when someone told her that the people had no bread, had joked, "Let them eat cake!" The story wasn't true, but it angered many. On a rainy October day hundreds of women—housewives, shop assistants, even "ladies with hats"—marched all the way from Paris to Versailles demanding bread. They stormed inside the palace and they almost caught and lynched the queen. The next day they marched back to Paris, proudly bringing the royal family with them, virtually as prisoners. They were sure these three would somehow put more bread on kitchen tables, and they chanted, "We're bringing the baker, the baker's wife, and the baker's lad to Paris!"

The fall of the Bastille and Louis's humiliation ended royal rule, although the king did not step down. After Louis had decamped from royal Versailles to rebel Paris, the National Assembly did the same. Now the deputies could safely go ahead and carry out reforms and write a constitution.

Overnight, they broke apart an ancient social system, as an able traffic cop dissolves a gridlock. Among many other acts, they sold the lands of nobles who had fled, proclaiming that "the properties of patriots are sacred, but the goods of conspirators are there for the unfortunate." They confiscated the land of the Catholic Church. They ended slavery in France's New World colonies. And what was most important, they wrote a constitution, as Americans had done. It gave the vote, the key to democracy, to the better-off taxpayers. Later, almost every Frenchman got the right to vote, but the revolutionaries made sure that only their supporters exercised the right.

If Louis had been wise enough to go along, who can say what would have happened? But he wasn't. The king and queen fled Paris, disguised as a valet and a governess, but an official recognized them, and soldiers took them back to Paris. A year and a half later revolutionaries tried the king for treason, convicted him, and chopped his head off. Eight months later they also killed the queen. They decapitated both with a guillotine, a new device that was said to slice off heads humanely.

In 1793 the Revolution reached its radical extreme, called the Reign of Terror. Other nations had been waging war with France to put a king back on the throne. Fearing enemies both in and outside France, revolutionaries handed almost all power to a twelve-man body called the Committee of Public Safety. It leader was Maximilien Robespierre, a lawyer who powdered his hair and dressed in knee breeches like the hated gentry, but who preached of terror. "Terror is nothing but justice, swift, severe, and inflexible. . . . Terror is the mainstay of a despotic government. . . . The government of the revolution is the despotism of liberty against tyranny."

Whatever that may mean, he meant it. Within a year, the guillotine and older methods took the lives of nearly twenty thousand "enemies of the people." Some of these were truly foes, who fought against the revolution sword in hand. Many priests and noblemen were guilty mainly of not fleeing France when they could have. As the months went by, many victims of the terror were moderate revolutionaries, guilty of such crimes as inspiring discouragement, misleading opinion, depraving morals, and harassing patriots.

In July of 1794 the moment came when everyone had witnessed too much terror, too much blood. Even those in power, close to Robespierre, were fearful they would be the next to lose their heads. Suddenly, with little talk, many changed their minds about the Terror. On July 27, Robespierre wanted to address the legislature. Shouts rang out: "Down with the tyrant!" The chairman refused to give him the floor, and the members ordered his arrest. On the following day the guillotine took Robespierre.

With his death the Terror stopped as if a cooling rain had swept through France. The Revolution now was near its end, and moderates took charge. They wrote a constitution that preserved the republic (that is, no king) but put it in the hands of wealthy men. Their biggest mission was to beat the enemies of France abroad.

The wars that followed ended France's revolution and, at the same time, spread it over Europe. In 1799 a brilliant young general, Napoleon Bonaparte, made himself the military dictator of France. He enacted codes of laws that partly carried through the Revolution and in parts denied it. Then he conquered most of mainland Europe. He introduced his Code in many countries and claimed he brought them France's proudest boast, equality. (He could hardly say he brought them liberty.) But in 1815 Prussia (Germany) and Britain crushed him on a battlefield in Belgium, and penned him on a lonely South Atlantic island. He and it were both extinct volcanoes.

In later decades, France had other revolutions, lesser versions of its big one. With many leaps and many backward falls it moved toward democratic government.

AS WE KNOW, back in the 1500s much of South America and a sizable chunk of southern North America fell to Spanish conquerors. Spain's America was vast; it stretched from California to the southern tip of Argentina. Spain ruled it as four provinces, whose capitals were in Mexico, Colombia, Peru, and Argentina.

After the conquests, the region drowsed for three hundred years. Spain sent only a sprinkling of its people to its colonies. Governors and troops went out to rule, priests to rescue souls, and businessmen and ranchers to get rich in sugar, coffee, timber, and tobacco. But the colonies were huge and the jungles dense. The Europeans touched the lives of Indians only on the seacoasts or where the soil was rich and worth exploiting.

Between the whites and Indians was a gulf as deep as the Andes were high. The whites had money, rifles, ships, and books; the Indians had ignorance and hunger. But the whites were also split, and this would be important in the Latin revolutions. On one hand were the Spanish governors, priests, and soldiers. They were transients;

they came from Spain and later would return there. They were loyal to their king in faraway Madrid and to old and settled ways.

On the other hand were "Creoles," the white men born and raised in Spain's New World. They loved the Old World well enough, but their destinies were in the New. For that reason, news of revolutions that had shaken France and North America excited them.

Independence fever in Spanish America started in the early 1800s. Across the ocean, Napoleon was conquering much of Europe. The king of Spain and Spanish America had weakly given up his throne and allowed Napoleon to name his own brother king of Spain. For white Spanish Americans, to have a Frenchman on the throne of Spain was shocking. Faced with this appalling fact, Creoles in cafés and universities talked with passion of democracy and independence.

Our tale begins in Mexico, the only place in Spanish America where Indians got deeply involved in a revolt. Even here it was a Creole, not an Indian, who lit the fire. In 1810 Miguel Hidalgo y Costilla, a well-read small-town priest, learned about Napoleon's takeover in Spain. Did he really care about that far-off event, which meant so little to the humble people of his parish? Most likely not, but he saw an opportunity. He urged the local Indians and mestizos (people partly Spanish, partly Indian) to expel their Spanish governor. Hidalgo had no program, no thought of a republic or a constitution, but his oratory reached a lot of ears. His hearers shouted, "Long live America! Death to foreign tyrants!"

Indians and mestizos everywhere in Mexico took up that cry and swarmed along the country roads. Soon Hidalgo had an "army" of 6,000, armed mostly with slings and clubs. They stormed the mountain town of Guanajuato (famous for its silver mines and handsome churches), massacred its Spanish guards, and sacked it. In other places also, violence occurred, and on both sides of the struggle priests were often leaders. In one town a pro-Spain priest sliced their ears off Indians and wore them on his sombrero.

Hidalgo's army grew to 80,000, and he led it to the gates of Mexico City. There he halted for a while, uncertain what to do, and so he lost his chance. Many of his troops went home, and the rest were crushed in battle. The Spaniards shot Hidalgo. In ravaged Guanajuato they displayed his head inside an iron cage.

José María Morelos, a mestizo priest, tried to carry on. Unlike his friend Hidalgo, Morelos had a program, an astonishing agenda. He insisted on equality of Indians and whites, confiscating huge (and mostly Spanish-owned) estates and sharing them among the peasants, and a law that every able-bodied adult must work. What a program! Morelos wanted not democracy alone, but something near socialism, whose day would only come a century later. For all the whites in Mexico, royalists and Creoles alike, his program was unthinkable. Spanish forces crushed his army. They defrocked the rebel priest and shot him.

As it happens, only a few years later Mexico was freed from Spanish rule. But independence came about through cynical opportunism, not because of a popular democratic movement. In the century that followed, wealthy despots ruled the country. The Mexican revolt had really ended when a Spanish firing squad shot Morelos in the back.

Everywhere else but Mexico, Indians played only a minor role in the Spanish American revolutions. Creoles instigated them, and Creoles largely carried them out. The greatest of these Creoles, part George Washington and part Napoleon, was Simón Bolívar.

This dazzling man was born to wealth in Venezuela, at the northern end of South America. While he still was young Bolívar (boe-LEE-var) pursued women, married one (who died quite soon), read a lot, and traveled everywhere. Legend has it that in Rome, standing on the top of Monte Sacro, Bolívar made a vow to free his native land. He sailed back home, and in his later twenties led a little force of Venezuelan rebels against the Spanish. He proved to be a dauntless, dashing leader—and a cruel one sometimes, capable of slaughtering his prisoners.

In 1819, while Spanish troops still held most of Venezuela, Bolívar formed a plan to free Colombia, Venezuela's next-door neighbor. Colombia was the center of what was now Spain's northernmost province (after losing Mexico). With 2,500 men, including British mercenaries, Bolívar began to cross the Andes. This was a terrifying challenge. Other Spaniards thought these mountains were impassable. Hannibal, two thousand years before, and Napoleon, quite recently, had crossed the Alps at the awesome height of 8,000 feet. Bolívar and his men dragged their cannons and themselves along the icy brinks of cliffs at heights of up to 12,000 feet.

Heights like those aren't made for humans; they give them nausea, driving headaches, and pounding hearts. The weary soldiers punched each other just to keep themselves awake. A hundred men and all the horses died, and when the survivors reached Colombia they were exhausted and many of them were sick.

But they recovered. After resting they fought and routed a Spanish army half again as big as theirs. The Spanish governor fled, and a rebel assembly named Bolívar president of a nation that combined three territories: Colombia, Venezuela, and Ecuador. However, Spaniards still held the latter two. Bolívar returned to Venezuela and beat the Spaniards there, and then he drove another Spanish army out of Ecuador.

While Bolívar "liberated" the north of South America, General José de San Martín had done the same for the south. Though born in Argentina, San Martín had lived in Spain since childhood. He studied at a Spanish military school and for twenty years commanded Spanish troops. But he had a change of heart (no one knows just why) and sailed back to his native land. There he joined some rebels who had thrown out Argentina's Spanish governor.

San Martín took on the next big task of liberation, the conquest of Peru. He quickly decided that for his army the pathway from Argentina to Peru led through Chile, on the other side of the Andes. Two years before Bolívar did it in the north, San Martín led a force across the mighty Andes. In bitter cold his army crossed the mountains at heights of 10,000 to 12,000 feet, with condors soaring overhead poised to pick their flesh. Once in Chile, they freed it from the Spanish. From Chile they sailed northward to Peru and drove the Spaniards out of Lima.

Now the general ran into problems. One was that the still imposing Spanish army waited for him in the nearby mountains. Another problem, even more disturbing, was how to rule Peru. San Martín was patriotic but he had no faith in representative government—not in Peru, with its small elite of noisy Creoles and multitudes of unschooled Indian peasants. He let the Creoles know what he believed: Peru required a king. "Every literate man who knows his country and desires order," he said, "will naturally prefer a monarchy to the continuation of disorder and confusion." But the Creoles disagreed. While they argued, San Martín was acting governor, and the Creoles found him distant and domineering. They called him "King José," and even his friends abandoned San Martín.

Burdened with these problems and depressed, he sailed to Ecuador to meet triumphant Bolívar. Probably he wanted men and arms. The meeting of the hero of the south with the hero of the north was a failure. Magnetic Bolívar and distant San Martín had nothing in common. Bolívar didn't want to help the general take Peru; he wanted the glory of the final triumph for himself. He refused even to let San Martín serve as officer under his own command.

They also disagreed about the shaping of Peru. As we mentioned, San Martín opposed a republican, or representive, government, while Bolívar, high-mindedly but unwisely, demanded one. "Once the idea of a republic has taken root," he said, "it cannot be extinguished." San Martín hurried back to Lima, where, sick and disappointed, he resigned. He later sailed for Europe, where he died in exile.

With his path to glory cleared, Bolívar went to Lima, where he gathered soldiers, mules, and ammunition. He climbed the mountains and his army won two major battles, after which the Spanish governor surrendered. Bolívar's forces later beat another Spanish army, the last in South America, in the center of the continent. This area was made a nation, and, in honor of the Liberator, it chose the name Bolivia.

But Bolívar's triumphs curdled, as had San Martín's. When he urged the former colonies to unite and form a single nation as long as one and a half continents (from California to Argentina), only four of them agreed, and even they allowed the plan to die. When he wanted the infant nations to appoint presidents who served for life, like kings, and grant the right to vote to only very few, ultra-democrats opposed him. It did no good for Bolívar to argue (as San Martín had done in Peru) that universal suffrage would never work in countries where the masses were so huge and poor and ignorant. Liberal enemies invaded his palace one night, and his mistress barely saved him from their daggers by stalling while Bolívar escaped through a window.

What was worst of all, the nations he had freed from Spain made war on one another. Sadly he concluded that his very presence threatened the peace of the people he had freed. He wrote a friend that "He who serves a revolution ploughs the sea." It was as if he, like San Martín, were one of Shakespeare's tragic heroes and he now had reached the final act.

Bolívar had tuberculosis. When his doctors told him he required a healthier climate, he decided he would go to Europe. When he reached his Venezuelan seaport, he learned that assassins had killed his finest general, whom Bolívar had trained as his successor. By this time he was very sick, and he called the trip to Europe off. A friend invited him to his estate, and there he died a few months later.

In the next century and a half, democracy did not take hold in Spanish America. It couldn't, as Bolívar and San Martín had realized, because the Indians were unschooled and very poor, and one simply couldn't reach them in the mountains and the jungles. When a revolutionary general entered Bolivia to proclaim the revolution, he spoke to a vast crowd of Indians about liberty, equality, and citizenship. They didn't make a sound, so desperately he asked them what they wanted. As one they shouted, "Brandy, *señor*!"

The majority of the better-educated Creoles had no knowledge of self-rule and no desire to try it. So ranchers, miners, businessmen, and generals ran the former colonies to suit themselves. Only in the last half of the twentieth century did democracy take root in much of Spanish America.

SINCE THE 1700S, democratic rule has slowly spread across the world. How this happened is a story full of bloodshed, sacrifices, fuller stomachs, demagoguery, parliamentary maneuvers, assassinations, spreading literacy, deals, elections, coups, revolts, hard-won victories, and many setbacks.

CHAPTER 14

We make more and live better.

IN THE 1980S and 1990s, a madman in America mailed bombs to businessmen and technologists. These killed three people and injured more than twenty. Later he released a manifesto explaining why he did this. His victims, he declared, were agents of the Industrial Revolution, which he blamed for every evil. "The Industrial Revolution and its consequences," he wrote, "have been a disaster for the human race." This chapter deals with just that subject: what the Industrial Revolution did for—and to—the human race.

FOR MANY THOUSAND years, nearly everybody lived in want. Humans got along on little, and were lucky if they lived long enough to see their hair turn white. The very poor were everywhere; we have met them in this book. In ancient Israel the needy, said the prophet Amos, were "sold for a pair of shoes." In China nearly all the Chu family starved to death, leaving just one son, who couldn't pay the cost of burying the others. In France the Cervel family lost the father and two out of three children to hunger and disease. Slaves from Africa were shipped to the Americas owning absolutely nothing.

Not everyone was poor, of course. In ancient Egypt a wealthy farmer might own 1,500 cows. An Egyptian prostitute built herself a pyramid, and later a prostitute in Constantinople owned a chamber pot of solid gold. A Chinese poet of the 700s says that the wealthy wined and dined their guests with "Red jade cups and rare dainty food on tables inlaid with green gems." At about that time, someone offered Tahya the Barmakid, a wealthy Arab, an enormous sum for a box adorned with pearls and precious stones; he wouldn't sell. Russian grandees who visited London in the 1600s were rich and filthy; they arrived in ballrooms "dripping pearls and vermin." In the 1700s a Hungarian noble told a wealthy Briton that he had more shepherds than the Englishman had sheep.

Nevertheless, before the Industrial Revolution the living standard everywhere was low. The major reason was that those who made the things one needed most—houses, textiles, tools, and food—produced so little. Not that craftsmen were unskilled—far from it. If you wander through an arts and crafts museum you'll marvel at the quality of old-time hinges, hammers, swords, and saddles. Their makers knew what they were doing, and they did it well.

But their tools were simple and they worked alone or in small groups, each worker doing all production tasks instead of focusing on one. Therefore they could not produce a lot. With such small outputs, almost everybody lived in want. Little pies mean tiny slices.

Take iron, for example. Iron was essential. Imagine reaping wheat, as prehistoric farmers did, with sickles made of flint, or hanging doors with leather hinges, or guarding horses' hooves with wooden shoes. Housewives needed iron scissors, cutlery, and pots, and warfare called for iron spears and swords and, later, iron guns and cannons. However, iron making called for lots of trouble. An ironmaster and his helpers roasted ore for hours in a furnace with a charcoal fire. The product was a cubic foot or two of costly iron.

Now consider how iron craftsmen, those who used that glob of iron, worked and lived. Four centuries ago the town of Sheffield,

England, was widely known for making scissors, knives, and blades for tools. Many of its people earned their living in those trades. They worked in shops inside their homes, using simple files and ancient methods. Their basic raw material was iron—costly iron.

Since they produced little, they earned little and lived badly. An official report describes the Sheffielders' lives in 1615, dividing them into three groups. At the top were a hundred families sufficiently well off to be able to help others in need. But even they were hardly rich. The report describes the heads of these households as "poor artificers." Ten of them owned a little land, enough to keep a cow and grow some food.

The second group was made up of 160 families who at least could feed themselves and didn't have to beg. They could not afford to help others, and if the money-earner in a family fell sick for a couple of weeks his wife and children would be forced to beg. According to the report, both of these first two groups, the "poor artificers" and the barely self-sufficient, "live of small wages, and are constrained to work sore to provide them necessaries." But then there was the bottom group, who made up a third of the total. They were called the "begging poor," who could not live without some charity from their neighbors. No less than a third of the town were "begging poor."

EVERYWHERE THE PROBLEM was the same: how could people *produce* more, so that everyone could *have* more? The answer was industrialization.

This lengthy word means, simply, raising your output by using aids to production. One of these aids, of course, is machines. Another is the use of power sources that have many times the strength of humans. We used to use the force of rushing streams to run machines for grinding flour, making lint for paper, or working bellows in an iron mill. Later on we used the power of steam, and later still electricity.

Other aids to production have to do with how the work is planned—that is, with organizing workers so that they produce the biggest output possible. The use of all these things—machines, and power, and organization—multiplies by many times the work one does without them.

The first country in the world to have an industrial revolution was England. Why England? One important reason is quite clear. In the 1700s the English, isolated from the wars of the continent, were prospering, and their number was rising. That increase caused a strong demand for basic goods, and manufacturers and inventors saw their chance.

It was in the textile trade that they moved the fastest, and there John Kay began it all. Before Kay's time a weaver, working on a loom at home, made woolen cloth with considerable effort. He "threw" a wooden shuttle that held a thread back and forth and in and out between two sets of lengthwise threads. In the 1730s Kay made the weaver's work much easier. With Kay's improved loom, the weaver, by simply pulling strings, worked two hammers that drove a "flying shuttle" back and forth on grooves. He wove much faster than before, and made a wider bolt of cloth. (However, weavers didn't like Kay's new machines at first because they feared the things would put them out of work. They were so hostile to Kay that he fled one textile city hidden in a woolsack.)

Kay's shuttles made more cloth, but they also raised a problem: his hungry looms required more thread. At this time a spinner made one thread at a time by twisting one fiber to another. But now the spinners couldn't keep up with Kay's productive looms; they couldn't spin as fast as weavers wove. James Hargreaves, a poor, uneducated spinner and weaver, solved the problem. In about 1764 he invented a machine he named a jenny (for his little daughter, who had accidentally given him the idea for it). With this device the spinner turned a wheel that wound thread onto sixteen spindles at once.

However, one man's muscles weren't enough if you wished to add even more rods and spin more threads. What cloth-making needed now was another source of power to run the jennies harder. Richard Arkwright, a barber, wig-maker, and business genius, invented a solution, or, more likely, stole it from another man. The key was using rivers as a source of power. He began to make machines, called "water frames," that did what workers up to then had done by hand.

A water frame could spin as many as eighty robust threads at once. Arkwright found some backers, and he built a mill that soon employed three hundred men, spinning night and day, and it wasn't long before he ran ten mills at once. The former barber made a fortune, bought a castle, and boasted he would soon be rich enough to pay the national debt.

Now a clergyman named Edmund Cartwright enters this story of machines and rising outputs. Up to this point in his life, Cartwright's main achievement was the writing of some elegant but frigid poems. But while he was on a vacation, he chatted with some textile men and learned of Arkwright's new machines and a problem they'd created. Now the old dilemma was reversed: weavers couldn't weave as fast as water frames could spin. Although Cartwright hadn't ever seen a weaver weave, he understood the need for faster looms. So with some help from craftsmen, he devised a mechanical loom that the owner could power either with a horse or with a waterwheel. By 1815 two of Cartwright's "power looms," with a child to oversee them, could do the work of fifteen old-style weavers.

Iron making also had a speeding-up. We saw above how masters once had smelted iron in little furnaces. English ironmasters started using more productive methods in the 1600s, often building furnaces into hillsides to make it easier for laborers to shovel in the ore and charcoal. But changes came on faster in the 1700s, and the stimulus was the need for cheaper fuel. England's forests were disappearing, so charcoal cost too much to use. When Abraham Darby found a way to smelt his ore by burning coal instead of charcoal, ironmasters could smelt more iron at lower cost.

Later, Darby's son devised a way to use a waterwheel to work a giant bellows and make a bigger, hotter fire. This made it possible to use a larger furnace, with a saving both of fuel and labor. Other men found ways to make an iron that was easier to work with, and while it was still white-hot they shaped it into bars and sheets by running it through rollers.

Because of these and other changes, in the century after 1770, British iron output increased eightfold. Now the country had sufficient iron for many newfound uses: looms and jennies, railroads, bridges, ships, even coffins, and—this became an English specialty—machines to make machines.

It was a Scot who created the English industrial revolution's puffing, pounding hero. Young James Watt made scientific instruments, and he was asked one day to fix a model of a simple steam-powered engine that was used for pumping water out of mines. He saw at once the defect in this kind of engine. The cylinder that contained the steam had to be heated so that steam could push a piston, and then cooled with injected cold water (to condense the steam and create a vacuum) for the piston's return stroke. These changes, hot to cold to hot, wasted fuel and steam.

While strolling on a Sunday afternoon, Watt conceived a better way to make an engine. If one added a separate condenser, the cylinder need not be cooled between strokes. This change would greatly increase efficiency. It took him only weeks to make a model of his scheme but years to translate that into a working engine. He had to find a backer to supply the research money, and craftsmen with the skill to make the valves and other complicated parts. Luckily, a manufacturer had just discovered how to bore a cannon with precision, and the method was exactly right for making cylinders. Watt began producing engines that were four times as efficient as the old ones they replaced. They proved of use in iron mills, and for pumping water out of mines and into breweries and reservoirs.

His new machines, however, had a drawback: their steam was used to drive a piston back and forth. This was fine for pumping water, but he wanted more than that. He needed to convert the motion *to and fro* into motion *in a circle,* since that was the movement needed for machines. He solved the problem in the 1780s with a series of inventions, each one simple and yet brilliant. Soon his engines had a movement that was circular and smooth enough for all machines.

For the whole world, this invention was a great event. Within decades, steam replaced our muscles and the waterwheel. It drove machines in mills, trains across a country, and ships around the earth. Watt's business partner once showed a visitor through the factory where they made their engines. Pointing to the busy scene, he said, "I sell here, Sir, what all the world desires to have—POWER."

Meanwhile, businessmen were finding ways to organize the *process* of production. These changes were as crucial as the new machines. For example, it was clear that textile workers could no longer work inside their houses; they could hardly find room for big machines. So manufacturers like Arkwright built large factories, and the workers lived nearby in company houses, boring boxes set in rows.

Not only where they worked, but how the workers lived had to be arranged. Six days a week, when business was good, the workers—adults, even little children—came to work at dawn, tended rows of clattering machines, and didn't leave till dusk. Many of them had to learn a discipline that they had never known on farms or shops. No longer could they drink on Sundays and then make Monday a holiday, "Saint Monday," and sleep it off.

Managers found better ways to organize the work itself. In 1776 an economist described what workmen did in a pin mill he had seen. It went about like this: Worker Number One heated iron and drew it into wire, and Worker Two straightened out the wire. Three chopped it into inch-long pieces, and Four sharpened one end of each piece. Five flattened the other ends, where the heads would go. Six and Seven made the heads. Eight fastened heads to pins, and Nine painted the heads white. Ten put the pins in holders.

When they pushed themselves these workmen turned out nearly fifty thousand pins a day. If, wrote the economist, "they had all wrought separately and independently . . . they could certainly not each of them make twenty, perhaps not one pin in a day."

Other makers carried this practice (having each worker do a different task) much farther. One was an American, Eli Whitney, who contracted in 1798 to make 10,000 muskets for the U.S. government. For the time, that number was enormous, and he promised to deliver them within a mere twenty-eight months.

Whitney knew he couldn't simply hire a lot of smiths to make the muskets in the usual way. A gunsmith was a skillful craftsman. When he made a musket lock (the part that makes the charge explode), he filed each wrought iron part until it fitted closely with the others. To make a few expensive muskets took a year.

Whitney didn't know a lot about muskets, but he did know what he called his "uniformity system." The way to make a lot of musket locks was to break the process down to its component tasks. He hired unskilled workmen and trained them each to cut and file a single part, always to a stipulated size, over and over again. In this way, each man made a lot of parts, and these were quickly put together and (if broken) easily replaced.

To convince U.S. officials that his system really worked, Whitney gave a demonstration. On a table before the president, vice president, and cabinet, he poured out piles of different parts, and he asked the men to pick up one of each and fit them all together. They did so, with amazement. Thomas Jefferson (the vice president) reported that one could "take a hundred locks to pieces and mingle their parts and the hundred locks may be put together as well by taking the first pieces which come to hand . . . good locks may be put together without employing a Smith."

Whitney went ahead and made the muskets. It took ten years, not two, and for a decade's work he made a profit of only $2,500. Still, he had proved what could be done with "uniformity." Other weapons makers later used his system with machines that turned out many weapons a great deal faster. In just one year (1863) a British firm produced 100,000, and three years later Frenchmen manufactured guns three times as fast as that.

Whatever other makers had achieved before him, Henry Ford did bigger and better. Ford was born in Michigan on a farm outside Detroit, and he studied in a one-room school. When he was sixteen he walked to Detroit to work in its machine shops, and he got to know a new machine, the internal combustion engine.

This power maker is quite different from Watt's steam engine. The fuel it uses, gasoline, burns inside the engine, hence the name "internal combustion." Expanding gases push against a piston that turns the wheels. When Ford began repairing them, thousands of these engines were already chugging in the factories and farms, powering pumps, machines, and saws.

Ford liked to tinker and experiment, and he decided that he'd build a gasoline-powered car. He built an engine in his spare time, and he set it on the frame (the chassis) of a one-horse carriage, atop four bicycle wheels. It worked, so he sold it and used the proceeds to finance the building of a better car. With what he made from that one he built a third, and so on. By this time, other inventors had also made self-powered cars, and firms were starting to produce them.

But Ford kept tinkering and testing. He wished to make a sturdy car that everybody could afford, what he called "a motor car for the great multitude." He would market them not to millionaires but to the many millions of families whose incomes now were rising as America industrialized. After a decade he finally had a model that was easily produced and easily repaired. He christened it the Model T.

To make his cars not only good but cheap Ford knew he had to standardize, to "make them come through the factory just alike." Model Ts were certainly alike. People joked that you could have them any color you wanted, as long it was black—but they bought them. In the nineteen years that he made the Model T (from 1908 to 1927), Ford sold 17 million of them, which was half the auto output of the world.

It wasn't only uniformity that made the Model T so cheap. It was also the planning of production, and especially the assembly line. Ford would say he got the concept of the line when he saw how meat producers hung their slaughtered steers on trolleys overhead to move them from one cutter to another. A Ford car started as a bare steel chassis. A conveyer, never stopping, moved the chassis down the line past one man, or group of men, after another, and each performed a task. One or two installed the engine, someone else the steering wheel, then others did the hood, the wheels, the seats, the lights. Others painted, others greased, and so on to the end.

But an assembly line was more complex than that. The parts that workmen on the line installed had just been assembled by other men on feeder lines. These parts on their conveyers reached the main line precisely at the place and time when they were needed. This system grew efficient to the point that the company produced two cars a minute.

Ford found another way to lower costs: "vertical integration." He did not invent this method but he showed what it could do. His costs, he knew, began the moment power shovels dug his iron from the earth, and continued till the cars rolled out the factory door. He could cut expenses if he built an empire in which he owned and ran the total process. By 1927 he had it all in place. Every morning a Ford freighter (one of many) reached the plant with ore from Ford iron mines on Lake Superior. Furnaces smelted the ore with coal brought in from Ford mines in Kentucky. A Ford factory had made the tires and a Ford glassworks the windows. A Ford sawmill cut the floorboards out of lumber from Ford's 1,100 square miles of timber.

Model Ts and other cars changed the lives of many millions of people. Ordinary people learned that for the first time they were mobile. A farmer hitched his "buggy" to a saw and cut his firewood, then used it to drive his family into town to see a movie. A salesman sold his city home and bought another in the suburbs. Gangsters used their cars for robbing banks, and adolescents for romance.

NOW WE'LL TAKE a closer look at how industrialization altered lives. We will focus on a village by a stream in Pennsylvania, on America's east coast. What happened in this setting is a capsule social history of mankind.

Gatherers of food begin the story. Before the 1700s Indians lived around French Creek. They trapped the beaver, caught the fish with nets, and hunted deer that came to drink. To this day people find their spearheads.

In the 1700s farmers settled in this place. The first had left his home in Germany to escape religious persecution. Another farmer bought some land along the northern bank, and battled with the wolves that fattened on his sheep. By 1800 a dozen farmers, a miller, and two or three black slaves were living in this place. The Indians had left to get away from these intruders. The farmers planted wheat and corn, and tended cattle, sheep, and geese. They lived in simple houses, dressed in homemade clothes, and worked the whole day long. But with their large and fertile fields they had enough to eat.

The future of this village (so tiny that as yet it had no name) would be connected to the water of the creek. The miller used its power to turn his wheel and grind the farmers' wheat. Later someone bought the mill and used the waterpower to turn a circle-saw and cut up timber. But the village reached its turning point in the early 1800s when outsiders bought the mill to use in making nails. At this time, workers manufactured nails by flattening a white-hot bar of iron and slitting it. At first the French Creek makers bought their iron from other firms, but soon they built a furnace and began to smelt it for themselves. They dug the ore from bluffs beyond the creek, and bought the coal from nearby mines.

One owner named the little firm the Phoenix Works, and the town that grew around it took the name of Phoenixville. The phoenix is a mythic bird of Arabia, which lives for half a thousand years, roasts itself to death, rises from the ashes, and begins another cycle. It nicely symbolizes the making of iron by roasting ore.

After several decades, bigger manufacturers bought the site and dropped the nails in favor of rails. By this time, railroads were a universal project. For making rails Phoenix needed much more coal and ore so they hauled them in from far away. They also bought surrounding farms, and soon their furnaces, mills, and workshops sprawled for several acres. They renamed the firm the Phoenix Iron Company.

Starting in the middle 1800s Phoenix had another owner, and this one switched from rails to more sophisticated products. These included heavy, quarter-circle iron planks which, when four were bolted to each other, made columns to support a factory's girders. The company also manufactured cannons, using a new method of squeezing crisscrossed white-hot rods together so they wouldn't blow apart. (Both sides used these cannons in the Civil War.)

But the major product of the firm was lofty railroad bridges. When a railroad builder placed an order for a bridge, a Phoenix engineer would travel to the site (a gorge or river anywhere) and plan the bridge. Then the many heavy iron parts—all standardized—were shipped from Phoenixville by train and "pinned" together on the site. Heavy though they were, from afar these bridges looked like strips of lace.

The company grew in part by gathering "fixed capital." By the 1870s it owned, along with many other things, a canal and railroad spur to haul in coal and ore, three blast furnaces, "puddling mills," a rolling mill, a foundry, a machine shop, and several rows of workmen's houses.

However, growth meant people too. Many men—but not a single woman—worked for Phoenix Iron. These included the owner (who lived in a hilltop mansion from which he could survey his mills), engineers, foremen, "puddlers" (who tended furnaces so hot their clothes could burst in flames), mold makers, machine tenders (who worked unprotected with white hot iron), shovelers, "lifters," sweepers, water carriers, clerks, and mule drivers. Most of them worked in constant danger fifty-five hours a week.

Many workmen had been born on nearby farms, but beginning in the 1850s a goodly number came from Ireland. These young men had always lived in huts, dressed in rags, and eaten little but potatoes. When blight had rotted the potatoes in the 1840s, they had left their homes and journeyed to America. Later, many equally impoverished workers came to Phoenixville from central Europe; and, later still, liberated slaves came up from the southern United States.

Clearly the Phoenix raised its workers' living standard as the company prospered. The clearest proof of this is the changes in their dwellings. Early in the century workmen and their families rented cabins that the company had built for seventy-five dollars apiece. Later on they lived in rows of little houses that the company had built on streets above the mills. Later still, they lived in houses of their own.

The workforce had been roughly twenty in the nail-making years; in the 1870s it rose to two thousand. For about a decade, Phoenix was perhaps the biggest iron producer in the world. But this was not to last, because iron makers were expanding everywhere, and other iron firms combined, forming giant corporations. During the 1800s, iron making all around the globe increased by fifty times.

But Phoenix Iron remained a midsized firm. It hung on for a hundred years, and then it shut its doors. The buildings are still there, so big it wouldn't pay to tear them down.

DESPITE THE DECLARATION of the bomber whom we quoted at the start—that the Industrial Revolution brought "disaster for the human race"—the fact is this: Where the Revolution has occurred, and where it still is taking place, it has been a blessing for many, many people. We live longer and are healthier and probably happier than those who went before us.

CHAPTER 15

The richer countries grab the poorer.

AT THE RISK of being teacherly let's be clear. Here we *won't* discuss how Europeans seized the two Americas and bits of Africa and Asia. We talked of that in earlier chapters. When revolutions ended European rule in much of the Americas, that age was past.

What is called the "New Imperialism" occurred mostly in the 1800s. Here's an overview. Once again, most of the empire builders were European nations, only they were richer now. For purposes of conquest they had all the things they needed: money, steamships, guns, and greed. In a patriotic frenzy, European empire builders gobbled poor and "backward" countries, those that couldn't save themselves. The richer countries wanted, yes, such useful things as cotton, rubber, jute, and rice, and markets for their products. But they also craved the thrill of ruling others and, perhaps, of teaching them to love the Lord and cover up their breasts.

Other nations joined these European empire builders. America was one of them. As the 1800s ended, the United States seized some islands in the tropics and the land on which to build a link between two oceans. (We'll say more on that in the next chapter.) The other novice empire-building nation was astonishing Japan.

INDIA, UNDER BRITAIN, was the classic case, the prime example of the way the wealthy northern countries seized and bossed the southern ones. Besides India, Britain owned a lot: Canada, Australia (recently acquired), bits of China, chunks of Africa, and more. These colonies lay all around the globe. But if it was often said that the sun never set on the British Empire, it was also said that if its empire was a crown, India was the crown jewel.

The high value of India was surprising. When English businessmen first entered it, long before, India was their second choice, not their first. On the last day of the year 1600 English merchants organized a firm they called the English East India Company. But the name is misleading. These merchants really wanted to trade in the East Indies *islands,* which today are Indonesia. One could make a fortune out there trading spices. But the Dutch already traded in those islands, and they kept their rivals out. So the English settled for India, as second best.

Until the 1750s the East India Company was nothing but a thousand British (mostly English) traders and their clerks, sweltering in seaside shops. They did a tidy trade in coffee, pepper, cottons, silks, and (later) opium. The Company was just a mole on India's vast body. However, then it happened that the shahs, or rulers, of much of India began to lose control. They were feeble leaders, and India was big and hard to rule. Princes and some would-be princes tried to fill the vacuum, fighting with each other and the shahs.

For the British businessmen this civil war brought thrilling prospects. The Company formed an army, disciplined and well equipped with guns. First they fought and beat their French business rivals and shut their company down. Then they faced the hostile Nawab of Bengal, who recently had locked 146 Britons in "the Black Hole of Calcutta" on a summer night, causing all but 23 to suffocate. With the help of regular British troops, the Company defeated him. And so, in 1764 British businessmen found themselves the rulers of Bengal, the richest state in India.

Problems soon arose. Back home in London, it was learned that Company officials in India were plundering and grafting. The British government was also troubled by the way the Company was meddling in wars of other Indian states. Britain therefore named a governor general to "regulate" the Company, that is, to govern British subjects in India. By the beginning of the 1800s, however, a governor general did far more than that. He ruled the British zone in India and everybody in it, British and Indians alike, as if he answered only to his chiefs in London.

By now, in the age of the *New* Imperialism, the Company wasn't what it had been, but British rule was doing well. One by one the British won the states around them, then the ones that lay beyond. Sometimes they annexed a state, but often they merely forced its prince to let them run his foreign policy, which amounted nearly to the same thing. In taking over, they were moved no doubt by love of power, but they often claimed to have no choice but to conquer or be conquered. (The ancient Romans used to say the same.)

By the time it finished, Britain—just a little nation far away—controlled them all, those Indian rulers with their tasty titles: the Nawab of Oudh, the Nabob of Jubbulpore, the Maharaja of Travancore, the Gaikwar of Baroda, the Sultan of Mysore, the Rani of Jhansi, the Begum of Bhopal, the Ahkoond of Swat. (One important conquest was the plains of Sind. After it surrendered, the British general sent his chief a message with a single Latin word, "*Peccavi*," which means "I have sinned.") By the middle 1800s the British ran all India, from the snowy Himalayas to the steaming south.

As they won the land, the British changed their views about themselves and those they ruled. In the early days the Company officials, who were only guests residing in a foreign land, adopted Indian ways. They mixed with Indians and learned their many tongues. They took to Indian food and dress, and some of them to Indian mistresses and wives.

In the 1800s, that all changed. Now the British were the masters, and they learned to scorn the Indians. Now a governor general could write, as one did in 1813, "The Hindoo appears a being nearly limited to mere animal functions and even in them indifferent. Their proficiency and skill in the several lines of occupation to which they are restricted are little more than the dexterity which any animal with similar conformation but with no higher intellect than a dog, an elephant, or a monkey, might be supposed to be capable of attaining."

With wonderful disdain, another Briton clarified why Indians needed Britain. The role of Britain was "the introduction of the essential parts of European civilization into a country densely peopled, grossly ignorant, steeped in idolatrous superstition, unenergetic, fatalistic, indifferent to most of what we regard as the evils of life and preferring the response of submitting to them to the trouble of encountering and trying to remove them." The British saw themselves as instruments of God, helping an inferior people. They decided it was best to live apart and treat the natives with a cool correctness. They didn't hesitate to throw Indians out of their "Europeans Only" railroad cars.

Did the British do more good or harm to Indians? They brought about these gains: they caught and hanged or jailed the Thugs, India's gangs of ritual murderers, who were worshippers of Kali, goddess of destruction. They constructed bridges, roads, and thousands of miles of the best railroads in Asia. (As a result, food supplies moved faster, and India's awful famines nearly vanished.) They invested in local industries, especially textiles. (Here British capital helped, but Indians did much themselves.) The British stamped out suttee, the Indian rite of burning widows alive in their deceased husbands' cremation fires. (When the British outlawed this, Indian religious leaders remonstrated to the governor general, "But, Your Excellency, it is our religious custom." He answered, "My nation also has a custom. When men burn women alive we hang them.")

The British also fostered schools, and by 1900 one Indian man in ten could read and write. This was then a lofty rate of literacy in Asia. However, only one Indian woman in 150 could read and write.

In 1857 Indians revolted, terrifyingly. What triggered their "Great

Mutiny" were the cartridges that the British army issued to its Indian troops. Hindu soldiers were aghast to find the cartridges were greased with fat from cows, which to them were sacred. Because they had to bite the ends of cartridges before they loaded them in guns, the fat could easily pollute them. Many of them mutinied, and rebellion spread through much of north and central India. Many Indian princes joined the mutiny, hoping to regain the power they'd earlier lost. The program of the mutineers was simply to return to India's pre-British past.

In putting down the mutiny the British had advantages: tauter discipline and better weapons. In two grim years of fighting Britain quelled the great revolt. Because the mutineers had massacred some British captives, the British struck back hard. At Delhi they drove everybody into open fields and executed thousands after token trials, or none at all.

After the Mutiny, Britain would rule India for another ninety years. But the British rulers trusted Indians less and were even more aloof than they had been before. As one Briton said, "We are not in India to be pleasant." They were careful always to station British troops with Indian ones, lest Indians dream again of governing themselves.

IN THE EARLY 1800S, most of the continent below them was a mystery to Europeans. They *did* know North Africa, which was nearest to them. But they knew little of the region south of the Sahara, which is nearly as big as North America. Their ships, it's true, had anchored on its coasts to pick up palm oil, ivory, and slaves. But Europeans feared the heat, the fevers, and the warriors of the interior, and they rarely tried to penetrate it.

Curiosity and other motives had their way at last. After about 1850, numerous explorers landed on the coasts and wandered inland, using local bearers. One of these was Dr. David Livingstone, a Scot, who wanted to bring medicine and Jesus into Africa and also open it for trade. Honest business, he believed, would undercut the still-surviving Arab trade in slaves. For thirty years he hiked through central Africa, preaching, trading, treating people for disease, and mapping lakes and rivers. Between safaris, he wrote books on Africa that many read, and he urged that other Britons carry on his work.

In 1871 Henry Morton Stanley sailed to Africa to "find" the legendary Livingstone, who was thought to be in trouble. Stanley was an English journalist, reporting for the *New York Herald.* With his caravan he trudged to Lake Tanganyika, where he found the doctor in a village. Walking up to him he asked the famous question, "Dr. Livingstone, I presume?" As expected, Livingstone was sick. Stanley gave him medicines and bullets, pots and kettles, and cloth and beads to trade with. Intrepid Livingstone refused to leave his work. When he died the British buried him in London's Westminster Abbey.

Stanley's role in the New Imperialism, however, had only started. He wrote a book about his great adventure, called *How I Found Livingstone,* and Britain's Queen Victoria received him and gave him a gold snuffbox. He became an explorer, and wrote more books about his doings, with such alluring titles as *In Darkest Africa.* When they read about his deeds and those of others, Europeans were entranced. Darkest Africa had everything that statemen, missionaries, businessmen, and medal-hungry generals could desire.

A "scramble for Africa" soon got under way. Britain, France, Belgium, Spain, Portugal, Germany, and Italy all bit pieces until, after only fifteen years, they had eaten all of Africa but Ethiopia. We can't tell all this lurid and exciting story, so we'll focus on the way a Belgian took a giant helping, and let it represent the whole.

In this Belgian venture—caper—Stanley played a major role. Stanley was a harder man than Livingstone; he dreamed of more than saving lives and souls and slaves. He saw that he could make a fortune, so he went to Europe seeking an investor. He found him in a businessman who was also a king.

Leopold II, king of the Belgians, was searching for a place where he and his business friends could make a lot of money. "The Universe lies in front of us," he wrote, "steam and electricity have made distances disappear, [and] all the unappropriated lands on the surface of the globe may become the field of our operations and of our successes." When he heard about the ivory, the palm oil, and the rubber trees in the Congo River basin he knew that this was just the fertile field he wanted. And Stanley was the man to plow it. With some backers, Leopold and Stanley formed a company to exploit the Congo. This business would be private, not governmental. The Belgian people were to have no part in it.

Stanley journeyed to the Congo, where he held a meeting with over five hundred tribal chieftains. He persuaded them to sign over to the king's company the land it needed. Perhaps his words lost a lot in translation. As the chieftains understood it, they were giving up only some distant lands of no importance. They thought they understood that they would keep their villages, the cultivated plots around them, and the nearby forests. It was in these forests that they gathered modest quantities of rubber, ivory (from tusks of elephants), and gum, which they floated down the Congo and sold to Europeans. But what the chiefs believed they understood was wrong, and in the 1880s they lost everything to the outsiders.

Leopold now was master of a portion of the earth eighty times as big as Belgium. As he would say, "My rights over the Congo are to be shared with none . . . the King [himself] was the founder of the state; he was its organizer, its owner, its absolute sovereign." All of it was his to use as he desired, and his timing was exactly right, because just then industrializing countries needed rubber for the tires of bicycles. The Congo was one of the few places in the world with forests of rubber trees. Ivory too was in demand—for billiard balls, piano keys, and knickknacks.

To make a fortune one merely had to get the Congolese to tap the trees and hunt the pachyderms. However, this turned out to be a problem. The Congolese were quite content to live as they had always done: farming, hunting, sometimes selling tusks and rubber. In the steamy heat of the equator, that was all they chose to do. Why sweat for white-skinned men from who knew where?

Leopold's managers solved this labor problem by forcing the Congolese to pay "taxes" in raw rubber, tusks, and food. (The food—

cassava bread, bananas, game—was for the managers and their Congolese enforcers.) If a village failed to pay its "tax" the managers took hostages and jailed the chief. If those measures didn't work they sent enforcers to arrest the tax delinquents and to flog them, using whips they cut from hippopotamus hide. The goons would shoot to death the worst offenders, chop a hand from each, and turn in basketfuls of severed hands. The hands were proof they hadn't wasted bullets. (However, some historians believe they also cut the hands from living Congolese.)

The wretched local people often fled, but some of them struck back. The warriors sang:

We cannot endure that our women and children are taken
And dealt with by white savages. . . .
We know that we shall die, but we want to die.
We want to die.

Word of how the whites were brutalizing Africans was slow to make its way from isolated Congo villages to Europe. Even Leopold, delighting in his palace or his villas in the south of France, cruising on his yacht, riding on his stately tricycle, strolling in his hothouses, romping with his buxom teenaged mistress, didn't know a lot about it. His company was making handsome profits, and no doubt he didn't ask his agents exactly how the system worked. When foreign missionaries told him what was happening in the Congo, he told them, hand on heart, that he was terribly disturbed. He met with aides and told them, "I will not allow myself to be spattered with blood or mud." But his agents in the Congo knew his righteous wrath would pass, and they kept on reaping "taxes."

To his credit, Leopold did build some roads and bridges in the Congo, and harbors and cities. His contractors braved the heat, the rain, and the jungle to build a railroad to the interior that bypassed cataracts on the Congo River. Formerly that trip had taken months; now it lasted only hours. On the maiden voyage, white-gloved Congolese served iced champagne to the railroad's European guests. But the king spent even more on public works in Belgium. He gave generously to education, and to building arches, avenues, palaces, and parks. What paid for them of course were tusks and rubber gathered by his Congo serfs.

When Britons heard reports about the Belgian ruler's arm's-length brutality, the British foreign office told its consul in the Congo to investigate. The consul traveled deep inside the forests and reported finding villages abandoned. He blamed this mainly on the dread disease called sleeping sickness, but he also claimed that many Congolese had fled the Belgian bosses and the brutal goons. Survivors asked him, "Are the white men never going home; is this to last forever?"

By the early 1900s, Leopold had made great profits but had also fallen into debt. He borrowed money from his kingdom, and agreed that Belgium should inherit the Congo on his death if he hadn't paid the debt. The Belgians were upset about his treatment of the Congolese and embarrassed by the outrage other countries voiced. In 1908 Belgium took the Congo over from the king, a year before his death. The settlement was complicated, but the king came out ahead. He claimed that he had always meant to give the Congo to his people.

After Belgium started governing the Congolese, their lives improved. Belgium did away with chop-hands serfdom, and fought malaria and sleeping sickness. Its practice was to treat the Congolese like children. It encouraged missionaries to give elementary schooling and to train some carpenters, gardeners, and cooks. But the Congo people had no voting rights and didn't know such rights existed elsewhere. The Belgians strove to keep them that way. It's said that when a Congolese sailor jumped ship in Europe, Belgium refused to let him return home for the rest of his life. Until it granted independence, in 1960, Belgium claimed the Congo was a happy place.

EUROPEANS WERE THE big imperialists in the 1800s, but not the only ones. Near the end of the century the United States joined them, and so did Japan, to everyone's surprise. No one would have thought Japan would play a major role in world events; the Japanese had always lived apart. Geography in part accounts for that (see map, page 69). A broad and foggy sea divides Japan from mainland Asia, and its islands rise abruptly from the water and appear to tell the world, "Let us be!"

In 1543 a tempest blew a Chinese junk to Japan. Aboard the ship were three Portuguese, and they may have been the first Europeans ever to set foot in Japan. Later they reported what they'd seen, and others—Portuguese, Spaniards, Dutch, and English merchants and missionaries—also came to Japan. These foreign contacts lasted for a century.

Then, however, 30,000 Japanese who had converted to Catholicism revolted against their local lords, who slaughtered them. The shogun, the general who ruled Japan in the emperor's name, decided to protect his country from pollution. He ordered foreigners to leave Japan, and warned them that if any should return he would execute them. Three years later a Portuguese vessel actually did come back. It bore no cargo, only foolish men who carried gifts and hoped the shogun would relent. He did not; he chopped the heads off all but thirteen men, whom he advised to tell their friends to "think no more of us; just as if we were no longer in the world."

For the next two centuries, the Japanese lived nearly in seclusion, like nuns behind a wall. They did admit some Dutch and Chinese merchants. (Apparently the shoguns thought that since the Dutch were Protestants they were not Christians, and therefore weren't as dangerous as Portuguese.) Through this contact with the Dutch the Japanese kept up with what the Western world was learning in such fields as medicine and astronomy. They also bought some western goods, such as telescopes and watches.

Even in these centuries of seclusion, the Japanese economy began to modernize. Large business firms grew up, among them the House of Mitsui, which still survives. Mitsui was so up-to-date in merchandising that on rainy days it gave its patrons free umbrellas that bore its trademark.

In 1853 the country reached another turning point in contacts with the world. On the other side of the earth, a bustling, bumptious nation wished Japan to open its doors for trade. Audaciously, the United States sent an expedition to persuade or force Japan to end its isolation. Commodore Matthew Perry led his two steamships and two sailing vessels directly into the fortified harbor of Uraga. The Japanese commanded him to leave, but Perry sent a message: if Japan did not accept a letter he was bringing from the U.S. president, he would deliver it by force. The Japanese knew that the defenses of their harbor were feeble, so after several days they took the letter. Perry told them he'd return in a year for the answer, this time with a larger fleet.

The shogun thought it over. He didn't want Americans to shell his capital, Edo (soon to be renamed Tokyo), or cut off Edo's food supplies, which reached the capital by sea. And he knew that many Japanese would like their country opened. Major business firms, down-on-their-luck landowners, and free-lance soldiers all believed they'd prosper in an open-door Japan. And so, when Perry returned in 1854—this time with nine ships—the shogun signed a trade agreement. He later did the same with several European nations.

The West had taught the Japanese a lesson they were not too proud to learn. Now they knew that if they didn't want some empire-building nation to devour them, they had better make Japan a modern nation. Perhaps they would have modernized in any case, even if the American bully hadn't paid his visit. The business moguls who had urged the opening would surely have demanded change.

Japan began to modernize—or westernize, which meant the same. This process happened fast, and it started with a major change in governance. Influential Japanese compelled the shogun to retire and seemingly "restored" the emperor to power. A new constitution proclaimed that Japan "shall be reigned over and governed by a line of Emperors unbroken for ages eternal." In practice, though, the emperor remained what he had been before, a symbol, and his elderly but energetic advisers were in charge.

Let's see how all of this concerns the New Imperialism. For one thing, Japan's rulers built an army. In doing so, they relied on French advisers till the Germans beat the French in war, when they switched to Germans. They formed a navy too, and naturally they modeled it on Britain's, since Britain ruled the waves. They set up schools, and made the Japanese the most literate people in Asia, surpassing even Indians. They would need to read and write if they were to modernize their country. As the emperor explained: "Knowledge shall be sought for all over the world and thus shall be strengthened the foundation of the imperial polity."

The Japanese transformed their economic lives. Five years after seeing Perry's steamships, they began to buy them for themselves. A dozen years later they had a railroad and a telegraph line between Tokyo and their major seaport, which had been a fishing village when Perry made his visit, and they built a giant steel mill and began producing world-class ships. They also built a textile industry, which thrived in part because many Japanese were starting to wear western clothes. Women switched from kimonos to skirts and men from skirts to trousers.

Foreign trade increased and so did the country's population, which rose by a fifth in only twenty years, from 1890 to 1910. That trade and population rose together was not surprising, since each one drove the other. The problem for the government was how to keep the economic system going. Only if Japan imported metals, rice, and coal and exported manufactures could its densely crowded people get along.

And that is why Japan began to do unto others as other modern nations were doing unto others. It was clear where they should start. Conveniently nearby was Korea, a peninsula that juts from Asia like a warning finger. Korea was the "land of the morning calm," as isolated from the world as Japan had been before the Perry visit. Forests, where tigers, bears, and leopards lived, still covered much of it. Korea looked to be of no great value, but it had three things that Japan greatly needed: markets for its products, coal to power its mills, and rice to feed its people. Korea also gave Japan an opportunity to show the world its military power.

Japan approached Korea as Perry had approached Japan: by threatening to strike it. Korea responded in 1876 by agreeing to exchange ambassadors and to open several harbors to the Japanese. This was not enough, but Japan could not push harder. Enormous China was the problem. For several thousand years the Middle Kingdom had been the Far East's major power, and Korea was its satellite.

As it happened, China was itself in trouble, thanks partly to the New Imperialism and its failure to modernize as Japan had done. For several decades European countries had been barging into China, claiming regions of the country as their spheres of influence. A Chinese leader once remarked, "The rest of mankind is the carving knife . . . while we are the fish and the meat." China was too feeble to expel these parasites. A knowledgeable Briton opined that "twenty-four determined men with revolvers and a sufficient number of cartridges might walk through China from one end to the other."

In 1894 Japan attacked the stranded whale. Although China was many times as big, the Japanese were well prepared. Japan won the war with little trouble, and it then made China recognize Korea's independence and give Japan the island of Taiwan and the Liaodong peninsula of Manchuria, in northeastern China. The world was shocked to see Japan's proficiency in bullying and war, since these were thought to be the special skills of Europeans.

Japan's aggressiveness alarmed the Russian Empire, which just then was completing a 6,000-mile-long trans-Siberia railroad from Moscow to a Russian port on the Sea of Japan. Manchuria, although it belonged to China, was vital to Russia's planning for the railroad's eastern end. So Russia won the help of Germany and France, and the three of them "advised" Japan to return the Liaodong peninsula to China. Outraged and indignant, Japan gave in, whereupon the Russians induced the tractable Chinese to lease *them* the peninsula. They also started angling for Korea, the special object of Japan's desire.

Japan considered what to do. Korea and Manchuria had the markets and the iron, coal, and rice it wanted, but Russia blocked its way to both these places. So why not go to war with Russia, both to win these lands and exercise its troops and ships? Suddenly, in 1904, Japan attacked and sank some Russian vessels off the Liaodong peninsula.

Could Japan defeat an empire that could draft more soldiers than Japan had fish? Japan and Russia sent large armies to Manchuria's treeless plains, and war began. In fighting near a major city, Mukden, two-thirds of a million men took part, more than the world had ever seen in any battle. The Japanese prevailed. Russia had not quite completed its long railroad, so it couldn't use it to send out more troops. Instead it sent a fleet of forty ships from the Baltic Sea to Manchuria, steaming around Europe, Africa, and much of Asia. When the ships at last arrived, Japan's new navy promptly sank or captured all but two.

Japan had trounced the giant, and now it got most of what it wanted. Russia handed over its leases from China of land in southern Manchuria, and recognized Japan's ascendancy in Korea. Japan sent a "resident general" to Korea, and when a Korean patriot killed him, it annexed the whole peninsula.

In Korea and Taiwan, Japan tried to benefit both the colonies and itself. It gave them railroads, factories, and schools, but it carried off such quantities of rice that it left the colonies underfed. The Japanese were often brutal, and the Koreans, who had ruled themselves (as China's vassals) for a thousand years, hated their oppressive rulers. When they rebelled, Japan struck back by burning villages.

A few years after Japan had acquired Korea, World War I gave Japanese imperialists a splendid opportunity. Naturally, the Japanese cared little about Europe, where most of "World" War I took place. However, they were wise enough to join the winning side. They declared war on Germany and snapped up German interests in China. They also forced the powerless Chinese to grant them rights (that Europeans had already) to exploit their country.

Japan had done so much so fast! In 1918, at the end of World War I, Japanese still living could remember Perry's visit. Now their country attended the peace conference in France as a recognized great power, one of the wealthy nations that ruled the world.

THE WEALTHY EMPIRE-GRABBING countries claimed that they were bringing God and other blessings to the "backward" peoples of the world. An American president, William McKinley, once told an audience that he had hesitated to annex the Philippine Islands, which the United States took from Spain in 1898. But, he said, he arose from a night of prayer convinced it must be done. The Filipinos were "unfit for self-government. . . . There was nothing left for us to do but to take them all, and to educate the Filipinos, and uplift and civilize and Christianize them [most of them were already Catholics], and by God's grace do the very best we could by them, as our fellow men for whom Christ also died."

At about the same time a British politician addressed an audience of professors and students at Glasgow University. Of the British Empire he exclaimed: "How marvelous it all is! Built not by saints and angels, but the work of men's hands; cemented with men's honest blood and with a world of tears, welded by the best brains of centuries past; not without the taint and reproach incidental to all human work, but constructed on the whole with pure and splendid purpose. Human, and yet not wholly human—for the most heedless and the most cynical must see the finger of the Divine."

We may well call those silly speeches laughable and insincere. It's true that empire-building nations often squeezed the poor ones, made heaps of money from them, crushed their governments, and prevented them from working out their destinies alone. On the other hand, they also brought them things that many needed: order, law courts, harbors, railroads, medicine, and schools.

CHAPTER 16

We multiply, and shrink the earth.

IN OUR FIRST 150,000 years on earth, we modern humans increased very, very slowly. Our number rose a little when times were good, and fell not quite as much when they were bad. Even as late as the 1600s (when Europeans settled in the New World, and the Manchus conquered China, and astronomers discovered where we are) all the people on earth numbered roughly half a billion.

In those 1600s we began not merely to add to but to multiply our number. From then to 1900 our number rose threefold. The population of the Americas and Australia multiplied especially fast because many Europeans left their homes and occupied these nearly empty continents. Despite that exodus, Europeans tripled. Asians nearly tripled, and even Africans increased, although slave ships carried off at least ten million of them. On the eve of World War I, humans numbered easily a billion and a half.

When humankind began to multiply, we didn't do so for the reason you might guess. Women did not start to bear more children than they had before. Married couples already had large broods, as they had to do since so many children died so young. Birthrates (births per 1,000 people) were already high, almost everywhere.

No, what made our number swell was not a rise in birthrates but a fall in death rates. That drop in deaths is what we must explain, and in order to explain it, we shall focus mostly on the Europeans. To do so makes good sense, for in the early centuries of the global increase Europe led the way. If we include in Europe's total not the people of that continent alone, but all the Europeans who settled in the other continents, and also their descendants, then Europeans multiplied by five.

While they multiplied, Europeans also industrialized. Surely, one would think, these facts must be related; industrialization must have made the numbers rise. That was the case, but the point is hard to prove. The biggest growth in people often occurred in the rural parts of Europe, far from mills and coal mines. It was often the death rate, not the birthrate, that was high where development was happening fast. In the grimmest parts of Manchester, the English textile town, a boy would be lucky to make it to twenty, whereas a boy in England's rural Surrey might survive to fifty.

Just the same, industrial revolutions did support the population rise. What they did was put more money in the hands of millions. Think of what the poor could now afford to buy, and how these things could help them to escape an early death. If they purchased more and better food, their bodies could resist disease. If they dwelled in dryer houses, and slept in several beds instead of crowded all in one, they might escape tuberculosis. If they purchased cotton clothes, which are easier to wash than woolens, they might escape the lice that carried typhus. If they ate on shiny dishes, mass-produced in mills, they could wash them cleaner than the old ones made of wood or earthenware, and banish dysentery.

Prosperity alone does not explain why Europeans multiplied. Obscure events, observed by no one, also cut the death rate. Here is an important example. As we saw (five chapters back), the sickness known as plague once slaughtered millions, but no one understood just why. They knew only that when plague appeared the best thing was to flee.

But then, quite quickly, plague left Europe. At the time, no one knew what made it vanish.

Centuries later, men and women using microscopes learned what plague is all about. Despite appearances, it is a disease not of humans but of rodents, chiefly rats, and the villains are bacteria. If a rat infected with plague should die, fleas will leave his cooling corpse and look for other rats. Many of these fleas will have drunk his blood, and therefore carry plague bacteria. If they can't find other rats to live on, then they *may* move onto human beings (reluctantly, no doubt; we aren't so furry). When they bite these human hosts, they infect them, and from them the disease may spread to other humans and kill them by the thousands. *But* this is only incidental to a rodent epidemic.

Historians have a theory on the disappearance of plague from Europe. It runs like this: In the centuries when plague was common, European towns were homes to *Rattus rattus,* the black rat. These rodents often lived quite close to humans, perhaps in roofs of straw, or rubble walls, or underneath the floorboards. Since these rats were such near neighbors, when they got sick with plague their fleas could easily move to humans. Rat epidemics caused human ones.

In the 1700s, Europe had a *rattus* revolution. Apparently, it started in the east. Russians noticed Asian rats, brown ones, swimming west across the Volga River, and in Europe these brown immigrants replaced the black natives. (They are known as *Rattus norvegicus;* it's not clear why Norway gets the credit for them.) But the newer rats were not at ease with humans, like the black ones; they lived apart, at greater distances. So (the theory goes) if brown rats suffered an epidemic of plague, they were far less likely to pass it on to humans. As a consequence, this lethal illness, scourge of cities, disappeared from Europe. (It lingered elsewhere until the nature of the disease was understood.)

Europeans clearly can't take credit for the rout of plague, but cholera is another story. Unlike plague bacteria, those of cholera are found in food or water. In a human body, they can swiftly lead to diarrhea, dehydration, and demise. In the 1800s six cholera pandemics (nearly global epidemics) led to many deaths.

A London doctor, John Snow, discovered in 1849 that in one neighborhood where many died of cholera, the source was in their drinking water. They drew their water from a surface well by means of a hand-operated pump. When Snow explained this to the local board of guardians they asked him what to do, and Snow replied, "Take the handle off the Broad Street pump." They did so, and cholera epidemics stopped.

In other places, too, experts found that dirty water led to cholera. Meanwhile, governments in several countries began to see that public health should be a public care. In many towns and cities, local governments laid pipes beneath the streets to bring in clean water, and other pipes to carry sewage out. Where this happened, cholera vanished.

Another English doctor, Edward Jenner, studied smallpox, which was once a leading cause of death. He had heard the widespread observation that a person who caught cowpox (Latin name: *vaccinia*), a harmless disease caught from cattle, did not catch smallpox. So Jenner, in a daring test, injected in a little boy some *vaccinia* "matter" from the lesions on a milkmaid's finger. The boy developed cowpox. Six weeks later Jenner broke his skin again, but this time he injected smallpox matter. The dread disease did not appear; somehow *vaccinia* had saved the boy from harm; it "vaccinated" him. In 1798 Jenner publicized what he'd learned, and knowledge of his findings spread.

Smallpox would remain a killer, however, until governments got active, as they had in fighting cholera. It took a war in 1870–71 to show the need of vaccinating one and all. The Germans vaccinated troops for smallpox but the French did not, whereupon the Germans lost 300 men to smallpox while the French lost more than 20,000. However, Germans on the home front were not vaccinated, and 130,000 of them died of smallpox. German rulers drew the obvious conclusion and required vaccination for everyone.

• • •

AS PROSPERITY INCREASED and medicine improved, populations rose. And as the hungry mouths increased, so did the need for food. Could farmers grow enough to feed the rising numbers?

One way to produce more food, of course, was to grow more food per acre. But a major problem was that fields could not be cultivated every year; if they were, they lost their fertility. The custom therefore was to use a field for crops a couple of years, then leave the field unused a year, and then plant again. That year of resting fallow let the field recover.

But farmers in the Netherlands, where land was scarce and mouths were many, hated leaving any field unused so long. With a lot of trial and error, they found a better way to use a field. If they rotated different crops, instead of always planting one, and added fertilizer, they didn't have to rest a field one year in three. A farmer might, perhaps, grow flax (for linen) for a year; then turnips; then oats; then clover. (Clover isn't only cattle fodder; it also fertilizes, taking nitrogen from air and "fixing" it in soil.) After this rotation, without resting his field at all, the farmer might begin again with flax. With such rotations, land no longer went unused one year in three, and so production rose. Crop rotation farming spread in Europe and throughout the world.

Another way to feed more people was to plant potatoes. Since ancient times, the staple in the European diet had been bread, baked with flour from wheat and other grains. But grains were always hard to grow in places where the soils were damp and days were short.

The Spaniards who came to the New World with Pizarro got to know potatoes, which the Indians grew (and grow today) among the Andes. The Spaniards brought them back to the Old World, along with beans, tomatoes, and other plants. At first, Europeans decided that potatoes were for pigs, not for humans, but with time they changed their minds. They learned how trouble-free potato farming was, and that the poor could raise them even if they had no plows or horses.

Using only spades, they dug some "lazy beds" and grew enough potatoes for themselves. An acre of potatoes feeds many more people than an acre of wheat.

Here is the good and bad that potatoes did to the Irish. In the 1700s, the potato plant became the center of the peasants' way of life. A boy and girl would marry in their teens, occupy a scrap of land, build a hut of earth, and raise potatoes. In time the two would have a family, and all would live—not well, it's true—largely on potatoes. The Irish people nearly tripled.

But, as everybody knows, disaster followed. Potato blight arrived in 1845 and 1846. With horror, Irish peasants found that their potatoes turned to mush. A million died of hunger, and a million others fled to England or America. But just the same, and not forgetting 1845 and '46, potatoes played a major role in Ireland's and in Europe's population growth.

Another way to feed the growing numbers was to find more land on which to grow more food. This happened in Australia and in North and South America, where settlers and railroads opened vast and virgin plains. For Europe these new lands were windfalls, providing grain, meat, and wool just when the demand for food and clothing rose.

Australia had its semi-arid plains, the Outback, where Aborigines had wandered no one knows how long ago. This is land so blank and empty (someone once observed) that it's like a newspaper that has been completely censored. When Britons settled here they found the plains just barely moist enough for cattle, but later on Australia nourished many millions of sheep.

In southeast South America lay another windfall, the Pampas. European settlers were amazed by the immensity of these treeless, grassy plains, which sprawl across a third of a million square miles. Cowboys soon were tending giant herds of livestock on them, and the Pampas, like the Outback, were to play a major role in feeding Europe.

But they were nothing when compared to North America's Midwest, where seas of grass flourished on a third of the continent.

Before settlers from the east coast of Canada and the United States first entered it, they feared this unknown land might be a desert. They were pleased instead to find sufficient rain, and rich and loamy soil. The grass was like a forest, high enough to hide a horse. Its roots were tough and thick, and when a farmer's plow "broke" through the virgin sod the roots would pop like wires. Except in regions where the rain was scarce, the new arrivals thrived, and the scale on which they farmed was huge. To harvest wheat they used enormous "combines" that could cut *and* thresh *and* clean the grain. Up to twenty horses pulled them.

On the dryer plains, farmers put together giant ranches. The land was simply endless, so cowboys couldn't tend the scattered cattle. Instead they used a new device, the barbed wire fence. Behind these fences wandered herds of cattle often numbering about 2,500. Every spring the cowboys drove the surplus cattle to the nearest railroad, which was often hundreds of miles away. Their beef would help to feed the multiplying millions in America and Europe.

THOSE WHO LIVED in modernizing nations knew that they were witnessing astounding changes. One such change was how their number multiplied, and another was the new machines and all the things they made. What was most amazing, and the symbol of the age, was speed.

When the 1800s began, humans couldn't travel very fast. A walking horse could carry someone on its back a little faster than a human walked: five miles an hour. The swiftest public transport was the stagecoach, which was pulled by teams of horses bred to do the work. Rested horses pulled a coach about ten miles per hour if the roads were good. However, roads were often bad, and people joked about the wooden-legged man hobbling down a road who refused a lift in a coach, saying, "No thanks, I'm in a hurry."

The English led the way to speed. Miners had long been hauling coal on carts with wheels, that men or horses pulled on iron rails. The rails, which lessened friction, were all right, but men and horses weren't efficient pullers, so mining engineers replaced them with locomotives equipped with steam engines. When this proved successful, someone saw that if engines could move coal, they could also move people. So engineers made railroad cars that looked like horse-drawn carriages on tracks. Their owners gave the locomotives winning names, such as *Catch-Me-Who-Can, Puffing Billy, Comet, Rocket, Novelty,* and *Royal George. Rocket* proved a great success, running an amazing thirty miles an hour, three times as fast as a stagecoach.

In 1830 British investors built a railroad from the textile town of Manchester to the port of Liverpool, and Frenchmen laid a railroad from *their* big textile town, Lyon, to nearby coal mines. In a few decades, trains were running forty, even fifty miles per hour in Europe, Asia, North America, Africa, and Australia. Canadians and Americans ran trains from sea to sea and, as we saw in chapter 15, Russia built the longest railroad in the world. The trans-Siberian ran from Moscow eastward to the Sea of Japan, an eight-day trip.

Railroads ran through regions of the world where no one lived as yet, though towns and cities soon would rise along the tracks. When the English writer Dickens rode a train through the American wilderness he wrote, "The train calls at stations in the wood, where the wild impossibility of anybody having the smallest reason to get out, is only to be equaled by the apparently desperate hopelessness of there being anybody to get in. It rushes across the turnpike road where there is no gate, no policeman, no signal: nothing but a rough wooden arch, on which is painted: 'WHEN THE BELL RINGS, LOOK OUT FOR THE LOCOMOTIVE.' "

Travel on the seas was also changing. In ancient times, when Rome was at its peak, it took two months to sail the length of the Mediterranean, and in the era of Columbus, it still took sixty days. But builders then improved the sailing ship, above all by adding sails. By the middle 1800s "clipper ships"—lofty, sharp-bowed vessels with many sails—could really move, especially when they carried Chinese tea to England. The captains knew that early-arriving cargoes of the yearly crop would bring high prices. In 1866, three "clippers" raced each other from Fu-chou to London. They reached the Thames on the same day, and docked on the same tide. In just below a hundred days, these ships had journeyed many times the length of the Mediterranean.

But steam did to sails what it also did to horses. An American had a steam-powered boat built for him in England, and shipped it home. In 1807 it puffed 150 miles upstream on the Hudson River, from New York to Albany, in under a day and a half. It looked, said someone, like the Devil in a sawmill, but it made a lot of money. It wasn't long till puffing boats with paddlewheels were hauling goods and people on the rivers of America and Russia.

Travelers distrusted the early steamboats, and they had good reason, since the boats were made of wood and were prone to smash. Within some decades, though, builders made their boats of iron, then of steel. Steam was introduced on oceangoing ships, like the two that Perry sailed on to Japan, and when the 1800s ended steam had all but conquered wind. The newer vessels traveled faster even than the clipper ships that once had raced with tea from China. At the century's end, the German liner *Deutschland* crossed the Atlantic Ocean at an average speed of twenty-five miles an hour.

As railroads bonded peoples, steamships hooked up continents. Just consider these travels of one British "tramp steamer." Leaving home in 1910, it carried railroad rails and other goods to Australia's western shore. There it picked up lumber and carried it to southeast Australia. With a load of farm machinery, it headed west to Argentina, where it loaded wheat. From there it sailed to India, and picked up jute (for making ropes and sacks). Next the freighter sailed to North America, and from New York it carried factory goods to Australia. It then returned to England with a load of sheep's wool, wheat, and lead. In about a year and a half the ship had traveled 72,000 miles and visited six continents.*

*John P. McKay, Bennett D. Hill, and John Buckler, *A History of Western Society* (4th ed., 1991), p. 826.

What the Irish famine was to the potato, the *Titanic* was to oceanic travel. This British liner started on her maiden voyage in April 1912, heading from Britain to New York. The *Titanic* was the biggest and most luxurious of ships, and was said to be an engineering triumph. She was powered not by one enormous engine but by two, and her hull was double-bottomed. The *Titanic* was, they said, "unsinkable." But two nights later, steaming much too fast off Canada, she grazed an iceberg. Her vaunted hull plates buckled, and the ocean flooded in. The captain urged his crew, "Be British!" Down went the *Titanic,* and two-thirds of the passengers and crew.

Two annoying bits of land were major obstacles for ships that sailed from sea to sea. One was the Isthmus of Suez, a piece of desert that blocked the way between the eastern Mediterranean and the northern end of the Red Sea (which leads to the Indian Ocean). Because of it, ships that sailed between Europe, Asia, and Australia had to sail completely around Africa, the second largest continent. The other nuisance was the Isthmus of Panama, which linked North and South America. If only ships could somehow sail right through this slender thread of land, they would save themselves long voyages around South America.

A French promoter formed a Suez Canal Company, and the firm set out to dig a channel through the desert, sea to sea. The canal would pass through marshy lakes, and making it was not a complicated job. Just the same, it took ten years to dig and dredge across a hundred miles of sand and marsh. When the task was finished in 1869, ships could quickly pass between the Asian and the European seas.

It was not a company but a nation that wished to make a channel through the other isthmus. The United States wanted to do this for both business and military reasons. As anyone could see, the place to cut the channel was in Panama, where (and this will twist your tongue) the isthmus was the thinnest. But Panama was a province of Colombia, in northern South America, and when the United States asked Colombia for consent, it refused the U.S. terms.

The U.S. president, Theodore Roosevelt, was not to be denied. (He inspired a palindrome: A MAN A PLAN A CANAL PANAMA, which reads the same way backward.) Disregarding another nation's rights, he sent a warship and some U.S. troops and encouraged the Panamanians to revolt. This they did, with the loss of the lives of a man and a mule. When Panama declared itself a nation, America was quick to recognize it, and the two agreed that the wealthy northern country would build a canal across the tiny Central American one.

Panama held steamy forests, mountain ranges, and tumbling rivers. These were guarded by tarantulas, crocodiles, deadly snakes, and mosquitoes bearing yellow fever and malaria. The Americans began by warring on mosquitoes, and nearly wiped out both diseases. Then U.S. Army engineers began to build the giant, complicated system. Among other things, they had to move a hundred million cubic yards of earth.

The canal, when finished, worked like this. Ships could enter from (let's say) the eastern side, rise in giant locks (sections that pumps could fill with water) to eighty-five feet above sea level, steam across a man-made lake, navigate a channel and a "cut" chopped through the mountains, descend a lock, cross another lake, and drop down two more locks to the Pacific. Roosevelt, never underspoken, would call the Panama Canal "the biggest thing that's ever been done."

With railroads, steamships, and canals, humans had discovered ways to move around the earth much faster than they ever had before. Jules Verne, a Frenchman, wrote a novel that conveys the sense of wonder at this feat. He named the book *Around the World in Eighty Days,* and around the world it had a huge success. Its hero is a wealthy Englishman named Phileas Fogg, who bets the other members of his London club that he can travel around the earth in eighty days. He sets out with his French valet, and journeys on a yacht, an ocean liner, coaches, sleighs, railroads, and an elephant. En route, in India he saves a pretty widow who would otherwise have been burned alive atop her husband's funeral pyre, survives two shipwrecks, and narrowly escapes from North American Indians. Does he win his bet? Read the book.

A young reporter for the *New York World* set out to beat Fogg's fictional eighty-day record. Starting near New York, Nellie Bly sailed to France, paid a call on Verne, and sped by train to Italy. (Every day she kept excited readers of the *World* informed.) From Italy she sailed (that is, steamed) to Suez, glided through the desert on the canal, and journeyed onward to Japan. At times she rode on donkeys, sampans, and rickshaws. After she had sailed across the Pacific to San Francisco, she had only three thousand miles of the United States to go. She raced across the country on a special train (as other trains were made to pull aside and let hers through), and she was met at major stops by fireworks, politicians, crowds, and bands. At Chicago, Bly switched to another train and traveled east until at last she reached her starting point. The trip had taken eight days less than Phileas Fogg's, but of course her editor demanded what had taken her so long.

WHILE MANY THOUSAND workers dug the Panama Canal, two brothers in America were working by themselves on a machine that would do more than trains, canals, and ships to shrink the world.

Wilbur and Orville Wright were a midwestern parson's sons. Early on they showed a gift for things mechanical, and when they reached their twenties they made bicycles and printing presses. Wilbur, though, began to dream of something grander when he read about a German who had flown in gliders, drifting on the rising flows of air. (This pioneer, Otto Lilienthal, had perished in a crash in 1896.)

Wilbur pondered how to build a craft that flew with power of its own. A major problem was controlling flight, that is, making a machine that could simultaneously tilt, climb, and steer to right or left. He and Orville noticed that a vulture shapes its movements while in flight by twisting both its wings. So the Wrights made gliders that had wings and rudders that the pilot could mechanically twist.

A week before Christmas, 1903, the brothers' first machine with power of its own was ready for its trial. They had already tested it at home, in Ohio, in a wind tunnel. Now they planned to really fly it, over sand dunes on a peninsula called Kitty Hawk, on the coast of North Carolina. With Orville at the plane's controls and Wilbur running by the side to stabilize the wing, "Flyer I" arose and flew twelve seconds. Orville telegraphed their father: "SUCCESS FOUR FLIGHTS THURSDAY MORNING . . . STARTED FROM LEVEL WITH ENGINE POWER ALONE AVERAGE SPEED THROUGH AIR THIRTY ONE MILES LONGEST 57 SECONDS INFORM PRESS HOME CHRISTMAS."

Soon the Wrights made more and better planes. In 1905 they turned out "Flyer V," which they could fly in figure eights and keep in motion more than half an hour. Wilbur later flew a plane in France two hours and twenty minutes. They went into production, selling planes in Europe and the America, but competitors in Europe drove them out of business. French inventors then improved the airplane. Many terms they used, such as *aileron* (a wing flap) and *fuselage,* are still in use, evidence of France's contribution.

By a generation after flight began, planes were crossing continents. This was risky work. Once a pilot tried to scout a route across the Andes, but he was forced to land his airplane on a two-mile-high plateau. For two days he and his mechanic hunted for a place from which to fly away, but the brinks of the cliffs were perilously close. Finally they climbed back in the plane, started up, raced it to the edge of a cliff, and plunged into space. At first the airplane dropped, but then it picked up speed and the pilot coaxed it past another mountain. Then the motor quit, but now they were above the plains. They crashed while landing, but survived.

In 1927, an American, Charles Lindbergh, took off from near New York in a single-engine airplane loaded down with gasoline. Never in his job as air mail pilot had he flown where land was not in sight, but he was bound for France. He hoped to win a prize of $25,000 for being first to fly alone, nonstop, across the ocean. For a day and a half he flew the plane and fought to stay awake, singing, bouncing up and down, and stamping on the floor. At last he crossed the British Isles, and soon he drew near Paris. Just at dusk, he glimpsed the small French airport, dimly lit. He landed at one end and taxied to the center, thinking there was no one there. Amazed, he found a cheering crowd of thousands.

In 1931 two pilots flew around the world in nine days. In 1932 a woman flew alone across the Atlantic; she would vanish five years later on a flight across the Pacific. And in 1938 a pilot crossed the Atlantic by mistake, or so he claimed. Starting from New York he flew without permission in an ancient single-engine plane, and he landed one day later in Ireland. He claimed that he had aimed for California but made a navigational mistake. Reporters nicknamed him "Wrong Way" Corrigan.

CHAPTER 17

We wage a war to end war.

IT LASTED MORE than four horrendous years, from 1914 to 1918, and the world had never seen its like. Those who lived it knew it as the Great War, or the World War. Now, alas, we call it World War I.

To understand the war, one needs to be acquainted with the four major European "powers" of the early 1900s, the nations that were strong enough to shape events. The four of them stretched in an arc across northern Eurasia, from the British Isles to the Bering Strait. They were: Britain, which ruled the oceans and many far-flung colonies; France, which Germany had crushed in 1870, but had a proud military tradition; Germany, rich and capable; and Russia, poor and backward but so populous that it could field huge armies. Two lesser powers were Italy, which was better known for music than for military medals, and Austria-Hungary, which was a huge but disunited empire.

In the early 1900s, tensions between these powers were the rule. They quarreled over colonies in Africa and Asia and over the complicated struggles of the mostly Slavic peoples in the Balkan Peninsula. In the spirit of the times, they longed to prove their greatness in a

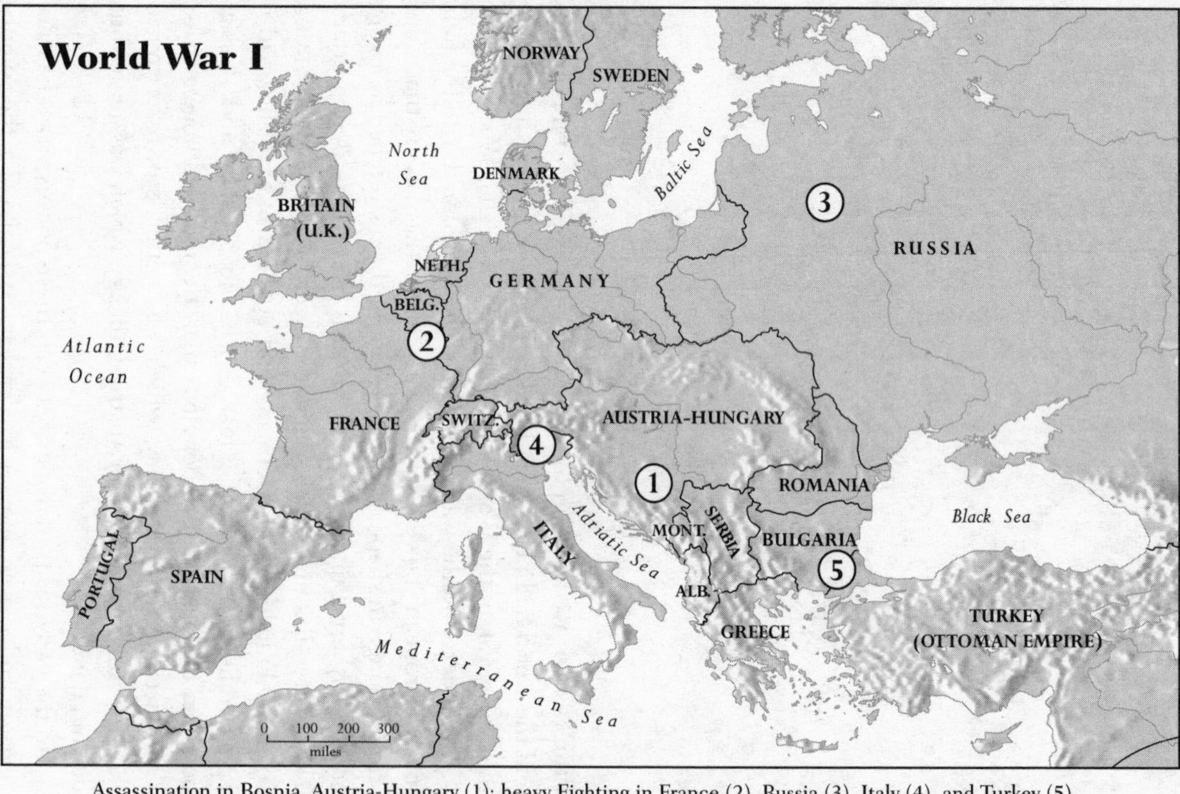

Assassination in Bosnia, Austria-Hungary (1); heavy Fighting in France (2), Russia (3), Italy (4), and Turkey (5).

war. Along their common border Germany and France erected lines of forts, and Germany raced to build a bigger navy than Britain's. Russia tried to build the railroads it would need to move its armies in the event of war. Germany, Austria-Hungary, and Italy, which together filled the center of Europe, had formed a "Triple Alliance" aimed at France and Russia. But fear of Germany resulted in a "Triple Entente" (or understanding) between France and Britain, to the Germans' west, and Russia to their east. Feeling encircled, Germany had drawn up a secret plan to knock out France and Russia, should a general war break out.

A general war was possible, and yet it seemed unlikely. Despite their quarrels, Europe's "powers" were usually friendly to each other. Several of the kings were cousins, and these lofty men were mostly on good terms. Smartly uniformed as generals and admirals they went to one another's weddings and funerals. When Victoria, queen of Britain, died a few days after the beginning of the twentieth century, two emperors, three other rulers, nine crown princes and heirs apparent, and forty princes and grand dukes attended the funeral and walked behind her coffin in procession. The German kaiser and his cousin-by-marriage, the Russian tsar, whose troops would soon be slaughtering each other, had for decades carried on a friendly correspondence. They wrote in English, calling each other Willie and Nicky.

THE IGNITION END of the fuse that led to war lay inside Austria-Hungary. This huge empire was like a wobbly Dagwood sandwich, made with layers of different peoples. Its army regulations recognized all of these languages: German, Hungarian, Czech, Slovakian, Slovenian, Serbo-Croatian, Ukrainian, Italian, Polish, and Romanian. Nation-fever had infected many of the groups who spoke these tongues. They wanted to leave the empire and live as independent nations.

On a sunny day in June 1914, a young man bitten by this separation fever sparked the First World War. He did this deed in Bosnia, a

Balkan province in the southern tip of Austria-Hungary. An Austro-Hungarian archduke, the heir to the empire's throne, had just arrived in Bosnia's capital, Sarajevo. Terrorists were waiting for him. When his open car came to a stop, the chauffeur having made a wrong turn, nineteen-year-old Gavrilo Princip jumped on the running board. With a shot from his pistol he killed the archduke, and with a second he killed the archduke's wife.

Princip and his friends were members of a secret group, the Black Hand. They aimed to rescue Bosnia from Austria-Hungary and join it to Serbia, a new nation east of Bosnia. They viewed Serbia as Bosnia's rightful home, since Bosnians and Serbians spoke the same Slavic language. The terrorists had trained in Serbia and got their weapons there.

Although the Austro-Hungarians had no proof, they were sure that Serbia had planned the murders. (They were right. The head of a semi-official Serbian secret society called Union or Death had plotted the assassination.) The Austro-Hungarians also feared that not just Bosnians but also other Slavic-language groups, the Croats and the Slovenes, wished to leave the empire and perhaps join with Serbia. So Austria-Hungary decided it must punish Serbia, and use it to teach all separatists (or nationalists) a lesson.

The danger of war was great. If a parliament of nations had existed then, it might have intervened and kept the peace. But no such league existed, and in a matter of days the fuse that Princip lit in Bosnia would burn through country after country toward the dynamite. Generations of historians have traced the fuse's sparking path and tried to fix the blame for what turned out to be a widespread war.

The Austro-Hungarians, as was said above, were allied with the Germans, their neighbors to the north and west. So now they checked to see if rich, dynamic Germany would support them. That help could be decisive.

The Germans' choice—to back their friend or not—was largely up to Kaiser William II. William had utter confidence, utterly unjustified, in his ability to make foreign policy. He dreamed of military glory, and he loved to dress in the uniform of an admiral or a guards officer, with a helmet crested by a waving plume and a golden eagle. So it was not surprising when he chose to stand behind the Austro-Hungarians if they went to war. Perhaps he hoped for war and glory, or he may have simply felt obliged to back his ally. Germany gave Austria-Hungary what has been called a "blank check," promising German backing if the greatest power of Eastern Europe, Russia, threatened to help Serbia fight Austria-Hungary. When the German envoy in Vienna telegraphed that Austria-Hungary might only mildly punish Serbia, William wrote in the margin of the telegram, "I hope not."

Sure of German backing, Austria-Hungary sent the Serbs an ultimatum. It made demands intended to provoke a war. Little Serbia, faced with doom, agreed to all demands but one (which dealt with trying the plotters). That one refusal was enough; it gave Austria-Hungary the pretext it wanted. The empire declared war on Serbia and began to bombard Belgrade, the country's capital.

Now what would Russia do? The Russians liked to play a leading role among the little Balkan states, including Serbia. Since they claimed, as Slavs, to be protectors of their little Slavic brothers, they couldn't simply watch as Austria-Hungary crushed the Serbs. And they had a bigger reason for concern: they feared that Austria-Hungary and Germany might subjugate the eastern Balkan peninsula. In that case, those powers might block Russia's water route to the Mediterranean, choking Russian foreign trade.

Russian officials saw that they must quickly act—or bluff. They had to get the jump on Austria-Hungary because, in their huge country with its horse-and-wagon roads, it took so long to raise an army and to move it to the front. So they began a limited mobilization. This was enough to alarm the Germans because, if it came to war, the Germans, like the Russians, wanted to be ready first. They knew that if the Russians should be first to put an army on the German eastern border, they would have an advantage if they immediately attacked.

Germany demanded that the Russians stop assembling forces.

Now the Russians could no longer bluff, if indeed they had been bluffing, for the Germans were in earnest. Since Germany might fight beside Austria-Hungary, the Russians had to act quickly against Germany as well. So they made the big decision to fully mobilize not only on their Austro-Hungarian frontier but on their German one as well. (In his diary that evening the none-too-brilliant tsar summed up his more important doings on this momentous day: "I went for a walk by myself. The weather was hot . . . had a delightful bathe in the sea.")

Germany now felt it had no choice, so in the Russian capital the German ambassador handed the Russian foreign minister a declaration of war. "What you are doing is criminal," the Russian told him. "The curses of the nations will be on you." "We are defending our honor," replied the ambassador. "Your honor was not involved," said the foreign minister. "You could have prevented the war by one word [to Austria-Hungary]. You didn't want to." The German burst into tears.

Germany also had to decide what to do about France, its neighbor to the west. The Germans were convinced that France would join the war on Russia's side. So Germany declared war on France as well as Russia, claiming that French armed forces had crossed the German border.

Germany had long ago foreseen the chance of war on two frontiers and had therefore made a master plan. Since France's German border was so strongly fortified, German forces first would dash straight west through little Belgium, which officially was neutral. Then they would wheel to the south, enter France, spread out, and encircle France's armies. In the meantime, German generals assumed, Russia would only slowly mobilize its armies, hampered by its meager roads and railroads. After crushing France, Germany would turn its troops around, shuttle them across the homeland on its first-rate trains, and crush the Russians at their leisure.

Now they put this plan in motion, but as they did they drew another power into war. When the Germans had declared war on France and Russia, they counted on Britain's staying out. This proved to be a big miscalculation. It's true that Britain was not bound by the terms of the Entente to go to war alongside France, but it was likely that, in the end, it would. Since many Britons wanted to stay out, their government needed an excuse to stand with France.

Germany, preparing its attack on France, demanded that the Belgians permit its armies to march across their land. Britain, when it learned of this, insisted that Germany not violate Belgian neutrality. Despite the British warning German troops invaded Belgium. Here was the pretext Britain needed. It promptly declared war on Germany and started sending troops across the Channel.

It was only five days since Austria-Hungary had declared war on Serbia.

Other countries soon declared war, and the opposing lineups were clear. On one side were the "Allies," chiefly Russia, France, and Britain. They would soon be joined by Italy (even though it had been a member of the Triple Alliance) and much later by America, and they would also have the help of French and British colonies throughout the world. On the other side were Germany and Austria-Hungary, soon joined by Turkey, which was then a large but rickety empire. Smaller countries joined the war on one side or the other.

NEARLY EVERYONE WAS sure the war would last only for weeks. The German emperor told departing troops in early August, "You will be home before the leaves have fallen from the trees." All through Europe young men rushed to join the armies, cheered by teachers, priests, politicos, and girlfriends. Recruits marched smiling down the flower-strewn city streets, and women darted out of cheering crowds to kiss them. This personal ad appeared in the *Times* of London: "PAULINE—Alas, it cannot be. But I will dash into the great venture with all that pride and spirit an ancient race has given me."

At first the German battle plan—shatter France, then conquer Russia at leisure—worked well. The Germans crashed through Belgium, wheeled south, and in big battles they pushed the French back almost all the way to Paris. But then they learned that the Russians had prepared for war much faster than expected; Russian armies had crossed the German eastern border. The Germans therefore had to pull two army corps from France and send them east, thus cutting down their strength in France.

The French were still in peril, and they hastily regrouped. Forces from Tunisia (part of France's African empire) had arrived in Paris, and two thousand Paris taxicabs rushed these troops and other reinforcements to the nearby battlefront. A month from when the war began, the French and British counterpunched and drove the Germans back. Germany gave up the idea of quickly crushing France.

Each side tried to get around the other, to outflank it in northwestern France and Belgium. As a result, the front line stretched from northeast France to the North Sea. And there it froze. All along the muddy, bloody front, the infantry of both sides dug trenches to shelter them from their enemies' big guns. This was not to be a war of movement on the western front, as Germany had thought. Not at all. From 1914 to the spring of 1918, the front between the Germans and the French and British would not move more than ten miles in either direction.

From time to time the armies on both sides sacrificed their soldiers by the many thousands to advance the front a mile or two. Typically they used these methods: artillery would pound the other side and try to blow up its machine guns. Then the infantry would climb up from their trenches. Bearing about sixty-five pounds of equipment they would lumber toward their enemies, crossing "no-man's-lands" of barbed wire, corpses, shattered trees, and giant craters made by shells. Enemy machine guns that had survived the shelling would fire six hundred rounds a minute at them. Poison gas might sicken them and blind them. Their enemies would try to drive them back with rifles and bayonets. If all went well, they won some trenches and held them for a while.

Most generals on the Allied side were portly seniors who had never been in battle. They lived behind the lines and telephoned their orders to the front. Those orders might result in many deaths, but prudent generals didn't ruminate on that. The French commanding general told his staff, after he had pinned a medal on a blinded soldier, "I mustn't be shown such spectacles again. . . . I would no longer have the courage to give the order to attack."

On the eastern front the fighting went more the way the Germans had expected. The western Allies had hoped the huge Russian armies would be a "steamroller," but it wasn't so. Russian and German armies struggled back and forth across the plains below the Baltic Sea, but the Germans had the better of it since their troops were better trained and better led. Russian generals had the bad habit of radioing each other "in clear," that is, not in code. German troops surrounded one whole Russian army and routed another. They captured 90,000 men and needed sixty trains to haul away the booty. The Russian general moaned to his chief of staff, "The tsar trusted me. How can I face him again after such a disaster?" In despair he shot himself.

Austria-Hungary had to punish Serbia; that had been the whole idea of going to war. In the summer of 1914 its forces crossed the Sava River. But the Serbs hit back, and after four months they forced the Austro-Hungarians to retreat. An Austrian spokesman explained it all: the invasion of Serbia had been only a "punitive expedition," and at the moment the empire's forces had to focus on the fight with Russia. The real drive against Serbia would take place on "a more favorable occasion."

In the second year of war the Austro-Hungarians and Germans pressed deeply into Russia. But soon they learned, as many have before and since, how hard it is to conquer Russia. Victories only drew them deeper, and the Russians had their endless plains in which to back away. The invaders captured, wounded, or killed many Russians—2 million in 1915 alone. But there were always more where those came from. In the following year the Russians hit the Austro-Hungarians so hard that their whole front collapsed, and the Russians captured a quarter of a million men. The Germans had to help their crumbling allies, so they hurried seven divisions to their aid. The Germans pushed the Russians back and killed or wounded a million of them.

Meanwhile, Italy had joined the Allied side, and it fought the Austro-Hungarians on their common border in the Alps. From the Allies' point of view, this fighting had the merit of tying down many Austro-Hungarians. After two and a half years of hard but inconclusive fighting the Austro-Hungarians, with German help, suddenly drove the Italians back across much of the north Italian plain. Their retreat is the background of Ernest Hemingway's famous novel, *A Farewell to Arms.*

Other countries fought on Europe's southeastern edge. In part because it was an ancient foe of Russia, Turkey joined the war on the German/Austro-Hungarian side. For the Allies this was bad because the Turks tied down a Russian army and also closed the Dardanelles, the Turkish strait through which vessels going to Russia have to pass. And so in 1915 Allied ships dropped nearly half a million British soldiers at the Dardanelles. The British quickly found themselves pinned down along a rocky shore where Turkish troops above could fire on them. The loss of life for both the British and the Turks was heavy. After half a year Britain pulled out its forces in the dead of night, having accomplished nothing.

Britain also fought the Turks in farther eastern Turkish lands: Syria, Palestine, Arabia, and far-off Mesopotamia. In the meanwhile, on the other side of the earth, the energetic Japanese had seen their chance. They joined the war as Britain's ally, and began to conquer northern China. The Chinese had recently deposed their Manchu rulers, but the country still was feeble and chaotic.

The outcome of the widespread war, however, would be settled on the front in France and Belgium, and in 1916 both sides tried to end the standoff. The Germans were the first to move. Early in that year they struck at the town of Verdun, where the French had built fortresses a hundred years before to guard a road from Germany to Paris. A victory at this famous place, the Germans thought, would break their enemies' morale. But the Germans' main goal was to pulverize as many French defenders as they could with cannon fire.

With all they had they shelled Verdun. Perhaps the French should simply have abandoned it; the forts were on an outward bulge in their line that they could have done without. But for patriotic reasons they chose to hold the place, and their general promised, "They shall not pass." The Germans had been hoping they could slaughter Frenchmen from a distance using cannon fire alone, but soon both sides were pouring in their troops. Four months of horror ensued. A French officer called it "a battle of madmen in a volcano." A third of a million Frenchmen were wounded or died and an equal number of Germans. This was not the kind of attrition that the Germans had wanted. Since they hadn't "passed," Verdun was called a victory for the French.

While the French were still repelling this attack, they and British forces launched their own strike farther west. They chose a site along the Somme River in northern France where French and British troops could battle side by side, something which was thought to be desirable. Since they were warned in ample time, the Germans were thoroughly prepared. They rolled out huge amounts of barbed wire and dug their bunkers fourteen yards in depth, safe from any shells the Allies' guns could hurl. Before the battle started 2,000 cannons carried on a shelling that was heard in London, 300 miles away.

On the first day of the battle, the British suffered 19,000 dead. A British soldier later told what it was like: "I see men arising and walking forward; and I go forward with them, in a glassy delirium wherein some seem to pause, with bowed heads, and sink carefully to their knees, and roll slowly over, and lie still. . . .

"And I go on with aching feet, up and down ground like a huge ruined honeycomb, and my wave melts away, and the second wave comes up, and also melts away, and then the third wave merges into the remnants of the first and second, and after a while the fourth blunders into the remains of the others, and we begin to run forward to catch up with the barrage, gasping and sweating, in bunches, anyhow, every bit of the months of drill and rehearsal forgotten, for who could have imagined that the 'Big Push' was going to be like this."*

The battle of the Somme would prove a turning point, not only in World War I but in the art of war itself. What British generals, trained in other times, still dreamed of was a glorious, victory-clinching charge by cavalry—gallant swordsmen riding noble steeds. At the Somme, the climax for the cavalry came late one day in mid-July after British infantry had overrun the Germans. Cavalry divisions were supposed to follow and assure a victory, but because of mud and craters it took them hours to reach the fighting. At seven in the evening, they trotted into battle through the waving grain with bugles blowing, lances gleaming. And then: machine guns mowed them down like hay.

This battle also launched the weapon that would replace the archaic horsemen. The British recently had started to produce what battle planners long had dreamed of: armored vehicles with guns. However, while the British at the Somme used forty-seven of these newfangled "tanks," they weren't enough, and the British used them badly. Tanks would later prove their value, lunging over trenches, flattening barbed wire, shielding men on foot, and squashing gun emplacements.

In three months' fighting at the Somme, the British and the French advanced a mere five miles. The four-month battle accomplished nothing for the Allies except to pull German forces from Verdun. The Allied killed and wounded numbered two-thirds of a million, and the Germans' casualties were something under half a million.

On battlefields in Belgium, British troops attacked in mud so deep it swallowed tanks. In one battle, they lost 300,000 killed and wounded, and the Germans 200,000. Next day, the British chief of staff surveyed the scene of battle—for the *first* time. He wept and cried, "Good God, did we really send men to fight in that?"

*Henry Williamson, *The Wet Flanders Plain* (1929), pp. 15–16, as quoted in Paul Fussell, *The Great War and Modern Memory* (1975), pp. 29–30.

After so much slaughter, and having long ago reached a stalemate, it was surely time to stop the war. Both sides claimed they fought to prevent future wars, but in fact they couldn't finish this one. Among the French, at least, the troops' morale became a problem. In 1917 the French began a drive that led to mutiny. While marching to the front a regiment bleated like sheep being led to slaughter. Fifty-four divisions refused to fight, and many thousands left the lines. The French commander had some soldiers shot, restoring order, but he promised France would wage no other big attacks.

Late in 1917, a shocking change appeared to favor Germany. As chapter 18 will relate, Russian revolutionaries overthrew the tsar and pulled their country from the war. Delightedly the Germans pulled their armies out of Russia and hauled them to the western front, where their generals prepared a killing blow. In early 1918 German forces shelled the French and British, poison-gassed them, and attacked. By the end of May they had pushed the French and British back to the point the Germans had reached at the start of the war, the Marne River, thirty-seven miles from Paris. It looked as if Germany and Austria-Hungary might win the war.

But now a partner joined the Allies who might make up in part for the loss of Russia. Until this point, America had sold supplies to France and Britain, but kept out of the fighting. But now the Germans made it easy for America to decide to join the Allies. To stop supplies from reaching Europe, German submarines began attacks on U.S. cargo ships. As a result, the U.S. president, Woodrow Wilson, went to Congress and obtained a rousing vote for war. However, America had almost no armed forces. To raise and train an army and to send it overseas would take at least a year. So a race got under way. Could U.S. forces reach the lines in France in time to help prevent a German victory?

When Germany began its major push in France, no U.S. troops had yet arrived. The Germans' hopes were high, and the British commanding general was anxious. He issued an alarming order of the day: "With our backs to the wall and believing in the justice of our cause each one of us must fight to the end." However, by the summer French and British troops had stopped the German drive and just begun to push the Germans back. Then nine American divisions reached the Allied line. It was still the Europeans' war, and even after years of it, they still were fighting hard. But U.S. troops began to play a useful role.

At the end of summer and in early fall, the Allies opened an offensive of their own. By now a quarter of a million Americans were arriving every month. Among these men was Corporal Alvin York, a mountaineer from Tennessee. York had first refused to fight, on moral grounds, then changed his mind. An expert marksman, he was used to rifle matches back home in Tennessee, where he shot at turkeys' bobbing heads. On one October day enemies surrounded his patrol. Single-handedly, York shot some 25, captured 132, and brought back 35 machine guns.

Early in 1918 it had looked as if the Austro-Hungarian/German side might win the war. As 1918 neared its end, however, that prospect had reversed, and the German High Command informed Berlin that Germany could not win the war. They didn't say that Germany might lose, but they obviously feared this, and many German people wanted only that the war should end. Parts of the country were on the brink of revolution. So Germany transformed itself. Its generals had gladly run a war but were just as glad to let civilians handle a defeat. They stepped back in the shadows. The kaiser appointed a liberal-minded prince as chancellor, and he in turn let the country become a republic. The kaiser fled to Holland in a special train, never again to see his native land.

The new regime was far from keen to lay down its arms since German armies were intact on foreign soil, and willing to fight on. But the civilian leaders feared a revolution, born of anger and frustration, like the one in Russia. So they consented to a truce, and in a French railway carriage they signed an armistice document. The truce amounted to surrender.

The war was over.

Firing halted on the western front on November 11 at eleven A.M.—the eleventh hour of the eleventh day of the eleventh month.

As the moment neared, a Scot recorded, "Officers had their watches in their hands, and the troops waited with the same grave composure with which they had fought. At two minutes to eleven, opposite the South African brigade, at the easternmost point reached by British armies, a German machine-gunner, after firing off a belt without pause, was seen to stand up beside his weapon, take off his helmet, bow, and then walk slowly to the rear."

At eleven, "There came a moment of expectant silence, and then a curious rippling sound, which observers from the front likened to the noise of a light wind. It was the sound of men cheering from the Vosges [in eastern France] to the sea."* But a German general bitterly informed his troops, "Firing has ceased. Undefeated . . . you are terminating the war in enemy country."

In the Allied countries work was halted for the day. Church bells rang, happy crowds filled streets and squares, many cheered and many wept. It's said that total strangers copulated in the streets.

What a war it was, and how much woe it caused! The world had never seen such loss of life. Nine million men in arms had died, and more than twice as many had been wounded, many of them maimed for life. Of all the Frenchmen aged from twenty to thirty-two in 1914, the war had killed half. Perhaps 5 million noncombatants had died of hunger and disease. But nature had an irony in store. In 1918, as the war was ending, an influenza epidemic spread throughout the world. It took the lives of 20 million humans. In India alone, it killed more people than the war had slain on all the battlefields.

TWO MONTHS AFTER fighting stopped, the victors gathered to arrange a peace. The leaders at this conference were the American president and the French and British prime ministers. Most of the talks were held near Paris at the Palace of Versailles, the splendid former home of kings of France.

*John Buchan, *The King's Grace* (1935), p. 203, as quoted in Martin Gilbert, *The First World War: A Complete History* (1994), p. 501.

Wilson, from America, was a complicated man, at odds with his own past. As president, he had bullied nearby Latin nations as if their peoples had no sovereign rights. But when he led his country in the First World War, Wilson had turned high-minded. A year before the conference, he had proposed a list of "Fourteen Points" that should, he thought, provide the basis for a peace.

As the fourteenth of his Fourteen Points, Wilson had made a great, imaginative proposal: the forming of "a general association of nations." Its goal would be to guarantee the peace and independence of "great and small States alike." So when he came to Paris Wilson pushed for such a body. The other leaders doubted that a league could keep the peace, but they finally agreed. The League was to rely on talks and treaties, not on force. Law-abiding countries, it was hoped, would rein in rogue ones with the threat of reprimand or sanctions.

Other countries duly formed the League, but the U.S. Senate then refused to let America join it. As we shall see below, in coming years the League would try to deal with great affairs. Among the things that hobbled it would be the absence of the very country whose president had pushed for its creation.

The victors' biggest problems at Versailles concerned the losers. At the very least they wanted to punish them, but better yet they hoped to break them into pieces, along their ethnic fault lines.

In the case of Austria-Hungary, this was not so hard. The huge empire was already crumbling into little nations, and the peacemakers did not so much order the breakup as take note of it. In the empire's north when the war ended, the Austro-Hungarian governor of Bohemia had simply telephoned the outlawed leaders of the Czech people's separatist movement. When they arrived at his castle, he handed them the keys and left, and thus was born the future Czechoslovakia. In the empire's south, where the war had started, Bosnians and others joined with Serbia to form Yugoslavia, "the land of southern Slavs." This would have pleased the assassin, Gavrilo Princip, who had spent the war in prison, but he had died of tuberculosis just before it ended.

Not at Versailles, but soon after, the allies carved up Turkey, the other complicated empire on the losing side. The Turks kept only Istanbul (the former Constantinople) in Europe, and most of Asia Minor. In the Middle East the oceans of sand that covered unplumbed seas of oil were broken into what would later be the Arab nations.

The biggest problem at Versailles was what to do with Germany. The Germans and their friends had nearly won the biggest war the world had known. The victors saw the Germans as a chronic threat to peace, like the hungry ogre in the castle in old folktales who from time to time drags off a village maiden.

One way to punish and weaken Germany was to lop off chunks of German land. But who should get the confiscated territory? To create a beneficiary, the victors had to improvise. More than a century before, a nation, Poland, had lain on Germany's eastern flank, but Prussia (Germany), Russia, and Austria-Hungary had divided and swallowed Poland. It had simply vanished from the map. Now the diplomats brought Poland back to life and gave it pieces of the same countries that had once devoured it.

What else to do with Germany? The victors took away its colonies in Africa and the Pacific, returned to France two provinces the Germans had taken fifty years before, and gave to France the use of German coal mines for fifteen years. They also fined the Germans heavily for the damage they had caused. Most important, though, they tried to weaken Germany. They limited its army to 100,000 men, and demanded that it surrender its fleet. (Instead, the German captains sank their ships off Scotland. Scottish children on a tugboat outing chanced to be there, and were thrilled to watch what they believed to be a show put on for them.)

Not surprisingly, the Germans hated to be treated as the losers of the war. They didn't feel defeated; they had no sense that they had lost the war. After all, they said, when the fighting had stopped they were "undefeated in the field" and willing still to fight. Not only did they not feel beaten, Germans also felt no guilt about starting the war. No one really knew who had caused it—if any single country had—because all the countries kept their diplomatic papers secret. Each side was free to blame the other.

To justify the harshness of the treaty the victors included in it a "war guilt" clause. If the Germans signed the treaty it meant that they "accepted the responsibility" for all the losses that resulted from the war. They also agreed that war had been "imposed" upon the Allies "by the aggression of Germany and her allies." The Germans found these words infuriating, and an insult to their honor. For that and other reasons, they at first refused to sign the treaty. They signed it only when the Allies threatened to resume the war.

In the years to come an army veteran would rant in front of German crowds about the heavy fine, the confiscated land, the army limit, and the war guilt clause. The humiliation of Versailles cried out for vengeance. As Adolf Hitler shouted, sneered, and threatened, thousands cheered.

CHAPTER 18

A utopia becomes a nightmare.

OUT OF THE MISERY of the First World War rose communism. And out of the Great Depression of the 1930s rose Nazism, which is the subject of the next chapter. They were both like noxious weeds that thrive amid the litter of a vacant lot, and their story is a grim one. Both promised a better world, even a Utopia, but they delivered something else.

THE MAN WHO planted the seeds of communism was born in 1818, a century before they germinated. Karl Marx grew up in western "Germany" (it wasn't yet a united country) in the early decades of the Industrial Revolution. At this time thousands of young men and women in Europe and North America were moving from the farms to the factory towns, and learning to run the clattering new machines.

Not Marx, however; as a prosperous lawyer's son he had better prospects. He studied law and philosophy at German universities and mixed with intellectuals, discussing atheism, politics, and reform. Some of these acquaintances were "socialists," who believed that all the people, not the wealthy only, ought to own the land, the mills, and the engines used in growing food and manufacturing.

In his early twenties Marx edited a radical journal in Paris. When he heard about him, young Friedrich Engels crossed the Channel from England to visit Marx. Engels was the son of a wealthy German textile manufacturer. His father owned a factory in England, and he had sent Friedrich there to learn the business and to run it.

Engels was utterly unlike the solemn Marx. He fenced and hunted, lived with a mistress, and welcomed friends for tea on Sundays. But he had another, intensely serious side. What he had seen in the back streets of the British manufacturing city of Manchester had stunned him. The filthy hovels shocked him, and so did the stunted workers, their hungry families, and the way they sought relief in gin and Jesus.

When Marx and Engels met, they were both in their middle twenties, and both were consumed with the idea of social reform. They talked for ten days and began an alliance that would last for forty years. Engels would bring to the team his indignation about the lot of the working class and his knack for quickly dashing off a sparkling tract. Marx, however, was the senior partner, the philosopher and prophet who drew lessons from the past and prophesied the future.

The two men called themselves "communists" rather than socialists. Communism meant something more drastic. Communists not only understood the oppression of the workers but were ready to carry out a workers' revolution. Marx and Engels joined a secret group of would-be revolutionaries and reshaped it into what they called the Communist League. The League had tiny subdivisions in London, Paris, and Brussels. The London group asked Marx to write a declaration of principles for them. So in early 1848, with help from Engels, he wrote a short and fiery pamphlet, *The Communist Manifesto*.

The pamphlet starts with words intended to shock: "A specter is haunting Europe—the specter of communism." It then explains the

Marx and Engels view of history: "The history of all hitherto existing society is the history of class struggles." In Europe in the Middle Ages, the rulers were the great landowners, since most people were farmers and the landowners held the chief means of production. But then, says Marx, wealthy men of the rising middle class overthrew the landowners. These victors were "capitalists," who owned ships, workshops, and enough money to lend to others at interest. In the time he was writing, Marx explains, the descendants of those triumphant capitalists were the powerful owners of banks and factories. They brutalized their workers and lived off their labor as parasites. Governments claimed to serve everyone, but they were really just committees of the middle class for the exploitation of the workers.

At any moment, however, fighting would begin, and the workers, guided by the communists, would have *their* revolution and overthrow the middle class. (The communists, Marx explained, might include a few enlightened members of the middle class who "cut themselves adrift and join the revolutionary class." He had in mind no others than Engels and himself.)

Once the revolution had succeeded, the workers (guided by the communists) would themselves be the ruling class, the dictators, for a while. They would confiscate all significant private property, such as land, banks, factories, and machines. After private property was gone, classes, and therefore war between the classes, would disappear. A classless society would emerge. Marx does not explain what life in such a classless world would be like, but it clearly would be better. Schools would be open to all, factories and farms would produce more, and poverty would vanish.

The *Manifesto* closes with a flourish. Communists, Marx says, "openly declare that their ends can be attained only by the forcible overthrow of all existing social conditions. Let the ruling classes tremble at a Communist revolution. The workers have nothing to lose but their chains. WORKING MEN OF THE WORLD, UNITE!"

The authorities in Belgium, where Marx composed the *Manifesto,* were arresting radicals, so he went to Paris and then to Germany. There he edited a newspaper until the government closed it down.

Marx printed the last edition in red ink (symbol of revolution) and left for England. For the rest of his life he would live there with his wife, three children, and their German maid, who stayed with them, unpaid, all their years. During one period of six years they lived in just two rooms, eating potatoes and bread. It was mostly Engels's gifts of money that kept them alive.

In his final decades Marx wrote out his theories of history and economics in a treatise that he named *Capital*. A volume of it appeared during his lifetime, and Engels edited two others after Marx's death. *Capital* would be communism's Bible. For a hundred years communists all over would study it and argue over every word. However, the fiery little *Manifesto* held Marx and Engels's basic message and reached many more readers.

As decades passed, however, the message began to look outdated. Workers did not revolt as Marx and Engels had predicted. On the contrary, in several countries they actually got the vote. As a result of their new influence governments ended the use of child labor, limited the hours of work, and provided public schools for all. Many workers' wages rose because of the general rise in productivity. If they didn't, workers organized in unions, and with the threat of strikes they compelled employers to increase their pay.

Marx was well aware of all these changes, and before he died he doubted that worker revolutions would ever occur. But many of his followers, who were known as "Marxists," disagreed. They argued with Marx and with each other, and for decades they would argue about what Marx had said, and how they ought to bring about the classless society. Marx was arrogant and certain of his views, and he wearied of the wrangling Marxists. He called them "rascals," "louts," and "bedbugs," and disgustedly he said, "All I know is I'm not a Marxist."

IN SPITE OF Marx's pessimism, communist revolutions did take place. The first began in 1917.

At that time the Russian empire covered two-fifths of Europe and Asia. It was so big that the United States, China, and India could all have been dropped into it. Europeans liked to call it "a thousand miles of mud," but it was five times as long as that.

Most Russians were peasants, sometimes called "the dark people." They lived in ignorance and isolation in shabby villages that dotted Russia's plains. Many of them worked small farms of their own, while others worked on the estates of landlords, who often lived in mansions in the towns and cities. Russia was still profoundly rural, but it had begun to industrialize. Already cities had their slums and hard-worked, poorly paid workers.

Petrograd (St. Petersburg), which was then the capital, lies in Russia's northwest corner. Tsar Nicholas II lived there in the Winter Palace or in his summer home outside the city. He was courteous and shy but a despot who was certain that democracy was "senseless and criminal." Russia had a feeble and conservative parliament that met in Petrograd but only when the tsar permitted.

Inside Russia (and outside, because many were in exile) small networks of radicals burned to lead the workers—and perhaps the peasants—in a revolution. Most of them aimed to oust the tsar and parliament and put the country in the hands of working people. Many of the rebels traced their principles to Marx, who had died about when they were born.

World War I gave the revolutionaries their chance. Germans had crushed the Russian armies and forced them to retreat deep into their own country. Millions of men had died, in part because the tsar and his ministers had grossly mismanaged the war. For example, the aged general who served as war minister had purchased too few machine guns and rapid-fire cannons because he saw them as newfangled weapons that only cowards used.

Late in 1916 the parliament found the courage to protest the government's bungling. Tsar Nicholas reacted predictably, and dismissed the parliament. As a result, many of its members, who until this time had been loyal to the tsar, concluded that they could save the country only by taking power.

However, it was not the parliament or middle class but working people who sparked an uprising. Because of the war, food in Petrograd was scarce and bellies were empty, but the fumbling government was slow to deal with the food problem. Early in 1917, workers began to strike and march in the streets. For the most part, they had no leaders. Riots broke out, and the crowds shouted "Down with the tsar!" Soldiers were ordered to fire on the rioters but refused to do so.

A revolt was turning into a revolution, but who was going to lead it? Two very different bodies took shape. One was an executive committee of the parliament, made up mostly of moderates who wanted a constitutional government. The other was a fiery Soviet, or council, of Workers' and Soldiers' Deputies, which socialists and communists led. In theory the Soviet was only a local body, whereas the parliamentary committee was national. In fact, however, the Soviet steered the committee. Under pressure from the Soviet, the parliamentary committee set up a provisional government.

Pushed by the Soviet, in March 1917 the provisional government telegraphed the tsar, who was at the war front, and called on him to leave the throne. He tried to return to Petrograd on his special train, but en route he learned that at a station near the capital rebel soldiers blocked the tracks. Clearly the army was starting to back the revolution, fatefully for Russia. For Nicholas all was lost, and in his silk-hung railroad car he signed an act of abdication.

FAR AWAY IN Switzerland news of the revolution reached Vladimir Lenin, a Russian revolutionary. Lenin, like so many other socialists and communists, came from Russia's slender middle class. His father had been an inspector of schools. However, his eldest brother had been a terrorist, and was hanged for joining a conspiracy to kill Nicholas's father with a bomb hidden in a book. Lenin also had become a revolutionary and the leader of the Bolsheviks, one of the factions of Russian communists. A fellow Marxist said of Lenin that he thought of revolution all day and dreamed of it at night.

The outbreak of a revolution was the moment he had long awaited. It was urgent for Lenin and other Bolsheviks to return from

Switzerland to Russia. But Germany, which was of course at war with Russia, lay across their path. They applied to German officials, who hated revolutionaries but hated the Russian government even more. The officials arranged for Lenin and his friends to ride through Germany in a sealed railway car, as if they were deadly germs. The Germans hoped, of course, that these bacteria would spread the disease of revolution in Russia.

Despite a common misconception, Lenin and his Bolsheviks didn't bring about the revolution. As we have seen, when they arrived in Petrograd in April 1917, the revolution had already begun. The tsar and his family were prisoners, and the army was melting like the snow that still lay on the fields and streets. The Provisional Government was shakily in charge, and workers' soviets like the one in Petrograd were forming everywhere. Lenin and his colleagues figured what they had to do. Their task was not to spark rebellion, but, like bandits climbing on a moving train, to seize a revolution that was under way.

Lenin waited half a year. By then the Bolsheviks had won majorities in all the provincial soviets, and he had the backing of the rebel soldiers in Petrograd. Now was the moment to snatch control of the revolution from the middle-of-the-road Provisional Government. To lead the coup Lenin chose the able Leon Trotsky, another middle-class revolutionary, the son of a well-off farmer. (The tsarist government had deported Trotsky during the war. When the revolution began, he had been even farther away than Lenin, in an apartment on 164th Street in New York City.)

On the night of November 6, 1917, Trotsky's men and some soldiers seized some power points in Petrograd: bridges, the main telegraph office, the central bank, railroad stations, and power plants. The coup was so quick and bloodless that the people of the city had no idea what was happening. The Bolsheviks quickly called into being a "Congress of Soviets," and it declared the Provisional Government dead. The congress then set up a new government and chose Lenin as its head.

Lenin's government held an election for a "Constituent Assembly" that was to draw up a constitution for Russia. But his own communist faction, the Bolsheviks, won only a quarter of the delegates. This would never do, since Lenin wanted a one-party government (his party, of course) to run the "dictatorship of the proletariat." He let the assembly meet just once. On the following day he sent armed men to shut it down.

The revolutionary government quickly announced the confiscation of the property of wealthy landowners, with no compensation to the former owners. This wasn't such a daring move, since peasants had begun to seize landowners' estates even before November 1917. But now the peasants took over millions of acres that had belonged to the wealthy gentry. Formerly the peasants had been dirt-poor; now at least the dirt was theirs. As in France in 1789, aristocrats began to flee from Russia.

Lenin and his colleagues faced enormous tasks. They had to arrange peace with Germany, and also end a civil war that broke out after the revolution and raged through much of Russia during three bad years. But the communists had an even bigger task. If they wanted to reshape the country, they had to get the Russian people under tight control. Lenin didn't mind this task; he was quite prepared to shove aside or murder anyone who blocked the path of history. He once declared, "the more members of the reactionary bourgeoisie [the middle class] and clergy we manage to shoot the better." He was known to attack even other socialists and communists as "ugly scum . . . blisters . . . pus."

The Bolsheviks exterminated every member of the former ruling family they could find. They shot and bayoneted the tsar and tsarina, their children, three servants, and their pet spaniel; cut them in pieces; doused them in acid; and dropped them into shallow, unmarked graves. They killed four grand dukes in prison, and they shot a nobleman and threw his body into a blast furnace. They beat some other aristocrats and dropped them, still alive, down a mineshaft.

Many once-rich landowners had already fled the country. Now many of the middle class did the same. So did many socialists and communists who had not been shrewd enough to join the Bolsheviks. But many didn't flee who should have. When sailors at a naval base revolted, demanding democratic reforms and the release of political prisoners, Trotsky used the new Red Army to quell the mutiny. Those who survived the army's merciless repression were shot to death.

Lenin and his governing committee moved the capital from Petrograd to Moscow, from which the tsars had once ruled Russia, and they governed from the Kremlin, which had been the fortress of the tsars. They gave Russia a new and bulky name: the Union of Soviet Socialist Republics. It would keep that name for seven decades, but many called it by its former name. To keep things simple we shall call it Russia.

Five years after taking power, Lenin had the first of several strokes. He died in 1924. On Moscow's Red Square an Immortalization Committee built a huge red granite tomb. They embalmed his corpse and placed it in the tomb in a glass coffin. There the faithful could file by silently and gaze on the hero of the revolution.

Now the Bolsheviks had to choose another chief. One might suppose the choice was easy; surely they would pick the brilliant Trotsky, leader of their coup and commanding general in the civil war. Not so obvious a possibility was Joseph Stalin, who had the weighty job of general secretary of the Central Committee of the Communist Party. "Stalin" was his revolutionary name; it means "the one of steel." Unlike Lenin, Trotsky, and many other Bolsheviks, Stalin really came from the working class. His father had been a shoemaker and his mother a washerwoman.

In public, Stalin posed as slow and placid, always puffing on a pipe. His rival Trotsky underestimated Stalin as "a gray and colorless mediocrity," but he was a hard-driving, head-bashing administrator. In private he was rude and crude. His daughter once witnessed this: while Stalin was pacing in his office, smoking and spitting on the floor as usual, his pet parrot imitated his spitting. Stalin pushed his pipe inside the cage and smashed the parrot's head.

Lenin had found Stalin useful, but in his final months, he lowered his opinion of him. Stalin had insulted Lenin's wife, shouting at her on the telephone that she was a "syphilitic whore." Lenin wrote a letter, to be opened after his death, in which he warned the Bolsheviks that Stalin was *grub,* meaning rude or coarse. When Stalin learned of this he cursed his one-time friend: "He shat on me and he shat on himself." Stalin kept Lenin's letter secret as long as he could.

Stalin was a skilled manipulator. As general secretary he had found jobs for many of the Bolsheviks, and they owed him favors and also feared him. As his friends and enemies were well aware, he had a special staff that maintained files on them. As time went by, he made alliances with other men of power (most of whom he later killed), and quietly won control of the party.

Then he disposed of his rival, Trotsky. First he eased him from his post as minister of war, and then he had him expelled from the Politburo, the group of Bolsheviks that made the policy decisions. Later a party congress led by Stalin expelled Trotsky from the party and exiled him within the country. He was dragged screaming from his apartment and hauled by train to a Russian town far to the east, near the Chinese border. The next year, Stalin exiled Trotsky from Russia. He went to Turkey, then France, then Norway, and finally Mexico. There he wrote attacks on Stalin that found many readers outside Russia and some inside.

Stalin soon regretted letting Trotsky get away. As we'll see, he never put him out of mind. He did obliterate him from Russian history books, as he did with many other fellow revolutionaries. Before allowing publication of old group photographs of Trotsky and other longtime Bolsheviks, Stalin's aides brushed out any men whom Stalin had later turned against and in some cases executed.

By 1930 Stalin was fully in charge. Where Lenin, in his few and hectic years in power, had scarcely started to remake the country, Stalin would have a quarter century in which to do it. He soon announced his goals. He would build a nation that was self-sufficient, strong, and up to date, and he would lay the foundations of a workers' society.

Those may sound like inconsistent goals: a modern nation *and* a communist utopia. But Stalin saw no contradiction, or if he did he smoothed it out with patriotic pride. "We are becoming a country of metal," he said, "a country of automobiles, a country of tractors." And when the people ride in cars and the peasants in tractors, he said, "we shall see which countries may then be 'classified' as backward and which as advanced."

If Russia was to be a nation of metal, millions of peasants would have to move to the cities and work in factories. At the same time the peasants who stayed on the land would have to grow more food, so that they could feed the growing millions who made the steel, the tractors, and the cars. That was the problem, how to get fewer farmhands here to grow more food for more factory hands over there. To Stalin, the way to raise more food was clear: Russia must "collectivize" at once.

FOR MANY MILLIONS of Russians, the real revolution in their lives was not the communist takeover in 1917 and '18, but collectivization, which began in 1929. In vast areas of Russia peasants now were suddenly forced to give up their little holdings, some of which they had taken over from the landowners only a decade before. Their scraps of land were joined to make big farms that averaged several miles in size, and here the communists reshaped the peasants' lives as if they were soft clay. The peasants were made to do what Marx would have wished: to work together, own no capital (that is, no land, animals, or machines), and exploit no laborers.

Not all peasants moved quickly to collective farms (or to factory jobs). In the Ukraine region southwest of Moscow, Russia's "breadbasket," the peasants resisted collectivization. Stalin decided to starve them to death. Two years in a row he demanded that the region supply the state with huge amounts of grain. Gangs of enforcers swooped down on farms, seized all the grain, and demanded more. Peasants caught eating a handful of their own wheat or rye were jailed or shot. Starving, they first ate dogs and cats. Then earthworms, rats, and ants, and a soup of dandelions and nettles. Then, sometimes, their children.

Naturally, in every Russian village some peasants owned a little more land than the others and sometimes hired their neighbors as laborers. The communists, pretending that these peasants were tough and grasping, called them *kulaks,* a Russian epithet meaning "fists." Stalin, who wanted their land for his collective farms, claimed that they were class enemies who had to be wiped out. "We must smash the kulaks," he said, "eliminate them as a class."

Exactly how to smash the kulaks wasn't clear. If the government simply shot them, it got no work from them. If it put them in its concentration camps, there was little work to make them do, and they had to be fed. Also, everybody knew that slave laborers were poor producers. Then a former wealthy man, himself in prison, wrote to Stalin to suggest a labor scheme. His basic concept, not very complicated, was to subordinate food intake to work output. Send the prisoners, he suggested, to some awful place where there was work to do, feed them very little, and squeeze perhaps six months of hard labor from them. And when they died of hunger, cold, and exhaustion, bring in others. For making this suggestion, the former millionaire was released from prison and appointed a director in the system he had proposed.

Men with guns stormed villages and rounded up the so-called kulaks, more than 15 million women, men, and children. They were jammed in railroad cars and hauled for days or weeks to lonely places in the Arctic north, and from there they marched across the tundra and through the forests. By the time they reached their camps, 15 or 20 percent, mainly children, had already died. Now they were told to dig a hole to live in, and soon they were told to grow their own food. Half starved, they were made to work until they died.

Some kulaks mined for gold in northeast Siberia, one of the coldest spots on earth. In some camps everybody died of cold and hunger, even the guards and their dogs. Other kulaks were made to dig a waterway from the Baltic Sea to the White Sea, using shovels, hammers, and chisels, and carrying rubble on their backs. Even in winter they were half-clad and housed in flimsy barracks. Perhaps a million of these prisoners died. (And then the waterway proved too narrow for the navy ships it had been intended for.)

How many peasants died in one way or another because of collectivization no one knows because, as one official said, "no one was counting." But this is fairly certain: 8 million more people died in Russia in 1933, the last year of collectivization, than died in 1934, when the worst was over. So many lives were lost that when the next census was taken, the government suppressed the results, and members of the census board were arrested and later "purged" for "treasonably exerting themselves to diminish the population."

As Marx would probably have wished, collectivization put all peasants (those it did not kill) on one "classless" level. Now they lived and worked without their brains, robbed of freedom to succeed or fail, and robbed of the right to leave the land at will.

It is true, however, that the collectives did, eventually, raise the farmers' living standards. After a decade had gone by, farmers had somewhat better housing and more to eat than they had had in the old precommunist days. They also raised more food for the cities, since the government provided the collectives with farming experts, and tractors and combines. By 1938, the eve of World War II, Russian farmers produced as much grain as a much larger number of farmers had done before World War I. So collectivization did advance industrialization, as Stalin had intended. Fewer hands were needed on the land, and 20 million peasants were plucked from farms to move to the cities and work in the new factories.

The "steel one" wanted a "country of metal," and he wanted it fast. His planners therefore ruled that a third of Russia's national income must be spent on building industries. Only two-thirds of everything produced was left for the people. This was a harsh decree in a land where most were poor. They were made to pay for the new plants and machines and railroads with heavy sales taxes, which the government hid from them by including the taxes in the prices of the things they bought. They also paid for industrialization with hard work at low wages. Labor was drafted from the villages and from the collective farms, which had to supply set numbers of factory workers. These workers usually lived in unheated barracks and ate one skimpy meal a day.

Russia did build up its heavy industry quickly. Huge steel mills rose in several places, and hydroelectric dams vastly increased electric power, and mines were opened, and industries arose even in the Asian grasslands, far to the east of Moscow. By the end of the 1930s Russia was the world's largest producer of farm tractors and railway locomotives. Only the United States and Germany were bigger industrial producers.

And it must be said that everyday life improved somewhat. By the late 1930s, stores carried more food and basic goods, and the state provided medical care. Everyone had a job, such as it was. Children went to school, and many of their parents, after working many hours, went to night schools and learned to read and write. In the most backward areas of Russia women had been emancipated.

BY THE 1930S Stalin loomed over Russia like a giant rearing bear. He was stronger and crueler than any tsar, and for the next twenty years he would have more control over more people and events than any other figure in the history of the world. But he had his enemies, of course, and they began to whisper and to plot. Some were simply jealous of his power, while others heeded Trotsky's cries from far away that Stalin was a monstrous traitor to the thoughts of Marx and Lenin.

Stalin was himself by profession a plotter, and by nature paranoid. He trusted no one, even Bolsheviks—no, especially Bolsheviks. When his spies reported seeing dirty looks and sullen moods, Stalin readily believed them. He knew his rivals, or he thought he did, and he had to wipe them out. He found a cunning way to carry out a purge.

One rival was Sergei Kirov, the communist boss of Leningrad (the former Petrograd, renamed). Like others, Kirov had expressed concern at Stalin's cruelty. What was just as bad, from Stalin's point of view, was that if Stalin should be ousted Kirov might be picked to take his place. So he was an enemy of Stalin and therefore an enemy of the people.

Hard proof does not exist, but it is clear enough that Stalin gave the order, "Kill Kirov." Someone hired a thug, who did the job, and Stalin rushed to Leningrad and personally ran an "inquiry." The hired killer was shot, of course, and so were his wife, ex-wife, sister-in-law, and a brother. Kirov's bodyguard, probably in on his boss's murder, was beaten to death with crowbars, and the chief of police, who may have hired the thug, was sentenced to a labor camp and later murdered. Stalin also seized the opportunity to shoot some rival politicians, claiming they were terrorists. At Kirov's funeral he helped to carry the coffin, and he named the Kirov Ballet for his late, lamented friend.

Using Kirov's murder as an excuse, he now began a purge, a Terror. He put to death more than a hundred of his enemies, real or fancied, and sent thousands more to labor camps. Then he put Bolsheviks on trial, men whom he had known for many years. Most of the accused were charged with murdering Kirov, plotting against Stalin, and taking orders from far-off Trotsky. Stalin appears to have watched their trials from an alcove where he could not be seen.

All around the world people closely followed these trials. To their amazement, the men accused stood up in court and regretfully confessed to killing Kirov, plotting against Stalin, and taking orders from far-off Trotsky. Most were quickly sentenced, quickly killed. Why, baffled people outside Russia asked, did they grovel and make these preposterous confessions? No one knows. Perhaps they did it because they had been tortured, or to save their families from death, or because they really believed that what the Party wanted from them must be just.

The public trials were only the visible part of Stalin's purge. Out of sight, many more Russians were arrested, often after a loud knock on the door in the dead of night. They were shot without a trial, or sent to the labor camps with a notation stamped on their files: "To be preserved [imprisoned] forever." Nine out of ten generals in the army were purged and about 15 percent of the officer corps. During his years in power Stalin approved the executions of 230,000 people, and lower officials approved many, many more. On just one day in December 1937 Stalin approved 3,167 deaths. Then he went to see a movie.

Stalin had not forgotten his old enemy, Trotsky. Three floors in the headquarters of his political police held files and operations rooms devoted solely to this man. Now Trotsky lived in Mexico, still writing against Stalin, who found him as maddening as a horsefly. In 1940, an employee of Stalin's secret police entered Trotsky's house in Mexico City and swung an ice ax into Trotsky's brain.

CHAPTER 19

A Leader tries to shape a master race.

WHILE LENIN AND STALIN steered Russia toward that ever-receding goal, the classless society, Adolf Hitler led Germany toward a different kind of Utopia. The means and the result were much the same: a state imposing nearly total power on human lives.

However, Hitler worked with different material. Whereas Russia had been torn apart by World War I, Germany emerged intact. Travelers who drove into the country from France right after the war were astonished, as soon as they had driven past the crumbling trenches, to find a smiling land of planted fields, well-cared-for houses, and decently fed people. Germany could still take pride in its modern industries, its famous universities, and its well-schooled and hardworking people. What's more, the Germans had reshaped their government at the end of the war. The kaiser had resigned, and Germany was, for the first time ever, a democracy.

But in the next fifteen years, the Germans had their troubles. For one thing, they believed they had been unjustly treated and humiliated. The Allies had forced them at Versailles to admit a guilt they didn't feel, and to agree to pay the victors compensation they could not afford. They had also been forced to give up the lands and peoples east of Germany that Lenin had conceded to them just a year before.

For fifteen years after the war the German economy wobbled through three crises. First, right after the war, many people, especially war veterans, were out of work. Then, after a recovery had begun, Germany had a terrible bout of inflation due to printing enormous amounts of paper money to pay war reparations. Money lost its value to the point where even a large bank account was not enough to buy a bunch of carrots or a pound of flour. Many Germans lost all their savings. And then, a decade later, the Great Depression overwhelmed Germany and most of the industrialized world. By the early 1930s 6 million Germans were out of work.

Other countries also had their problems, and they dealt with them as best they could. Couldn't able Germans do even better? Yes! shouted rabble-rousing Adolf Hitler. He knew just what to do, not merely to revive the nation but to make the Germans once again a master race.

Strangely enough, this Hitler, who would play a giant role in Germany, had not been raised there. His first home was just across the border in Austria, which at this time was the German-speaking heartland of the Austro-Hungarian Empire. His father was a retired customs officer, a hard, short-tempered man, and his mother a former housemaid. Both of them died while Adolf was in his teens. At sixteen he dropped out of high school, and two years later he went to Vienna, the capital of the empire, to study art.

The Viennese Academy of Fine Arts found his test drawings "unsatisfactory" and rejected him. But Hitler stayed on in the city for five years. He slept in public hostels, and after he ran out of money he did odd jobs such as shoveling snow and carrying travelers' suitcases at a train station. He also earned a little by painting postcards and making advertising posters for products such as Teddy's Perspiration Powder. A couple of kindly Jewish dealers sold his mediocre postcards. Hitler was a strange young man, lazy and moody. He wore a greasy derby hat and a well-worn black overcoat given to him by an old-clothes dealer, a Hungarian Jew named Neumann.

Meanwhile, Hitler was becoming Hitler. He was reading widely in a public library, peering at the world, and deciding on his loves and hates. He loved the German people of the distant past (as he dreamed they were), hardy forest dwellers, worshippers of warrior gods, and fighters even Romans couldn't beat. He loved the operas of Germany's Richard Wagner, with their tales of northern gods and heroes, demons, dragons, blood feuds, and even the twilight of the gods, when a princess seeking vengeance set the hall of fallen warriors on fire, and all went up in flames.

His hates were many. He hated the Hapsburg family, who ruled the Austro-Hungarian Empire; and he hated the aristocrats in their carriages who glided past him, a shabby ambler on the sidewalk; and he hated businessmen, socialists, and communists. Above all, like many Europeans of his time, he hated Jews and put the blame on them for every ill. He argued about his loves and hates with the other down-and-outers in the hostels, ranting and waving his arms.

A year before the start of World War I, Hitler moved from Austria to the southern German city of Munich. And so it was that, when the war began, he enlisted not in the Austrian army but in the German one. In the next few years, the war transformed him. In peacetime he had been a nothing, but now he felt caught up in something big and thrilling. He served as a "runner" who carried messages along the front lines, and was wounded in a leg. Later a British mustard gas attack temporarily blinded him. Twice he won the Iron Cross for bravery. The officer who recommended him for the second cross was Hugo Gutmann, a Jew.

For two years after the war Hitler stayed on active duty. The army sent him to Munich, his former home, and told him to keep patriotism alive in army veterans and workers. At this time Munich swarmed with secret clubs and bands of angry, jobless men bent on fighting pacifism, democracy, and communism.

On orders, Hitler joined the German Workers, which was part political party and part debating club. When he attended a meeting of its executive committee, he saw how insignificant the party was.

The meeting took place in a small back room in a beer hall, where (he later wrote), "Under the dim light shed by a grimy gas-lamp I could see four people sitting around a table." Two members must have been absent, because when Hitler joined the committee he became its seventh member. He was thirty years old.

He left the army and threw himself into party work, and soon he was the German Workers' leader. Although his offstage manner was harsh and jerky, he turned into a good speaker. His looks were unimpressive, with his smidgen of mustache, but he carried a heavy riding whip made of hippopotamus hide. Another member of the little party, Ernst Röhm, had a private force of brown-shirted Storm Troopers who served as the party's fist. They guarded Hitler and beat up communists and anyone else who needed it.

Meanwhile, the German Workers renamed themselves the National Socialist German Workers' Party. The name is misleading, since the party was not national, nor socialist, nor made up mainly of blue-collar workers. The name was also too long, but that problem was soon resolved. From the German word for *national* came the nickname everybody called the party: the Nazis.

In 1923 Hitler, an immigrant, former drifter, ex-corporal, leader of a minor party, decided to seize power, not merely in Munich but in all of Germany. He was counting on support from the army, and on nerve and bluff. On November 8 he entered a beer hall where Munich's political leaders were meeting. With him, inside and outside the building, were several hundred Storm Troopers.

He fired a pistol at the ceiling and pushed his way onto the platform, where he shouted, "The national revolution has begun!" Then he led a march toward the center of the city. But policemen fired at the Nazis and they scattered—even Hitler. He was captured and, after a sensational trial that every German followed, the court sentenced Hitler to five years in prison.

His jailers treated Hitler as the celebrity he was, and they kept him in prison only nine months, instead of five years. While he was there, flattering politicians and admiring women paid him visits, and he strolled in the prison garden with other Nazi prisoners. He also found time to begin to write a book, which soon was published in two volumes with the name *Mein Kampf* (*My Struggle*).

IN *MEIN KAMPF* Hitler told the world what he thought and what he planned to do. Life, he wrote, guided by his misreading of Darwin and others, is an endless struggle. The strong will flourish and the weak die out, as they deserve to do. Humans have climbed to mastery over other animals, and among humans one "race" has trampled on the others and emerged on top. This noble folk, this "master race," are the "Aryans" of northern Europe, especially the Germans.

However, Germans recently had lost their ancient force. To be a master race, said Hitler, Germans must again, as in the past, put their fate in strong and steady hands. His new Germany would have no "democratic nonsense." The ruler would be a leader, a man of action, a dictator. In short, Hitler.

The reason for the Germans' loss of mastery, he wrote, wasn't only that they had lacked a strongman. They had also shared their land, and even mixed their blood, with people of the lower races, especially Slavs and Jews. He put the blame on Jews for Germany's defeat in war, the humiliation of Versailles, greedy capitalism, communism, pornography, and trafficking in prostitution. The master race must eliminate the "lower peoples." "All who are not of good race in this world are chaff," he wrote. The Germans "must call eternal wrath upon the head of the foul enemy of mankind."

For the master race he had a master plan. Germans must, he said, venture from their corner of the world, show again their ancient valor, and conquer "living space" for themselves. The first step would be to destroy France, their ancient enemy to the southwest, so as to prevent the French from interfering with what was to follow. That done, Germany must drive to the east. It would annex the other Germanic peoples in Austria and Czechoslovakia and unite them once again with Germany. After that would come the climax, the crushing of the hated Slavs, victory over communist Russia, and the conquest of living space for the German people on the plains of Poland and southwest Russia.

Although he had been famous during 1923 and 1924, after he was freed from prison everyone had nearly forgotten Hitler. For several years the German economy was strong, and for a rabble-rouser times that are good are bad. People scarcely mentioned Hitler and the Nazis now except as butts of jokes, and his fizzled grab for power now was called the Beer Hall Coup. Even many of his longtime comrades believed that he was finished. In 1929 a scholar was editing the memoirs of a British ambassador to Germany. He wrote a footnote mentioning Hitler's jailing and added: "He was finally released after six [*sic*] months . . . thereafter fading into oblivion."

After 1929 the Depression gave him his chance. Shops and factories closed, and millions, who only a few years before had lost their savings to inflation, now lost their jobs. Many Germans lost all faith in their capitalist economy and in democracy, which they had known for only a decade. Some turned to the left and saw salvation in communism, with its promise of a classless world with food for all. But a larger number, especially the middle class, saw communism as certain death. They knew a little—but that was plenty—about Russia under Stalin. Who or what, they asked, could save them from hunger and from communism?

Hitler had the answer. In speech after speech he hypnotized the Germans, thrilled them with his wrath, and made them dream of what they could become. He denounced the Treaty of Versailles as an insult to a noble race, scorned the German democratic state as weak and futile, lashed the parties of the left, and swore to navigate the country out of the Depression.

Aided by hard times and fiery words, Hitler and the Nazis started on their climb to power. This time, unlike 1923, they did it legally. Their share of the seats in the Reichstag or parliament rose steadily, and by 1932 they were Germany's largest single party and entitled to lead the country. At the time the president was Paul von Hindenburg, a crusty aristocrat and former general. It was his task to name the chancellor (or chief executive). He distrusted Hitler, but he thought the Nazis had earned a chance to show that they could govern. In 1933 he named Hitler chancellor.

As Hitler had made clear in *Mein Kampf*, "democratic nonsense" was not for him. He had no intention to cooperate with other parties. He wanted total power, and what he needed was a pretext to secure it. At just this time an unhinged Dutchman set the Reichstag building afire, burning it to the ground. For Hitler, who may have secretly ordered it, the fire was opportune. He exulted, "Now I have them!"

He blamed the fire on communists and issued a decree that gave him all-embracing powers. But legally he needed to amend the constitution, so he convened the parliament. (Having lost their building, they gathered in an opera house.) By now the Nazis ran the parliament, since they had thrown in prison many Communist deputies and about a dozen Social Democrats.

Hitler urged them to approve a bill empowering him to write the laws without the parliament's approval, and if necessary to ignore the constitution. When the leader of the Social Democrats bravely stated that his party would oppose the bill, Hitler told him, "I don't want your votes. . . . The star of Germany is in the ascendant; yours is about to disappear. Your death knell has sounded." Overwhelmingly the parliament approved the bill, and in the square outside a mass of Germans roared approval.

Now the Leader could begin to make them all a master race, fused by hatred, fear, and pride.

At rousing such emotions Hitler shone. He was now a splendid speaker and had learned to stage a rally as Wagner did an opera. A packed crowd waiting in a stadium would watch in awe as columns of Nazis wearing jackboots strode inside, by the light of torches, to the beat of patriotic marches. A thousand men in uniform would mass in squares before the stage. Nazi banners rippled, and searchlights would create a dome of beams.

Finally the Leader would appear, saluting in the Nazi style, and step up to the microphone. From here his words would reach not just the crowd before him but, thanks to radio, all the Fatherland. Now the stadium would darken, except for one bright beam focused on the Leader. When at last he spoke he moved the crowd with words of fire and rage—and maybe madness. The crowd would roar as one, "*Sieg heil!*" ("Hail victory!") "*Sieg heil! Sieg heil! Sieg heil!*"

Soon after taking power, Hitler decided to eliminate his comrade Röhm and cut the number of Storm Troopers. They were a raucous lot, fit for helping him take power but not for holding it or making war. For his master plan of conquest he would need the regular army and its able generals. So Hitler prepared a "blood purge" for a weekend in June. (Throughout his career he liked to hit his enemies on weekends, when they were off their guard.)

Röhm and a number of his followers were staying at a resort hotel. Hitler went to Röhm's hotel room, woke him up, and had him shot. At the same time the Nazis murdered many other Storm Troopers and inconvenient people, perhaps a thousand of them. They shot two troublesome generals on the doorsteps of their homes. Nazis murdered a well-known music critic named Willi Schmidt by mistake; he had the same name as a man on their murder list. Just for fun some soldiers killed a group of Jews. The Nazis later killed a number of Communist leaders. Those who escaped fled to Russia, where Stalin killed them.

Hitler called his purge "the night of the long knives." When it was over he proclaimed in the parliament that he was the "supreme judge" of the German people. Far away in Moscow, when Stalin learned about the purge, he asked someone, "Have you heard what's happened in Germany? Hitler, what a lad! Knows how to deal with political opponents." Hitler's example may have influenced Stalin when he decided shortly after to conduct a purge.

Like Stalin, Hitler and the Nazis were totalitarians, which means they wanted to control the life of every German. They wanted Germans to love the Leader absolutely and to sacrifice their personhood. The Leader told the Hitler Youth, "We have to learn our lesson: one will must dominate us; we must form a single unity; one discipline must weld us together; one obedience, one subordination must fill us all, for above us stands the nation."

Borrowing the idea from the Russians, the Nazis set up prison camps for those who opposed them in the slightest ways. Prisoners were badly beaten, sometimes killed. The security service used 100,000 part-time spies to snoop on every German and report disloyal remarks. One never knew if his or her own secretary, son, or friend was an informer. Woe to him who dared to call the Leader the "rat-catcher." The remark might lead to a visit from Nazi thugs, then to torture and death.

To think, to preach, to lead, was dangerous. The Nazis discharged hostile newsmen, labor leaders, pastors, and professors. From the walls of art museums they pulled down "decadent" paintings by Van Gogh, Picasso, Grosz, and Chagall. A few months after Hitler was named chancellor, thousands of students marched by torchlight to a square beside the University of Berlin. Using torches they set fire to a huge pile of books, and as the flames took hold they threw on more until twenty thousand books were burning. The same event took place in other cities. The students stated that they burned any book that "acts subversively on our future or strikes at the root of German thought, the German home, and the driving forces of our people."

Totalitarian control takes more than concentration camps and bonfires. Equally important was the shaping of the coming generation. Children joined the Hitler Youth when they were six, and at the age of ten each child would take this oath: "In the presence of this blood banner, which represents our Leader, I swear to devote all my energies and my strength to the savior of our country, Adolf Hitler. I am willing and ready to give up my life for him, so help me God." Until they reached eighteen the Hitler Youth were trained in camping, sports, and soldiering, and learned the story of the master race.

HITLER HAD PROMISED in *Mein Kampf* to purify the German people, and as soon as he had taken power he set about it.

Jews were less than 1 percent of Germany's population. The Nazis drove them from their jobs in government, the schools, and universities. They decreed that Jews had no citizenship rights, and they forbade marriages or even sexual relations between Jews and others. Hitler genuinely hated Jews. (He once described the U.S. president, Franklin Roosevelt, as a "pettifogging Jew" married to a woman with a "completely negroid face" that showed that she was "half-caste." He was wrong on every count.) But Jews also were for Hitler what the middle class or "capitalists" were to Lenin and Stalin, a scapegoat and a tool for unifying a people by giving them a common enemy.

In 1938 a Jewish boy, half crazed because of Nazi mistreatment of his parents, shot and killed a German official. The Nazis answered by unleashing the Storm Troopers, who carried out a swift campaign of persecution called "the Night of Broken Glass." All over Germany they looted, smashed, and burned the shops and offices of Jews, as well as synagogues. They beat up several thousand Jews, raped or murdered unknown numbers, and rounded up some thirty thousand for the concentration camps. Nazi prosecutors ignored the murders, since the killers were more or less following orders. Rapists, on the other hand, were sternly punished because they had violated the laws forbidding sexual intercourse with Jews. The government later confiscated moneys that insurance companies paid to Jews whose businesses the Nazis had wrecked. It fined the Jews collectively a billion marks for provoking the attack on them.

Despite the horrors of the new regime, many Germans were apparently content. What if they *had* lost their freedom? They didn't know or didn't care. As Hitler started to rearm, preparing for a war, the Depression in Germany came to an end and almost everybody found a job. They joked that at least under Hitler you didn't have freedom to starve. Under a program called "Strength through Joy" the government provided workers with dirt-cheap cruises on the Mediterranean, vacations on the Baltic Sea, skiing in the Alps, adult education, and low-cost theater tickets.

Other Germans knew the truth about their rulers, and they hated it. Of course, that's not surprising, since Germany had given the world many men and women of principle. One man who dared to speak his mind was Martin Niemöller, the pastor of a church in a stylish suburb of Berlin. Niemöller had captained a German submarine in World War I and was a well-known military hero. In the early 1930s he supported Hitler but he soon grew disillusioned. He opposed the efforts of the Nazis to control the German churches. Although Nazis arrested him in 1937 for his open opposition to Hitler, and put him in concentration camps, he would later blame himself for having done too little.

After World War II Niemöller gave many speeches (about his pacifist beliefs), and he often ended with these words: "They came first for the Communists and I didn't speak up because I wasn't a Communist. Then they came for the Jews and I didn't speak up because I wasn't a Jew. Then they came for the trade unionists and I didn't speak up because I wasn't a trade unionist. Then they came for the Catholics and I didn't speak up because I was a Protestant. Then they came for me—and by that time no one was left to speak up." He was accusing many, not himself alone.

The takeover, the killings, and the warping of a people are only part of the Nazi story in the 1930s. In *Mein Kampf* the gangster boss had promised war and conquests. As we'll see in chapter 20, even as the Nazis reshaped Germany in the 1930s, they also started seizing nearby lands whose people spoke in German. They too were members of the master race.

When Hitler wasn't in Berlin, he lived in a chalet in the German Alps. On an elevator cut through solid rock he could ascend to his private retreat on top of Obersaltzberg mountain. There he could gaze on other mountain peaks and ask himself what next for Germany and Europe. A Nazi slogan promised, "Today Germany, tomorrow the world."

Now and then he must have thought about that other ruler, even more dominant than he was. Often in his speeches he had ranted about Stalin, but watching Stalin from afar he had learned some things, as Stalin also learned from him. Both of them knew the uses of giant lies, scapegoats, work camps, stomping boots, censors, hatred, purges, and terror. One demagogued on race and one on class, but what mattered most for both was power, complete control of humans in the mass. Stalin had said admiringly of Hitler, "What a lad!" Hitler would say of Stalin, "He's a beast, but he's a beast on the grand scale." And "In his own way, he is a hell of a fellow!"

CHAPTER 20

We wage a wider, crueler war.

IN THOSE AWFUL YEARS, 1914–1918, many had imagined that the First World War would also be the last. It was "the war to end war." But in the 1930s Europe, Africa, and the Far East lurched and skidded toward another war. Contempt for peace and world opinion spread like gasoline on water.

Trouble started on the eastern edge of Asia. In 1931, Japanese army units stationed in the southern corner of Manchuria took over all of this northern province of China. The order for this aggression hadn't come from Tokyo; the army units acted on their own. The land grab nonetheless pleased many Japanese, and their government accepted Manchuria as an "independent" state. What was almost as bad, other nations reacted to the incident with a mildness as alarming as Japan's aggressiveness. The League of Nations should have acted, but it didn't want a war so it made only a tactful protest. Even that sufficed to make Japan flounce hotly out the League's front door.

Only worse would follow. Japan desired to feed upon the trade of China, its huge and feeble neighbor. War between the two looked likely. A Japanese general remarked that any talk of not invading China was "like telling a man not to get involved with a woman when she is already pregnant by him." In 1937 Japan invaded China. As before, the League condemned Japan, and it asked the Japanese to meet with other powers. Japan of course refused, and went on fighting. One by one it conquered China's railroads and much of its fertile land and biggest cities. It ate up China, a western statesman said, "like an artichoke, leaf by leaf." At Nanjing Japanese soldiers, under orders, slaughtered more than half the population.

A European country, meanwhile, confirmed the truth of what Japan was teaching: that the League could not preserve the peace. Italy was governed now by a flashy strongman, Benito Mussolini, who wore black shirts and white spats. He declared that in Italy he had "buried the putrid corpse of liberty," and he boasted of imposing greatness on a people who loved art and life too much and warfare not enough. To be great, he said, a nation needs an empire. In 1935 he ordered that Italian troops invade what now is Ethiopia, in northeast Africa.

So the League now faced another test, and here again it failed. The members merely voted economic penalties for Italy, and even these lacked bite. (When Mussolini said that yes he *would* mind having his oil supply cut off, the League did as he wished!) Like the Japanese, Mussolini flounced—no, strutted—from the League. He proceeded with his war, using bombs and gas against a people mainly armed with spears, and when he won he trumpeted the founding of another Roman Empire.

Of all the awful portents of the great disaster that was soon to come, Hitler was the worst. Even in the 1920s he had made his plans quite clear. How fast he moved when he had taken power! In 1933, he pulled his country from the League of Nations. In 1935 he startled everyone by announcing that despite the Versailles Treaty, Germany would rearm. Britain, France, and even Italy protested, but Hitler knew what he could get away with. He built an army, and the biggest air force in the world, and he signed agreements with Japan and Italy that linked this outlaw trio in a union called the Axis.

Once, twice, thrice, he grabbed. He started with the Rhineland, a German province that the Versailles peace accords had made a disarmed buffer region shielding France from Germany. In 1936 Hitler boldly sent in troops—what few he had as yet—and took the Rhineland. Naturally, the major powers waxed indignant. France alone could easily have stopped him at this point but lacked the will. When the League of Nations council asked for talks, Hitler blandly answered that he wanted only peace and had "no territorial claims."

That statement proved a lie. Hitler badly wanted Austria. For one thing he had been born there, and for another Austria was German-speaking, so its rightful place, he said, was with Germany, the nation of the master race. When Austrians discovered that a Nazi underground had planned a coup, intending to unite their country with Germany, the chancellor of Austria went to see the Leader. Far from offering excuses, Hitler shouted at the man. He told the Austrian to sign a list of Hitler's demands, or "I will order the march into Austria."

He signed, but Hitler sent his soldiers to the Austro-German border just the same. The Austrian leader asked Britain, France, and Italy to back him, but they wouldn't. When the Austrian telephoned Mussolini, who by now was Hitler's lackey, he refused to come to the telephone. Hitler coolly moved his troops across the border into Austria. He ran a plebiscite in Austria on joining Germany, and 99.75 percent voted yes.

Soon it was another neighbor's turn to be embraced by Hitler. Small and democratic Czechoslovakia lay in the very heart of Europe. Now, after Germany took Austria, it all but encircled western Czechoslovakia. The people of the little country were mostly Slavs, who spoke Slovakian or Czech, but German-speaking people lived in Sudetenland, some hills along the German border. Hitler now began to clamor that the Slavic majority terrorized this German minority and had killed them by the thousands. He demanded that Czechoslovakia cede Sudetenland to Germany.

This demand produced a crisis, for a nation's life was in the balance. Sudetenland held vital forts and mountain passes; if those were seized, the Czechs would be defenseless on that border. And losing land would be a gross humiliation. Then Hitler toughened his demands. He seemed to threaten to take over not Sudetenland alone but all of Czechoslovakia.

Now France and Britain faced this question: should they go to war to stymie Hitler? They didn't want to go to war—not for Czechs, and certainly not then. British generals said that fighting Germany before you strengthened Britain's air force would be like charging a tiger before you load your rifle.

With tension at its peak, Hitler, Mussolini, and the French and British leaders met in Munich. Showing what a decent chap he was, Hitler said all right, he'd take only Sudetenland, not all of Czechoslovakia. Happy to avoid a war, the French and British leaders signed the peace accord. To save their faces, though, they guaranteed support of what was left of Czechoslovakia. When they got back home to France and Britain they were met with cheers. The British prime minister, Neville Chamberlain, told an English crowd that he had brought "peace with honor . . . peace for our time."

So Hitler gulped Sudetenland, while two neighbors, Hungary and Poland, snapped chunks of eastern Czechoslovakia that he tossed them. It looked as if the rest of the little nation might be safe. Not so. In early 1939 German troops marched in and seized the rest of Czechoslovakia. Despite their promise, France and Britain stood aside and watched.

It now was clear to all that Hitler was a mortal threat to those around him. But no one knew, not even he, that all-out war was only half a year away.

Late in August 1939, Hitler and the Russian ruler, Stalin, shocked the world. In sight of all they signed a nonaggression pact. It looked as if the two, the leaders of the ultraright and ultraleft, had turned from foes to friends. Of course, they were not friends at all, and Hitler still was planning to demolish Russia, as had long been his dream. But he wanted Stalin to stay neutral for the moment while he crushed another neighbor, one whose luck it was to live between these brutes. Stalin, being half afraid of Hitler, was glad to come to terms.

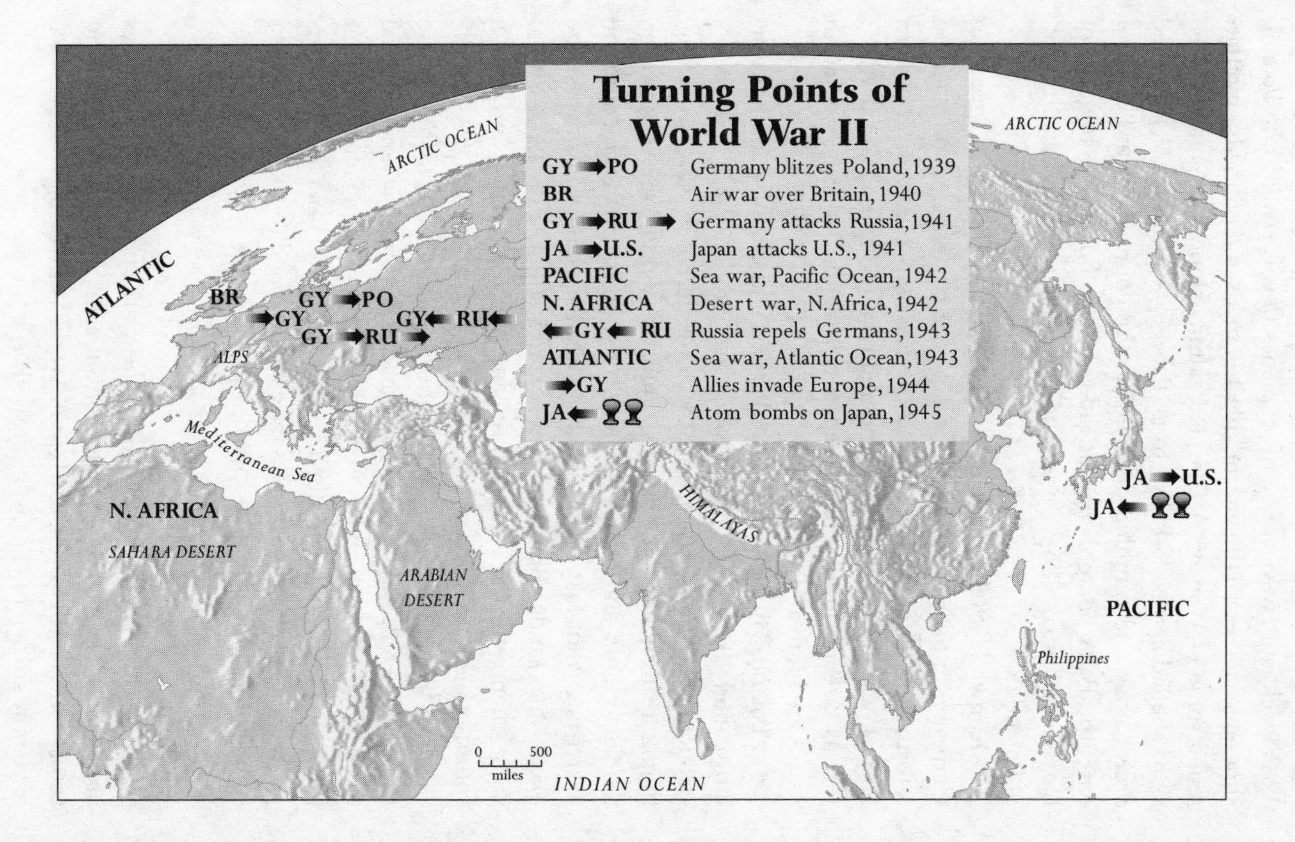
Turning Points of World War II
GY ➡PO Germany blitzes Poland, 1939
BR Air war over Britain, 1940
GY ➡RU ➡ Germany attacks Russia, 1941
JA ➡U.S. Japan attacks U.S., 1941
PACIFIC Sea war, Pacific Ocean, 1942
N. AFRICA Desert war, N. Africa, 1942
⬅GY⬅RU Russia repels Germans, 1943
ATLANTIC Sea war, Atlantic Ocean, 1943
➡GY Allies invade Europe, 1944
JA⬅ Atom bombs on Japan, 1945
ARCTIC OCEAN
ARCTIC OCEAN
ATLANTIC
BR
GY ➡PO
➡GY
GY⬅ RU⬅
GY ➡RU ➡
ALPS
Mediterranean Sea
N. AFRICA
SAHARA DESERT
ARABIAN DESERT
HIMALAYAS
JA ➡U.S.
JA⬅
PACIFIC
Philippines
0 500
miles
INDIAN OCEAN

A week later, waves of German planes flew eastward into Poland, shot up airplanes on the ground, blasted roads and railroads, and flattened some of Warsaw. A German training ship on a "goodwill mission" shelled a seaside Polish fort. "Panzer" (Panther) tanks stormed Polish lines and rumbled deep inside the country. Absurdly out of date, the Poles had only one brigade of tanks but twelve brigades of horsemen, armed with swords. The Germans quickly conquered half of Poland.

Meanwhile, Stalin made no protest. He and Hitler had agreed on Germany's attack in secret clauses of their nonaggression pact. In return, the Germans now allowed the Russians to take over eastern Poland, which Russia had surrendered near the end of World War I. Just as Czechoslovakia had disappeared, so did Poland, once again partitioned into air.

In Poland, Hitler had employed a new and crushing kind of warfare called "blitzkrieg," or lightning war. World War I had been a stalemate on the western front because defenses were so strong, thanks to trenches and machine guns. In World War II, however, thanks to tanks and planes, armies on the offense had the edge, and everything went fast. Tanks would race a thousand miles across the plains and sands. Bomber planes would streak across the sea at night, then return in time for breakfast. Between dawn and dusk a country might change hands.

Until his Polish onslaught, whenever Hitler bullied, France and Britain dithered. Not this time. Within two days from when he blitzed the Poles, the British and the French declared war on Germany. That this time they would fight him may have come to Hitler as a rude surprise. To this day no one knows if he had planned to take them on, or simply blundered into widespread war.

AND NOW THAT war, the worst the world has ever known, was under way. The fighting would continue for six years, and spread across much more of the earth than World War I had done. Here we have only sufficient canvas to let us sketch the great events.

For seven months in 1939–40 no all-out war took place. France and Britain were not ready. Germany was busy still in Poland and glad to spend the winter making plans and training troops. Someone named these months of waiting in suspense the "phony war."

In spring, however, Hitler sent his armies into battle. Wanting northern navy bases, the Germans conquered Norway and Denmark. Soon after this, the House of Commons, Britain's legislature, held tense debates. Britons wanted something more than caution, so as leader of the government they chose the bold and able Winston Churchill.

Now the Germans struck again, and hard. In May of 1940 German armies hit the Netherlands and Belgium. As in World War I, they were really aiming at France. British, French, Dutch, and Belgian armies waited for the Germans. However, they were ill prepared for blitzkrieg, and Hitler's tanks and bombers stunned them. The Germans split the Allied front in two and surrounded Allied armies in the north. The Dutch and many French and Belgian troops surrendered.

The British troops, however, and some French and Belgian forces retreated. They reached the English Channel at a French seaport, Dunkerque (or Dunkirk), from which Britain was only fifty miles away. Could Britain somehow save them? Strangely, Hitler held his tanks back, leaving only planes to block his enemies' escape. From England, tugboats, naval vessels, motorboats, paddle steamers, fishing boats, and yachts set out to save the troops. For a week these little boats, defended by the British air force, crossed the Channel many times. They rescued a third of a million men.

With Belgium and the Netherlands in hand, and Britain shoved aside, Germany could concentrate on France. Back in the early 1930s France's generals had prepared for a defensive war, the kind they'd known in World War I. For nearly ninety miles along their German border they had built a massive chain of forts, the Maginot Line. As the *Titanic* had been thought unsinkable, these forts were thought impregnable. Just north and west of the Line were Belgium's wooded Ardennes hills. The French saw these as a part of their defenses and were sure that German tanks could not get through them.

However, just as one might guess, instead of butting at the line of forts the Germans skirted it. Tanks and all, they plunged right through the Ardennes. The French fought poorly, and this time Paris toppled into German hands. Only then did the Germans storm and take the vaunted Maginot Line, but *from the rear.* On July 10, the French gave in. Hitler made them sign the papers of surrender on the very spot in France where, at the end of World War I, the defeated Germans had surrendered. The French collapse amazed the world, which wondered how, in but a month, so great a power had fallen.

At the end of World War I, Hitler had been a corporal. Since taking power, however, his army rank had risen. When his minister of war married a prostitute, Hitler fired him. He also fired his commanding general, falsely claiming he was gay. Then he named himself commander in chief. And so, in World War II, it was Hitler who made the big decisions for the Germans. He was quick to fire or even shoot a cautious general, and with utter confidence he made brilliant moves and ghastly blunders.

A year from when the war began, Britain was his only opposition and he had to swat that irritating fly. Only then could he destroy the Jews, subdue the Slavs, and conquer "living room" for Germany. All that lay between his troops and Britain was the narrow Channel, but if he was going to cross it he would need control of the air. His air force chief assured him that his planes could wipe out Britain's in about five weeks. In August 1940 German planes began the air assault.

Germany had nearly twice as many planes, but Britain's planes were better. Britain also had a new detection system, radar. By measuring the time it took for a radio wave to travel to an object and then bounce back, radar enabled the British to "sight" approaching German planes. The "Battle of Britain" lasted from August through October. Several thousand British and British Empire pilots, with a scattering of Poles, Czechs, French, and Belgians, more than held their own against the Germans. But German bombs gave British towns, especially London, a pounding worse than anything the world had ever seen.

Hitler hoped to shatter British spirits, but their morale held firm. On a day when German bombs shook London for six hours, the London *Times* reported the appearance in an air raid shelter of a crested grebe, an uncommon bird. Raucous British soldiers everywhere sang this ditty about Hitler and his leading colleagues, to the tune of "The Colonel Bogey March": "HITler has only got one ball / GOERing has two but very small / HIMMler has something sim'lar / As for Goebbels he's no balls at all."

In the end, Hitler reckoned that he couldn't win the battle over Britain, and he canceled the invasion.

Meanwhile, Allied spies performed a useful feat. For highest-level messages, the Germans used a code machine they called Enigma. Poles and Britons got their hands on one and smuggled it to England. They discovered how it worked, and learned to read the code. Decoding called for brilliant math and cleverness in making use of small mistakes the Germans made when putting words in code. Unknowingly, the German typists helped decoders by starting messages with "Heil Hitler!" For the Allies, knowing German plans was priceless.

Hitler's partner, Mussolini, proved less than useful. Only when he was sure that Germany had conquered France did Mussolini, hoping to make hay while the sun shone, send in bumbling and unneeded troops. He then invaded Greece, across the Adriatic Sea from Italy. Looking for an easy win, he found only disaster. The Greeks attacked his forces in their rugged hills and drove them out, and Hitler had to send in German troops to clean up Mussolini's mess. Shortly after this fiasco, British planes surprised the Italian fleet at its naval base and sank three battleships at their moorings.

Also thanks to Italy, the war now reached another continent. In North Africa Italy owned a colony, Libya, and Britain treated Egypt as a military base. Side by side these countries stretched for endless miles across the desert, making them a perfect battleground for tanks. From Libya, Italian forces struck at Egypt, whereupon the

British, using troops from India, pushed them back and captured 130,000 men. So Hitler had to help his ally once again. He sent down panzer tanks and his ablest general, Erwin Rommel. The British (and Australians and Poles) were waiting for him. Since Enigma had been cracked, they had read the Germans' coded orders. But Rommel pushed them nearly back to Egypt just the same.

With France defeated and Britain off balance, Hitler was prepared for Russia. He readied a campaign and named it Barbarossa for a red-haired German warrior emperor in the Middle Ages who, according to legend, would one day wake and fight again for Germany. In 1941, as summer started, Hitler sent 3 million men to fight in Russia on a front 2,000 miles across. He knew the risk. Napoleon had done the same, 130 years before, and lost half a million men to hunger, cold, and wounds. But Hitler counted on a rapid win, before the cold began. He confidently told a German general, "You have only to kick the door in."

This sudden blow took Stalin by surprise, though why it did so no one, to this day, can understand. Surely he had known that Hitler would break their pact when he wished to. Why, then, had he ignored the warnings that he got from Winston Churchill; from Russians watching on the German border; from Germany's ambassador in Moscow, who even told him the day of the invasion; and from a German deserter who revealed not just the day but the hour (and whom Stalin ordered shot)?

The Russians were surprised not only by the fact of the invasion but by its method. Like the Poles and the western Europeans before them, they were ill prepared for blitzkrieg. By fall, the Germans had encircled Leningrad in northwest Russia. In the center they were fighting in the Moscow suburbs. In the south they had conquered half the wheat lands of Ukraine.

As in World War I, the Germans captured Russian armies whole. During all of World War II they captured nearly 6 million Russians. Hitler had decreed that war in Russia not be waged in "knightly fashion," and more than half these captives died.

A German witness watched one time as captive Russians dragged themselves to German prison camps. "Suddenly we saw a broad, earth-brown crocodile slowly shuffling down the road towards us. From it came a subdued hum, like that from a beehive. . . . We hastened out of the foul cloud that surrounded them. Then what we saw transfixed us where we stood and we forgot our nausea. Were these really human beings, these gray-brown figures, these shadows lurching toward us, stumbling and staggering . . . creatures whom only some last flicker of will to live enabled to obey the order to march?"[1]

The Russians had already, for decades, suffered chaos, civil war, and Stalin's terror. Now on top of that they all, not only the soldiers, suffered greatly in the war. The people of Leningrad (now St. Petersburg) had it worse than most; they underwent a German siege that lasted eighteen months. Their daily ration was a scrap of bread. It's said that dogs and cats were first to vanish, then the crows. Exhausted, famished people dropped dead in the streets. One evening at the Pushkin Theater people pulled a dying actor off the stage; the show went on. Hunger, cold, disease, and German cannons killed a third of the city's million people.

This is sacrifice: Dmitry Ivanov had charge of rice at Leningrad's seed bank, where grains with genes the world might someday need were carefully preserved. After Ivanov had died of hunger at his post, someone found that he had faithfully preserved several thousand packs of rice. The specialist in peanuts, and the woman who preserved the oats, and half a dozen others also perished at their desks. Instead of eating priceless seeds they starved to death.

For the Axis side, the summer and the autumn months of 1942 were good. In west and central Europe, Germans now were ruling the very lands Napoleon once had held. In Russia, they had pushed far south to where they could seize grain and oil, and they smashed the major town of Stalingrad. (Hitler claimed that it was "firmly" in his hands, but we

[1]John Keegan, *The Second World War* (1989), p. 196.

shall see that he was wrong.) On the North Atlantic, German submarines were sinking British ships that carried vital goods from North America. In Africa, Rommel, called the "Desert Fox," felt so sure of taking Egypt that he chose a horse (a snow white stallion) on which to ride in triumph into Cairo.

NOW THE WAR, which so far had been fought in Europe and Africa, suddenly enlarged; it spread across the world.

The reason was Japan's unending drive for empire. Japan already held Korea, much of China, and Taiwan, but on the eve of World War II its leaders wanted more. Japan lacked raw materials, and they dreamed of building what they called a Greater East Asia Co-prosperity Sphere. If only they could take the Asian colonies of Britain, France, and the Netherlands, they'd have abundant rubber, metals, coal, and oil. The colonies they wanted included what now are Vietnam, Malaysia, Singapore, Myanmar (or Burma), and Indonesia.

The aloof and revered emperor of Japan, an amateur biologist, was more at ease with fish and fungi than with men. It was Prime Minister (and general) Hideki Tojo who shaped events. Tojo was a friend of Japan's treaty partner, Germany, and he led a cabinet of army and navy officers said to have "smelled of gunpowder." To them, the world in 1941 presented just the opening they needed. The European powers had either been subdued or were struggling for their lives. Their colonies begged for taking, like pies that a cook sets out to cool in a kitchen window.

The United States, however, blocked the road. America held the Philippines, which lay on the sea route from Japan to the European colonies it wanted. America was sure to keep Japan from grabbing colonies and getting oil and other goods that it could use in war. So the Japanese resolved to throw a sucker punch. If they did, they may have hoped, the soft Americans would never spend their blood and gold to get revenge.

Before it threw that punch, Japan sent forces down from China to occupy France's Indochina (now Kampuchea, Vietnam, and Laos).

Because of this and other matters, in the fall of 1941 Japanese and U.S. diplomats in Washington held discussions. By now Americans had cracked the Japanese version of the Germans' Enigma, and could read Japan's coded messages. They knew that even as the talks went on Japan was readying a blow against America. Washington sent warnings to U.S. commanders in Hawaii and the Philippines.

Early on a Sunday morning in December 1941, Japanese airplanes hit Hawaii. They caught the Americans by surprise, and sank or grounded warships that, despite the warning sent from Washington, were anchored in the harbor. (Most of them were later raised, repaired, and sent to war.) Luckily for America, none of its aircraft carriers was in port. The Japanese pilots found U.S. airplanes on the ground, bunched wing-to-wing, and easily destroyed them.

Franklin Roosevelt, the U.S. president, denounced this catlike pounce as "infamy," as if the Japanese should have told him they were planning to attack. America of course declared war on Japan, and so did Britain. Hitler and Mussolini backed their far-off Axis friend and declared war on the United States, which in turn declared war on them. Here the German and Italian leaders made a great mistake. It's true, the Axis pact obliged them to support Japan, but in the past Hitler had never felt compelled to keep his word. If he had just kept out, America might well have thrown its weight largely against Japan. Instead, he fused two almost unconnected wars, and egged America to fight him.

The Japanese had much to do before America had time to arm itself and hit them back. They quickly seized the European colonies they wanted and the Philippines. They even pondered taking India and Australia.

Only a few months after the surprise attacks, the United States won a battle at sea. This is how: In the spring of 1942 Americans discovered, from decoded messages, that Japanese ships were steaming across the Pacific. Their fleet included four carriers, each of which could serve as a base for seventy airplanes. But where exactly, on an ocean covering half the world, was this armada headed? The coded Japanese signals that the Americans were reading called the target

"AF." An American code analyst guessed that AF was Midway Island, a lonely Pacific atoll that served the United States as an air base. He tricked the Japanese into confirming (in a coded message to other Japanese) that he was right.

So U.S. warships, fewer than Japan's, hurried to the island. Over many hundred miles of ocean the airplanes of the fleets engaged. The battle reached a climax when U.S. bombers attacked the Japanese carriers at a time when their planes were loading and refueling. The decks were strewn with bombs and fuel, making each ship as vulnerable as a fireworks factory. The bombers plunged. They sank two carriers; disabled one, which was sunk at noon; and routed another, which they chased and later sank. This all took place between 10:25 A.M. and 10:30. In five minutes the Japanese had lost their dominance at sea.

AS 1942 RAN into 1943, the Allies, east and west, had more than Midway to cheer about. Factories in America were almost fully geared for war and were turning out supplies for all the Allies, including Russia. They would soon be turning out a ship a day and an airplane in five minutes. Meanwhile, on the North Atlantic, the Allies started to destroy the German subs that preyed on transport ships carrying goods to Europe.

In Africa, the British held the Germans back from Cairo, and Americans joined the fight. The Allies crushed the Germans and Italians, capturing a quarter of a million men. Allied armies next invaded Italy and slowly pushed a German army north.

The crucial zone of World War II, however, was on the Russian steppes. On those vast plains the battles were the biggest, and the suffering the greatest. And there, like other conquerors before them, German troops ran into troubles. One of these was "General Winter." The Germans were not ready for the cold, or hardened to it. Their engine oil turned first to paste and then to glue. While scrawny Russian horses seemed to thrive by nibbling straw from roofs, cold and hungry German ones collapsed. And while the Russian boots had warm felt soles, the German ones had leather soles held on by tacks. Warmth flowed down the metal tacks, and Germans by the tens of thousands lost some toes.

The greatest German problem was the abundance of Russians. Stalin scooped up new recruits as fast as his troops died in battle. As nature can replace an octopus's missing arm, so Stalin, if he lost an army, put another in its place. A country fights with what it has. When the war was over, Russian officers were shocked to learn that other countries' armies cleared a path across a minefield by blasting it with shells. Such a waste of shells! The Russians said their method was to form a column and order "Forward march!"

Although they had more men than shells, it's also true that the Russians manufactured most of what they needed. By rights, when Germans took their towns the Russians should have lost the factories that made their clothing, tanks, and guns. Instead the Russians hauled whole factories over mountains, and rebuilt them in the east. And what the Russians didn't make America provided.

For both sides, the central Russian city of Stalingrad was crucial. To control it was essential since, if the Germans took this sprawling river city, they could block the oil boats on the Volga, crippling Russia. And with its name, Stalingrad was a symbol. Hitler ordered that it be taken, and Stalin warned his generals, "Not one step backward!"

For months the soldiers battled hand to hand in cellars, sewers, and alleys, and from loft to loft in battered factories, up to their hips in rubble. A German soldier wrote that Stalingrad was not a town but "an enormous cloud of burning, blinding smoke . . . a vast furnace lit by the reflection of the flames. . . . Dogs plunge into the Volga and swim desperately to gain the other bank."

In early 1943 the Germans nearly won the city. But then two rested Russian armies dashed to Stalingrad from north and south across the snowy steppes. They circled it and trapped the Germans. The German commanding general wished to fight his way outside the city, but Hitler radioed him to remain. When a German army tried to bring relief, the Russians threw it back. The Germans' food ran out, and they died of hunger, cold, and sickness. Once they had numbered a third of a million; now they were down to 100,000. The Russians also suffered heavy casualties. They lost more men at Stalingrad than America would lose in all the war.

Finally, disobeying Hitler, the German general surrendered. Hitler had expected him to kill himself and rise to "national immortality." Instead, said Hitler bitterly, "he prefers to go to Moscow." In Moscow, Russians rang the Kremlin bells.

Russian armies followed up the victory. Using horses, trucks, and sleighs with camels, they struck the shaken Germans at other points along the front. In just two months they won back all the land they'd lost in 1942. Only once, at Kursk in the Ukraine in the summer of 1943, the Germans halted the pursuing Russians. In a giant battle lasting fifty days, the German forces conquered back a meager twenty miles. When they lost momentum, the Russians counterattacked and won back what they had briefly lost. The battle cost the Germans 1,400 planes, 3,000 tanks, and 40,000 men. After Kursk, the Russians rumbled on and swept the Germans back as far as Poland.

While the Russians, in the east, drove the Germans back, the western Allies were preparing for an invasion on the other side. Their supreme commander was a U.S. general, Dwight Eisenhower, a gifted organizer. In Britain he assembled an armada. When ready, it comprised two million British, Canadian, and American soldiers; 80 warships; 5,000 other ships; 1,500 tanks; 12,000 airplanes; 4,000 landing craft; and two huge artificial floating harbors. The Allies planned to cross the English Channel, storm the western coast of France, emplace the harbors, land the tanks and trucks, and battle eastward.

Before the dawn on June 6, 1944, the armada, so vast it reached to the horizon, crossed the Channel. This was but the first of many crossings on that day. (The planning had been so careful that every soldier carried not just seasickness pills but what the military labeled "Bag, vomit, one.") The Germans were indeed expecting an invasion, but they thought it would be at Calais, where France is nearest to

England. Instead, the Allies landed farther south at beaches on the coast of Normandy.

With heavy packs and rifles, men jumped off their landing craft in water full of swimming tanks, floating boxes, shattered boats, and severed limbs and heads. On several beaches they came under heavy German fire from atop the cliffs. Many died. Along one beach a colonel with a bloody cloth wrapped around his injured wrist strode among the wounded and the frightened shouting, "They're murdering us here! Let's move inland and get murdered!" They crossed the sand, and climbed the cliffs.

As they started inland, Hitler eased their task a little. Believing that this landing was a feint and a bigger landing would soon follow farther up the coast, he held back his reserves. The Allied troops fanned out, and launched a drive through France and toward Berlin. Later, other Allied forces landed in the south of France, hurried north, and joined the drive.

After the Germans had slowly retreated for half a year, Hitler hit the Allies back. The place was that familiar battleground, the Belgian forest of Ardennes, and the time was late fall, season of fog and snow. (A German planner had poetically named this German operation Autumn Mist. After the Germans had punched a salient in their front, the Allies named it the Battle of the Bulge.) At first the German counterblow succeeded, but then the Germans slowed, held back both by Allied bombing and the fact that they had, literally, run out of gas. The Allies pushed them back, and when they reached the Rhine they found a bridge the Germans hadn't wrecked. They poured across the river into Germany.

One has to wonder why the Germans kept on fighting as their foes pressed in from east and west. They hadn't enough men to fill their ranks, and fuel became so scarce they harnessed steers to drag their tanks to battle. The leading Nazis would not quit because they knew the Allies would refuse to parley and planned to punish them. But it wasn't only fear of punishment that kept the Germans fighting. For many of them, surrender was unthinkable. Better far to fight until the last, and then, as in the ancient German legend of the twilight of the gods, perish in the flames.

However, certain Germans did want a negotiated peace. Believing one could be arranged if only Hitler were removed, a group of officers conspired to kill him. One of these was Klaus Philip Schenk, Count von Stauffenberg, formerly a tank commander. Battle wounds had already cost him his left eye, his right hand, and two fingers of the other hand. But Stauffenberg agreed to carry out the risky task. At a meeting held at Hitler's headquarters, he left a time bomb in a briefcase next to Hitler's chair. But just before the bomb exploded, someone moved the case, or maybe Hitler moved. He survived, wounded, angry. He killed Stauffenberg, some other plotters, and five thousand others who might have known about the plot. General Rommel, the "Desert Fox" and hero of two wars, was marginally involved. The Nazis offered him a choice, suicide or trial, and Rommel swallowed poison.

In the meantime the Russians had fought their way through Poland and were the first to reach Berlin. They battled through the suburbs to the center of the city. As they fought with diehard Germans, shells exploded, buildings fell, and liberated prisoners and drunken Russians raped and plundered.

In a bunker far below the ground, Hitler carried on. He tried to motivate his broken armies with radio speeches, and wrote a "will" enjoining Nazis not to quit the fight with Communists and Jews. He learned that the Allies had defeated the German army in Italy. (Italian partisans had murdered Mussolini and his mistress and hung them upside down like slaughtered pigs in a public square.) Hitler knew the end was near. He ordered his cherished dog and puppies poisoned, wed his longtime mistress, and said farewells. While the Russians waved a banner from atop the Reichstag, in the bunker Hitler and his wife took poison and Hitler shot himself.

When the Russian forces learned of Hitler's death a general telephoned to Moscow and told the Russian leader. Stalin said, "So, that's the end of the bastard."

• • •

IN THE PACIFIC, the war went on. Already it had lasted three-plus years, in part because the British and Americans had put the war with Hitler first. In Burma, British forces kept three armies of Japan tied down, and the United States used the time to manufacture airplanes, ships, and landing craft, and capture Japanese island bases.

To understand the Asian war zone, picture one huge swath of ocean, diamond-shaped, dotted here and there with islands. Japan is at the top. In 1942 America had lacked soldiers, ships, and long-range bombers, so it couldn't strike directly at Japan. Instead, it had to start with Japanese-held islands at the bottom of the diamond, which were easier to reach. From there the U.S. forces had to battle toward the top, island by island. Many islands were only spits of sand or rings of coral, but the United States needed many of them as bases for its ships and planes. Its forces didn't hit each island. Instead, they hopped from one key island to another, leaving Japanese defenders stranded on unneeded islands in between.

The Japanese fought back with courage, prepared to give their lives for emperor and country. When they were defeated, officers would often lead a final, hopeless charge in order to be killed and thus escape the shame of being captured. On Iwo Jima island over 20,000 Japanese fought fiercely to defend what was really just a dead volcano. Almost to a man they died.

By 1944 Americans had pushed to several hundred miles from Japan, close enough for bombers to attack it. Air raids shattered factories, sank the remnants of the Japanese navy, and triggered fires that burned Japanese wood-and-paper homes as if they were oily rags. A firestorm burned sixteen square miles of Tokyo, boiled the water in canals, and killed 89,000 people.

In April 1945, while Allied armies neared Berlin, Americans struck the island of Okinawa, 325 miles south of Japan. The enemies fought bitterly on hills and ridges honeycombed with cannon ports and tunnels. Using flamethrowers and explosives, Americans killed countless soldiers and civilians in the tunnels. Offshore, young Japanese pilots deliberately crashed their planes on U.S. ships. One of them composed this haiku:

If only we might fall
Like cherry blossoms in the spring—
So pure and radiant.

After nearly ninety days the U.S. forces won Okinawa. All the senior Japanese commanders killed themselves.

The climax of the war, it seemed quite clear, would be a struggle for Japan itself. The Germans had by now surrendered, so U.S. troops in Europe would no doubt sail around the world to join the fight. Perhaps the Russians and the British, who had recently recaptured Burma, would take part. Recently conquered Okinawa, no doubt, would serve as launching site. Presumably the conquest of Japan would cost the lives of many people, since the Japanese would surely fight as hard as they had on the other islands. Half a million, maybe a million U.S. troops would die, and far more Japanese, both soldiers and civilians.

Happily, both sides escaped that fate. To make clear why that was so, we'll look at what had happened on another front.

LONG BEFORE THE WAR, physicists had learned that atoms are more complicated than they once believed. The atom isn't, as they had thought before, the smallest piece of matter. Inside it there is much, much more. At the center is a nucleus with particles called protons and neutrons. Usually, a force holds the neutrons together. However, it was found, the neutrons of one atom can be made to break the nucleus of another, and this will free a huge amount of the energy that holds the nucleus together. To illustrate: if all the energy in the atoms in an airplane ticket were released, it could power a plane for several thousand trips around the earth.

Just before the start of World War II, scientists were pondering that force. Some among them theorized that if they could find an element whose atom's nucleus they could split, *and* whose nucleus would shed more neutrons than it absorbed, they might produce a "chain reaction." They theorized that atoms of uranium—a radioactive metal—might do the job. Neutrons shed from split uranium nuclei might be made to shatter other nuclei, thus releasing yet more neutrons that would split yet other nuclei, and so on.

The energy released would be immense, like nothing ever known on earth. One might convert it into cheap electric power, which would be a boon for humankind. On the other hand, if one could free that energy in just an instant he'd have a bomb so strong it could destroy a city.

World War II provided an opportunity to test the theory. Early in the war four physicists—all Jews, born in Europe but living in America, and well aware of Hitler's plans for Jews—sent a spokesman to President Roosevelt. They urged that America make an atom (or nuclear) bomb before the Germans did. Roosevelt decided that it would do so, secretly and fast. Centers for the project rose up overnight, first in universities and then in lonely valleys and empty deserts. Other countries' experts joined the project, and by 1945 the work involved 120,000 men and women.

Problem number one was this: were chain reactions only theory or could you make them happen? A team directed by an Italian physicist, Enrico Fermi, did the dangerous research. They worked in secret in a squash court underneath the University of Chicago football stadium. Finally, in a scary test, they brought about a self-sustaining chain reaction without making the Windy City any windier.

In mid-July of 1945, technicians detonated an experimental atom bomb. It exploded on a steel tower in the desert of the southwest United States. Dazzling light inflamed the sky, and a shock wave roared and echoed off the hills. The heat transformed the tower to vapor, and it fused the desert sand to glass. A ball of flame arose, and then a giant, surging, bluish cloud. From twenty miles away the research leader,

Robert Oppenheimer, watched it all through welder's glasses. The sight reminded him of words from Hindu scriptures: "Now I am become Death, destroyer of worlds."

Here the story changes from science to its application. Roosevelt had died in early spring, and Harry Truman was now president. It fell to him to make the choice to use the atom bomb or not. He knew that if he did, many guiltless Japanese would die. However (this is clear from what he later said), he also thought the Japanese would quit the war if they were to witness what the bomb could do. Using it to force surrender, rather than invading, might save the lives of many Americans and a great many more Japanese.

Truman had another reason to use the bomb: the simple fact that he possessed it. Behind it lay two billion dollars and three years of work. Could he allow all that to go to waste? When he ordered that the fearsome thing be dropped on Japan, he had not so much decided to employ it as not to not employ it.

In early August 1945, airmen put an atom bomb aboard a B-29 "super-fortress" plane. Technicians called it Little Boy, but it looked like a long, black trash can with fins. The pilot flew the airplane 1,500 miles to south Japan, and there he dropped the bomb on Hiroshima, a seaport and an army headquarters. It flattened four miles of the city's center, and left 80,000 people dead or dying.

For a teacher and her pupils on the city's outskirts, the calamity occurred like this. The teacher cried, "Oh, there's a B!" meaning a B-29, and the children all looked up. They saw a brilliant flash and felt a wave of awful heat. Then, a pupil says, "pitch darkness [comes]; [and] from the depths of the gloom, bright red flames rise crackling, and spread moment by moment. The faces of my friends . . . are now burned and blistered, their clothes torn to rags; to what shall I liken their trembling appearance as they stagger about? Our teacher is holding her pupils close to her like a mother hen protecting her chicks, and like baby chicks paralyzed with terror, the pupils were thrusting their heads under her arms."*

*This and the following quotation are from Richard Rhodes, *The Making of the Atomic Bomb* (1986), pp. 716, 728.

Another teacher climbed a nearby hill. "I saw that Hiroshima had disappeared," he later said, ". . . I was shocked. . . . Of course I saw many dreadful scenes after that—but that experience, looking down and finding nothing left of Hiroshima—was so shocking that I simply can't express what I felt. . . . Hiroshima didn't exist—that was mainly what I saw—Hiroshima just didn't exist."

The United States called upon Japan to quit the war or to "expect a rain of ruin from the air." When Japan did not respond, the United States dropped another bomb, this time on the port of Nagasaki. This bomb, nicknamed Fat Man, was mightier than Little Boy. Nagasaki's hills confined the blast, and fewer people died at once than had at Hiroshima. But in the next few years wounds and radiation sickness killed far more.

Ever since Hiroshima, people have asked if it was right to use these bombs. That is not a historical question, but it does involve some matters of historical fact. Some objectors say Japan was ready to make peace; no bombs were needed to persuade it. But official Japanese internal messages sent at the time make clear that the Japanese had not decided yet to quit the war. Some objectors say the United States should have dropped the first bomb on a desert island to demonstrate the weapon's awful power. But if this distant demonstration had not impressed the Japanese enough, what then? Bomb a city? And then another? That could not be done, because America had only Little Boy and Fat Man. No other bomb was ready.

Some say the use of atom bombs was racist, that the United States would never have dropped them on Caucasians. In fact, the United States had intended them for Hitler's Caucasian Germans, but Germany had left the war before the bombs were ready. The Japanese suffered Hitler's beating.

Defenders of the using of the bombs agree with Truman that by sparing both sides an invasion the bombs saved far more lives than they destroyed. Defenders also make the point that ordinary bombs in other places took more lives than did the atom bombs at Hiroshima and Nagasaki. Allied air raids in Germany (all together) killed between one-third and two-thirds of a million. The pilots on both sides discovered in the war that, with "luck" a bombing raid could cause a giant firestorm. When this happened, everything in a city burned that could burn. Humans died for lack of air, then shriveled down to small black bundles. As we said above, such a firestorm killed 89,000 people in Tokyo. That was more than died at Hiroshima.

After Nagasaki, Japan gave up at once. The emperor broadcast the news to his subjects, who had never heard his voice before. He told them that Japan would endure the unendurable and suffer the insufferable. The Japanese surrendered on the U.S. battleship *Missouri,* which was anchored in Tokyo Bay.

Now the war was over, everywhere.

Among those present at the signing of surrender was the secretary of the Japanese foreign minister. He later wrote a report for the emperor. He said that while aboard the battleship he noticed many little rising suns, symbols of the flag of Japan, painted on a bulkhead. They represented airplanes, ships, and submarines the U.S. battleship had sunk. He tried to count the suns, he wrote, but "a lump rose in my throat and tears quickly gathered in my eyes, flooding them. I could hardly bear the sight. Heroes of unwritten stories, these were young boys who defied death gaily and gallantly. . . . They were like cherry blossoms, emblems of our national character, swiftly blooming into riotous beauty and falling just as quickly."*

ONLY AS THE fighting ended did the whole world learn about the Nazis' foulest deed, apart from the war itself. When Allied soldiers, fighting toward Berlin from east and west, burst through the barbed wire of the Nazi camps, they found the proofs.

Hitler had for long preached hate of "under-men"—meaning flea-bitten Slavs, "half-ape" blacks, and, most of all, satanic Jews. As soon as he won power, Nazis persecuted German Jews, and on the eve of war in 1938 they started here and there to murder them. But World War II gave the Nazis a free hand, not only in Germany but in all the

*William Manchester, *American Caesar: Douglas MacArthur, 1880–1964* (1978), p. 534.

lands they conquered. Now was the time for what Hitler called the "final solution to the Jewish question." Nazi leaders told subordinates to murder all the Jews in Europe. If we let a few survive, they said, one day they will seek revenge.

When killing Jews, the Nazis found, the hardest problems were logistic. Germany held only half a million Jews, but in Europe as a whole they numbered eleven million. By far the largest number lived in the Polish and Russian lands the Germans took in 1941 and '42. How best to kill so many was the problem.

Was it more efficient to bring death to victims, or victims to their deaths? The Nazis tried both ways. At first they sent their death squads to the towns where Jews resided, marched them to the nearby fields, and mowed them down with guns. They also sent out special mobile execution vans. Soldiers crowded up to sixty victims in a van and piped in fumes from the van's exhaust pipe. Early in the war, a satisfied inspector claimed that three such vans had "processed 97,000 without any evidence of mechanical defects."

Eventually the Nazis found it best to bring the victims to their deaths. They collected Jews, crammed them into freight cars, shipped them to transit ghettos or labor camps, housed them in filth, overworked and starved them, and let them freeze. They hauled the Jews who failed to die, and many others, to death camps, where they shot or gassed them. (Their gas of choice was an insecticide that had the merit of posing no risk to soldiers who released it.) Then they burned the bodies.

Those who did the killing were ordinary soldiers and civilians. They saw themselves as decent Germans doing hard but needed work. They never used such words as "haul" and "gassed," and they liked to gild their deeds with words like "special treatment," "resettlement," and "liquidation." They murdered Gypsies, the disabled, and 5 to 6 million Jews.

Did the Germans know what was happening in the death camps? Everybody knew. The Nazis sometimes boasted of the murders. In 1942 Hitler told a crowd, "At one time the Jews of Germany laughed at my prophecies. I do not know whether they are still laughing or whether they have lost all desire to laugh. . . . They will stop laughing everywhere." All over Hitler's Europe people watched as Germans rounded up Jews. German soldiers in the camps took photographs and told their families what they saw and did. The British learned about the killings from decoded German messages. The leader of the Catholic Church, Pope Pius XII, also knew, though he had little to say about it.

Surely, one would think, when Hitler came to power Jews everywhere in Europe should have fled. They must have learned about his persecutions, so why didn't they leave while they could? Not many did. Flight meant leaving all they owned, their families and friends, their homelands. One day, they believed, this misery would end. So most of them were caught.

Where they could, the Jews resisted, as they did in Poland's Warsaw Ghetto. The Germans herded nearly half a million Jews inside this section of the city and walled them in. From the ghetto, every day, they hauled about 5,000 Jews to a secret death camp in the countryside. They pretended those who were deported merely went to labor camps, but after they had nearly emptied the ghetto some Jews who had escaped the death camp brought back the awful news.

The Jews still living in the ghetto began to fight. Using homemade bombs and captured German weapons, they drove the Germans from the ghetto, but their enemies returned with tanks and cannons, gas and flamethrowers. For a month the two sides fought, around and through the shattered houses. Finally, the Jews ran out of ammunition, and many of their leaders killed themselves. It had cost the Germans several hundred soldiers to deport or kill some 55,000 Jews. They dynamited Warsaw's Great Synagogue, and their general named the report he wrote for Hitler, "The Warsaw Ghetto Is No More."

Among the Jews who fled from Germany in the 1930s were the family of Otto Frank. A Jewish banker's son, Frank grew up in comfort. He had a year of business training in America, and in World War I he served as an officer in the German army. A few years later he married. When Hitler came to power, Otto and Edith Frank sensed the danger, and with their little daughters, Margot and Anne, they moved to the Netherlands.

The Franks had not escaped. As World War II began, Hitler seized the Netherlands. The Franks and other Jews were made to mark their clothes with yellow stars, and authorities expelled the girls from school. When the Nazis started hauling Jews from Holland to the death camps, the Franks and several other Jews went into hiding in some rooms in a house that Otto had used for business. Here they lived for two years. At the risk of their own lives, non-Jewish friends brought them what food they could find, mostly lettuce and spoiled potatoes. In the daytime, office workers were near them, so the hiders spoke in whispers. For ten hours at a time they didn't flush the toilet.

Anne Frank, now fourteen years old, began to keep a record of their life in hiding.* She knew that Germans soldiers were arresting friends and other Jews. "We assume that most of them are being murdered," she wrote. "The English radio says they're being gassed." And later: "In the evenings when it's dark, I often see long lines of good, innocent people accompanied by crying children, walking on and on, ordered about by a handful of men who bully and beat them until they drop." In July 1944 she wrote: "I see the world gradually being turned into a wilderness, I hear the ever-approaching thunder, which will destroy us too."

An informer told the Germans about these Jews in hiding, and a German officer and four Dutch policemen went and took the little group away. (The Nazis paid the Dutch informer a dollar for each person he'd betrayed.) Later on that day, a good friend of the Franks climbed the stairs to the hiding place. On the floor she found what she recognized as Anne's papers. She took them home and put them in a desk unread.

German official records tell us that a train hauled the Franks and more than a thousand other "pieces" to Auschwitz, Poland, the most infamous of death camps. Their jailers separated Otto Frank from his wife and daughters. Because they found the Franks were strong enough to work, they spared them from the gas chamber. Edith died

*Anne Frank, *The Diary of a Young Girl: The Definitive Edition,* trans. Susan Massotty (1991).

of hunger and exhaustion after several months. Otto would survive the camp and later oversee the publishing of Anne's diary.

Late in 1944 the Russians were approaching Auschwitz, so the Germans in the camp began to hide the evidence of slaughter. They shipped 4,000 women, among them Margot (now seventeen) and Anne (fifteen), west to Germany, to the barren mud fields of the Bergen-Belsen camp. There, in winter months, they lay on sodden straw in crowded tents surrounded by ditches serving as latrines. In a storm the wind tore off their tent. Typhus, borne by lice, swept through the camp, killing 50,000 men and women. Presently Margot, overrun with lice, died and dropped to earth from the board she lay on. Anne, brokenhearted, now a naked skeleton beneath a rag, died shortly after. Their jailers probably dumped their bodies in the giant pits of sprawling corpses found by British soldiers when they entered Bergen-Belsen one month later.

WORLD WAR II had maimed the bodies and destroyed the lives of hundreds of millions. It took 60 million lives. Of these, millions were stillborn to famished mothers, crisped by flamethrowers, roasted alive in burning tanks, baked in firestorms, vaporized by nuclear bombs, felled by radiation, blown apart by shells or mines or misfires, crushed by falling buildings, smashed by tanks and trucks, worked or beaten to death, shot, gassed in vans and death camps, hung in group reprisals, dropped alive in grottoes, shot down in planes, or trapped and drowned in sinking ships.

Millions died of wounds, infections, dysentery, cholera, or typhus. Millions drank themselves to death, fell from mountains, froze, took their lives, or died of despair.

CHAPTER 21

The Asian giants try to feed their poor.

INDIA AND CHINA are the giants of the earth, big in hills and plains, but even more in people. It's as if the earth, our globe, were set in such a way that India and China were at the bottom and people tumbled down and gathered there. In the last half of the twentieth century China had the largest population in the world, and India was second and catching up. (Their neighbors, too, were big: Indonesia was fifth. Pakistan, seventh. Bangladesh, ninth.)

Anything that happened to the people of those two enormous countries happened to a third of all the humans on the earth. Their common problem was their poverty, but these giants followed different pathways as they tried to cope with it.

AT THE END of World War II, China was, as it had been for a long time, a tragic land. For a century and a half it had suffered from hunger, misrule, opium addiction, bullying Europeans, corruption, civil war, and Japanese invaders. For its many millions, existence was a daily war with violence and hunger.

The armies of Japan had left the country, but China once again was racked by civil war. On one side were the Guomindang, or Nationalists. Ever since the Chinese deposed the last Manchu ruler back in 1912, the Guomindang had tried to win control and run the country. But they were harsh and bungling, and their leader, Generalissimo Chiang Kai-shek, could not beat the Communists. A U.S. senator asked about him, "If he's a generalissimo, why doesn't he generalize?"

On the other side was the Chinese Communist Party, which we will call the CCP. Influenced by Russia's revolution, Chinese revolutionaries had founded the party in the early 1920s. The CCP and Nationalists had fought each other in the later 1920s and the '30s, but they fought together against Japan after 1937, although always hostile to each other. When World War II was over they resumed their civil war. In the next four years the CCP did well and the Nationalists lost ground.

The leader of the CCP was battle-seasoned Mao Zedong. Mao was born in 1893, in a village in the midst of China. His father was a rice merchant who owned a spacious house. (Mao would later say that, had his father lived until the Communists had taken power, "he would have been classified a rich peasant, and would have been struggled against," by which Mao meant reduced to poverty or killed.) In his early years, Mao had been a teacher, library assistant, laundry worker, writer, guerrilla, and army leader.

In autumn 1949 his armies won the bitter fight at last. In Peking's Tiananmen Square, Mao Zedong proclaimed the founding of the People's Republic of China. Chiang withdrew his battered forces to the island of Taiwan.

After a century and a half bedeviled China had a strong regime. Mao would lead the CCP for a quarter of a century. He didn't have the total power of Stalin, but as the ablest of the party chiefs he usually had his way. He cared, or said he did, about the poorest peasant, but he easily shut his eyes to suffering. He was full of life and full of words and liked to shake things up, then calm them down. Like the emperors before him he had many concubines. From time to time he used to demonstrate his vigor for photographers, swimming in the much polluted rivers, unperturbed by floating globs of excrement.

Back in 1946, when Mao was still a gaunt guerrilla leader, fighting Chiang and living in a cave, a well-known scholar came to see him. He asked Mao what would happen if the CCP should win the civil war and rule the country. "Dynasties," the scholar said, "begin with a surge of vigor, and then decay and fall apart. Has the Communist Party found a way to break this vicious cycle?" Mao assured him, "We have found a way. It's called democracy."

He must have lost that way by 1949. Although his party called their country a "republic," the CCP and Mao directed every phase of government. They had their hands on every source of information, every way of shaping thought. For this they used the schools, the media, and neighborhood and village meetings. As the Russian Communists had done, in their early years in power they liquidated anyone they feared and hated. Most of these were former Nationalists. Some years later, Mao discussed these killings. "Basically no errors happened; that group of people should have been killed. In all, how many were killed? Seven hundred thousand were killed, [and] after that time probably more than 70,000 more may have been killed. But less than 80,000."

Marxist dogma held that if you want to build a socialist state you have to start by seizing the means of production. In largely rural China what that meant was taking land, and so, besides killing Nationalists, the CCP destroyed the landlord class. The usual weapon was a "people's court" in which the peasants of a village would destroy the local landlord.

This is how they did it at a village in the west of China: With CCP members present, the villagers stood up, faced the landlord, and denounced him. Even Li Lao's wife—a woman so pitiable she hardly dared look anyone in the face—shook her fist before his nose and cried, "Once I went to glean wheat on your land. But you cursed me and drove me away. Why did you curse and beat me!" The landlord, Sheng Jinghe, bowed his head and then admitted that the charges all were true. The villagers decided that he owed them four hundred bags of grain. They beat him several times, and took his grain. When he saw them heating iron rods with which to torture him, he told them where he had buried his money, and they took it. They also found that Sheng had prepared a New Year's feast of shrimp, and dumplings stuffed with pork and peppers. So "everybody ate his fill and didn't even notice the cold."* Sheng no doubt lost his land, and possibly his life.

Historians believe the CCP and peasants put to death at least a million landlords. Mao explained, "A revolution is not a dinner party, or writing an essay, or painting a picture, or doing embroidery; it cannot be so refined, so leisurely and gentle, so temperate, kind, courteous, restrained, and magnanimous. A revolution is an insurrection, an act of violence by which one class overthrows another."

Anyone who used his brains to make his living was a problem for the CCP. But what to do about them? The Chinese had a deep respect for learning, and Mao himself had read widely and liked to write. But China's intellectuals, like Mao, often came from well-off families, and some had earned degrees abroad. So they were "feudal" or "reactionary"—therefore dangerous. In 1950–51, the CCP made tens of thousands of them go to "revolutionary colleges" where teachers lectured them incessantly about the thought of Mao and Marx, Lenin and Stalin. The worn-down intellectuals groveled, condemned themselves, and wrote their "autobiographies."

These memoirs were, of course, confessions. They had to be convincing, since the party disallowed confessions that it said were insincere. A well-known teacher of philosophy, trained at Harvard, wrote eleven pages on his failings. He criticized his easy life within his "bureaucratic landlord family." He condemned his "crust of selfishness," his earlier desire to stay above mere politics, and his taste for "bourgeois" thought. Then he hailed his newborn sense of purpose, for which he was indebted to the CCP.

*Jonathan D. Spence, *The Search for Modern China* (1990), pp. 492–93.

After seven years in power Mao and others thought the CCP had gone too far, and repression was itself a danger. In a speech to party leaders Mao discussed his new idea: to let "a hundred flowers bloom" in the field of culture, and "a hundred schools of thought contend" in science. Later on he gave a speech about the "Correct Handling of Contradictions among the People" and he said, "Never has our country been as united as it is today. . . . The days of national disunity and turmoil, which the people hate, are gone forever."

It took a while, but intellectuals grew convinced that they could safely air complaints. In 1957 flowers began to bloom. People railed against their poverty and the CCP's repression and corruption. "Party members," they complained, "enjoy many privileges that make them a race apart." And, "It seems as if an invisible pressure forces people to say nothing." And, "It is not true that all peasants consciously want to join the cooperatives." At Beijing University the students had a "Democratic Wall" and covered it with posters censuring the CCP. In many places students rioted, ransacked files, and called for changes.

CCP hardliners soon had had enough. They made it known that they opposed the blooming flowers and contending schools of thought. Mao himself observed how things were going and changed his message. He and other leaders made it known that the riots were anticommunist. They began to mow the flowers.

Officials labeled 300,000 intellectuals "rightist." Their careers and often their lives were over. Bright and useful men and women went to jail, or labor camps, or exile in the countryside. Many students and professors killed themselves. In the presence of 10,000 people, soldiers or policemen shot three student leaders who had led a protest against the way the CCP ran their school.

Killing critics wasn't hard. The CCP's big task was changing the economy. This meant raising much more food and vastly raising factory production. They had to do this the communist way, but without starving and slaughtering peasants as Stalin had done in Russia. The CCP began by giving peasants plots of land, but in the middle 1950s they tried to persuade them—and often forced them—to pool their land and form "cooperatives." Each cooperative was told to raise a specified amount of food. The CCP also began a Five-Year Plan to build up heavy industry. (They knew that their communist model, Russia, had done this so well that it was able to repel the Germans in World War II.) China's plan for industry went well, and output rose dramatically.

In theory, a regime as strong as Mao's could do what others couldn't. He once made war on sparrows, calling them a pest and nuisance. By the millions Chinese stood outside their doors, banging woks to scare the birds. The purpose was to make them fly until they perished of exhaustion. And they did. However, caterpillars, safe from sparrows, multiplied, ate the crops, covered trees, and defecated on passersby. The CCP halted the campaign.

The CCP decided how you dressed (like Mao), where you lived, what you did for a living, whom you were allowed to wed, and how much rice you ate. It even gave advice on frequency of sex: newlyweds: "very often"; others: every week or two. The party banished China's splendid poetry and novels. People sang such songs as "Socialism Is Good" and "Night Soil [feces] Collectors Coming Down the Mountain."*

The CCP decreed reforms but didn't always follow through. It outlawed old abuses of women, such as child marriage and concubinage, but gave most government jobs to men. The government claimed that it had wiped out illiteracy, but a census showed it hadn't.

By 1957 the leaders of the CCP were much concerned about the country's economic path. Industry was doing well enough, but not the new cooperative farms. Food production was increasing only 4 percent a year. That was not enough to feed the growing hordes of factory workers, and to pay back loans from Russia.

The CCP debated what to do. Some planners took a soft approach.

*For this and much other information about China's recent decades, I am indebted to Nicholas D. Kristof and Sheryl Wudunn, *China Wakes: The Struggle for the Soul of a Rising Power* (1995).

They thought the peasants would produce more food if only they received incentives and could purchase more consumer goods. They also needed farm machines and fertilizers.

Mao disagreed. He trusted, as he always did, the power of the masses and what he thought of as the heroism of the human will. He explained his thinking to the party's leaders: the vital point regarding China's many million people was that they were "poor and blank." This might look bad but in fact was good. "Poor people want change, want to do things, want revolution. A blank sheet of paper has no blotches, and so the newest and most beautiful pictures can be painted on it." He believed the way to raise production was to mobilize the peasants and inspire them. He wanted to make half a billion people find fulfillment in their never-ending work.

And so in 1957 and 1958 he launched a "Great Leap Forward." The CCP combined the new cooperatives into giant units known as communes. Their purpose was to grow more food and also to develop rural factories. A commune might contain as many as 10,000 households, and had communal kitchens, nurseries, and boarding schools. These arrangements largely took the place of families, freeing women so that they could work on farms and in the rural factories.

The slogan for the Great Leap Forward was "More, faster, better, cheaper." Communes struggled to fulfill their grain production quotas. Worker armies dragged in from the farms and cities dug enormous irrigation projects. A million peasants smelted iron in little furnaces behind their huts. According to the government's reporters, everybody worked with zeal. At an iron mill a reporter wrote, "The air is filled with the high-pitched melodies of local operas pouring through an amplifier above the site and accompanied by the hum of blowers, the panting of gasoline engines, the honking of heavy-laden trucks, and the bellowing of oxen hauling ore and coal."

Mao and colleagues claimed the Great Leap Forward was a triumph. Certainly the grain-production figures that the rural managers furnished were astounding. They showed production doubling, rising tenfold, rising "scores of times."

But in fact the Great Leap was a great disaster. The grain-production figures were all lies. The peasants had resisted doing what was ordered, and the crops were poor. A long and dreadful famine followed. Starving peasants tried to live on ground-up corncobs, husks of rice, and powdered tree bark. From 1959 to 1962 the famine took the lives of 20 million people.

What was Mao to say and do? He first denied the failure of the Leap, denied that there were peasants starving. Then he quit as head of state, implicitly admitting failure, but he soon returned. He decided to scale back the Leap, then abruptly changed his mind. At a party leaders meeting, he responded to his critics by defending the Great Leap and the communes, and insisting they continue. He told the CCP that Marx and Lenin also made mistakes, and he said that he would take some blame and so must they, the party leaders. Charmingly he told them, "If you have to shit, shit! If you have to fart, fart!"

In the end, the party moderates restrained their leader's zeal. They kept the giant communes, but they halted much of Mao's experiment in human engineering. As the Russians also sometimes did to raise production, they let the farmers sell a portion of the food they raised.

While the communes kept on failing, industries did somewhat better. Although steel was crucial for development, until the communists took over China hadn't ever produced a million tons of steel a year. After just a decade of the new regime, China's yearly steel production may have risen to nearly 20 million tons, although the quality was low. Its industries were poor and yet, like Russia, when it wanted to, China could do wonders. By 1964 its scientists had made and tested nuclear bombs. A decade later Chinese satellites were orbiting the earth.

In 1966, when Mao was seventy-three, he worried that the revolution he had led was sputtering. It was failing its ideals, its dreams. The party leaders were complacent, and they wouldn't always let him have his revolutionary way. Mao believed they were evolving into just another greedy ruling class, like the former one that they had put an end to. When he died, Mao asked, would the revolution carry on? It was time, he said, to shake the nation.

So Mao unleashed the Great Proletarian Cultural Revolution. He demanded purges of the highest ranks within the party; anyone who lacked the zeal for revolution had to go. He gathered regiments of students and young workers to impel the revolution forward. Waving books of Mao quotations, they denounced the lifestyles of the "middle class," demolished temples, and attacked and sometimes killed administrators and their teachers. To avoid harassment by the Red Guards, intellectuals destroyed their books and art collections. They often killed themselves.

When it looked as if the country had descended into rule by mobs and, to make things worse, the mobs were fighting one another in the streets, army leaders intervened. They talked with Mao, who authorized them to bring back order. They suppressed the zealots or dispatched them to the countryside. Inside their offices, however, party members carried on the Cultural Revolution. They intrigued against each other, and they sent 3 million teachers, students—even party workers—to labor camps or prisons, or to work on farms. To make things worse, with all the chaos, the economy broke down.

By this time, Mao was in his eighties and unwell. He had led a revolution that transformed the biggest country in the world. With brutality and mass murders he, and others in the CCP, had brought a hectic, muddled order to chaotic China. At least on paper, Chinese peasants now were just as good as anybody else, and women had rights they never had before. However, ordinary people lived like rookies in a boot camp, and many million peasants still were hungry,

When the "Great Helmsman" died in 1976, radicals and middle-of-the-roaders struggled to succeed him. From behind the scenes emerged a longtime party leader, Deng Xiaoping, then seventy-two. Deng had been a leader of the party moderates and had twice been in Mao's or the CCP's disfavor. During the Cultural Revolution, Red Guards had so tormented his oldest son that he leaped from a window and could never walk again. Deng now became the leader of the country, although nominally he was vice chairman of the CCP. He placed reform-minded practical men in top positions.

Deng proposed a startling program of reform and won the backing of the CCP. Gorbachev in Russia was doing much the same thing at the same time. Without admitting he was doing it, Deng largely threw out communism and focused instead on economic growth. He dismantled the communes, which had never been successful; private farmers got their land back. He encouraged private enterprise—yes, capitalism—and making money. (Amazingly, one of the government's slogans was "To get rich is glorious.") He pushed for more consumer goods, and factories now began to make the things that people wanted, such as washing machines and motorcycles.

Deng's program got results. Farmers now produced much more, and China even sold some food abroad. The gross national product had risen slowly under Mao, but in the 1980s it rose on average 9 percent a year. In parts of China living standards rose to heights the Chinese never knew before. By the early 1990s China was the world's biggest producer of coal, cement, and grain, and cotton, meat, and fish.

China also tried to curb its growing numbers. Like many other countries, especially the poor ones, China had a swiftly growing population. When the CCP took over the Chinese had numbered over half a billion, but by 1982 there were twice as many. The growing numbers threatened to outweigh any economic gains. So China took a daring step and set a limit of one child per family. Those who broke the law were fined and lost their rights to housing and to education. For a while this rigid limit worked and helped to slow the country's population growth. Later, though, many families broke the law, and, in effect, it died.

China's economic boom time had its downside. While many people lived much better than they ever had before, above all in the cities, the majority of peasants still were very poor. Meanwhile, the government didn't move as swiftly with reforms as it could have. The CCP refused to liquidate the state-owned industries, even though they made no money. And as in many of the poorer countries, corruption was widespread. Government leaders were closely tied to businesses, both the state-owned and the private ones, and could easily take bribes.

Despite the widespread economic gains, discontent arose. Intellectuals and younger educated people were aware that in other communist countries, such as Gorbachev's Russia, liberalization was under way. Many people now had short-wave radios, and they heard the British BBC, French Radio, and the Voice of America in Chinese. They learned how people lived in other nations, and they urged the CCP to loosen its controls. They claimed the right to call attention to corruption and to defects in the ruling system.

But Deng and other leaders of the Party would have none of it. They had freed up the economy, and that made sense. But they were firmly set against democracy, which they thought could only lead to chaos. It might also cause what was truly unthinkable: the downfall of the CCP.

In 1989 students demonstrated in Beijing in the enormous square called Tiananmen, or Gate of Heavenly Peace. Here Mao had declared the founding of the People's Republic forty years before, and from his giant portrait far above them Mao smiled blandly on the activists. They demanded change and set in place a plaster statue of a "Goddess of Democracy." Ordinary citizens and other students joined them, till a million people may have crowded in and near the square. As the world looked on via television, the demonstrations grew intense.

How would Deng respond? He discussed the crisis secretly with other party elders. A transcript of their talk, which is probably authentic, includes these comments:

DENG XIAOPING: We all feel Beijing just can't go on like this; we have to have martial law.

CHEN YUN: We have to stick with our principles no matter what. . . . It seems to me that if we can't even hold to these principles, then what we're doing is destroying our People's

> Republic—which we won from decades of battle, with the blood of thousands of revolutionary martyrs—all in a single day.
>
> WANG ZHEN: Give 'em no mercy! The students are nuts if they think this handful of people can overthrow our Party and our government! These kids don't know how good they've got it. . . . If the students don't leave Tiananmen on their own, the PLA [People's Liberation Army] should go in and carry them out.*

After a month and a half the party leaders sent in soldiers, tanks, and armored cars. The soldiers opened fire, and over several days they killed hundreds of Chinese. (The number is disputed.) At one point a man stepped bravely in front of a column of advancing tanks. He wore slacks and a white shirt, and held what looked like a shopping bag. He addressed the crew of the lead tank, probably urging restraint. The tank commander rose to the moral challenge by not mowing down the man, whom others pulled away to safety.

A shaggy-haired young student shouted at a foreign reporter, "Tell the world!" He grabbed his shoulder, screaming, "You've got to tell the world what's happening, because otherwise all this counts for nothing. So tell the world!"

The world did hear about the demonstrations and was shocked by how the CCP had dealt with them. But the freedom movement led, at least at first, to very little. Deng continued to oppose democracy. He pretended that his country's economic system was still communist, which meant that the protestors were counterrevolutionaries who "attempted to install capitalism in China." Of course installing capitalism was just what he himself was doing.

After Deng died, China carried on with his contradictory system of capitalism under autocratic rule. The government continued to free the economy. Before Deng, the state-owned businesses had produced more than three-quarters of the country's output; by the time he died, they produced less than a third. And in spite of many prob-

*_The Tiananmen Papers,_ compiled Zhang Liang, ed. Andrew J. Nathan and Perry Link (2001), pp. 204, 207, 208.

lems, such as government corruption, business thrived and many Chinese lived much better than they had ever dreamed they would. It was said that under Mao everyone had longed to have the "Four Musts": a bicycle, a radio, a watch, and a sewing machine. Under Deng they wanted the "Eight Bigs": a color television, a refrigerator, a stereo, a camera, a motorcycle, a suite of furniture, a washing machine, and an electric fan.

China, as we said, was no democracy; it was a single-party despotism. But although the party still commanded, it was not as rigid as before. For the first time, newspapers dared to tell about a man who actually sued the police, or how a little boy had been abandoned by his parents and ignored by the authorities. Talk radio appeared, and listeners called in to give their views on garbage pick-up, factory smoke, whether girls could keep their bras on during medical checkups at their schools, or what to do about a boss (in private business) who forced his employees to work sixteen hours a day.

It's often said that capitalism cannot function in an authoritarian system. But China was trying to make it work.

INDIA, THE OTHER GIANT, had a history unlike China's. Back when Britain joined India to its empire, in the early 1800s, India hadn't been one single country, and the British never tried to glue the parts together. At the end of World War II, when our story starts, the people of "India" still were not a single country.

However, many Indians had dreams and plans. They wanted independence, and they wanted to unite and form a single nation. Chief among these men—for nearly all were men—was old Mohandas Gandhi, who had for decades taught and fought (if that's the word) for his convictions. He wanted India's independence from the British, but he believed the way to reach it (and reach any other goal) was with nonviolent means. He once had caught the world's attention by leading villagers on a march to the sea to protest a British monopoly on making salt. The monopoly hit the poorest Indians the hardest. As they walked, admirers sprinkled water on the dusty road in front of Gandhi and his marchers and covered it with leaves and flowers. Gandhi won his goal; the British granted anyone the right to make salt for personal use.

As the salt march shows, Gandhi was a man of symbols and public drama. In order to encourage Indians to make their cloth, and not to buy it from abroad, he had himself photographed while spinning cotton thread (ineptly) on a wheel. When he went to tea at the palace of the king of England, India's ruler, he wore nothing on his skinny body but a loincloth. When India's Hindus and Muslims killed each other, Gandhi fasted nearly to death until they stopped. Many called him the Mahatma, or Great Soul, but he deplored this. "They say I am a saint trying to be a politician," he said, "but I am only a politician trying to be a saint."

Gandhi's leading follower, much younger than he, was Jawaharlal Nehru. His father had been a well-known leader of India's independence movement. Nehru was a moody, thoughtful politician. Although he was a sometimes imperious Brahman aristocrat, India's poverty deeply troubled him. The ordinary Indian understood this, and loved him. Even though in former decades India's British governors had jailed him nine times, he admired British democracy and wanted India to have it after the British left. At the same time, however, he saw Russia as a better social/economic model for his country than Britain. He believed the Russians had demonstrated how socialism could enrich a poor and backward country.

During World War II, Indians fought beside the British, but the New Imperialism was near its end. It was clear that when the fighting ended, troubled consciences and outside pressures would impel the Europeans to release their colonies. Knowing that the time was ripe, Gandhi, Nehru, and the Indian National Congress party demanded in the wartime years that the British "quit India." Because it needed India's help in waging war, Britain promised independence when the war was over.

When the war had ended and they had to carry out that promise, the British faced this problem: what about the Muslims? Although they were fewer than the Hindus, more Muslims lived in India than in any other country in the world. A body called the Muslim League was their spokesman, and the League insisted that the Muslims have a country of their own. They feared that if the Muslims stayed in India, the Hindu majority would not respect their civil and religious rights. The Congress party on the other hand opposed the notion of permitting Muslim areas to leave what had always been considered "India." They wanted a united India in which, they said, Hindus, Muslims, Sikhs, and other sects would worship as they pleased.

Reluctantly the British chose to do what the Muslim League demanded. Out of ancient "India" they carved two nations. The smaller Muslim part would be the Islamic Republic of Pakistan. (Incredibly, defying common sense, Pakistan would have two sections, separated from each other by a thousand miles of northern India. Later, after a civil war, the eastern section became the independent nation Bangladesh.)

All the rest of traditional India would be the huge Republic of India. Its population then numbered a third of a billion. The vast majority of these were Hindus, but about a tenth (even after Pakistan had broken off) were Muslims, and a considerable number were Sikhs. The Indians were divided not only by religion but by culture, and they spoke fifteen officially recognized languages. One thing they had in common was that most of them were very poor.

Independence was declared in August 1947. As the hour approached, Nehru spoke in the constituent assembly. He said, "At the stroke of the midnight hour, when the world sleeps, India will wake to life and freedom. A moment comes, which comes but rarely in history, when we step out from the old to the new, when an age ends and when the soul of a nation, long suppressed, finds utterance."

However, freedom carried with it tragedy as Hindus and Muslims immediately rioted against each other. At least 10 million Hindus, Sikhs, and Muslims fled their homes on both sides of the India-Pakistan border, all fearing to be trapped inside a nation hostile to their faith. As they tried to reach a land where they could live in safety, religious enemies slaughtered a million of them.

Five months later, a Hindu fanatic lay in wait for Gandhi, whom he blamed for favoring the dividing of India, which Gandhi had in fact deplored. He shot the old man as he walked to evening prayer in a garden in New Delhi. In a broadcast to the nation, Nehru said, "The light has gone out of our lives and darkness is everywhere."

When one thinks about its history, the fact that India had chosen to have a democratic government is stunning. Most other newborn nations at this time were falling under the rule of tyrants. In China, for example, the CCP would soon (in 1949) impose its iron grip. Could a democracy cope with India's giant problems? Let us briefly list them. Most Indians lived in poverty, and its hungry people were increasing every year by five million. Religious hatred could lead at any time to massacres, and India's neighbors, Pakistan and China, were often hostile. And three Indians out of four couldn't read and were perhaps incapable of taking part in a democracy.

For Nehru and the upper classes, the easy way to rule the country would have been with force. Instead, they chose the governmental system of Great Britain, the very country that had ruled India for a century and a half. Nehru, and many other educated Indians, even though they had struggled against British rule, believed in democracy. Therefore he and the parliament built what proved to be a staunchly democratic state. It soon turned out that in elections everybody willingly voted, even if they couldn't read and even if the candidates had nothing to offer. An Indian politician once observed that an Indian election was like sheep choosing a shepherd.

Nehru was a socialist—but not a communist—at heart. He believed that the British Empire, which had ruled his country for so long, was itself the expression of capitalism. Under British rule and British capitalism, Indians had lived in poverty. Therefore capitalism was an evil, even if British democracy was a good. Moreover, India could prosper by itself; it had no need of foreign, capitalist assistance. Like Stalin's Russia, India could improve its economy alone. The way to do this was with Five Year Plans, as Russia had done.

As with China, India's biggest problem was (and is) its poverty. In its huge and crowded cities, in its half a million villages, nearly everyone was poor. Millions slept in the streets. In parts of the state of Uttar Pradesh, in the north, it was common to eat one meal a day. In one place there the "untouchable" farm laborers were so poor that they often ate the seeds of grain that they picked from animal feces and then washed.

Families had too many children, but for a reason. Among the rural poor, around the world, children can be assets. They help to work the family's land, and they care for their parents when they're old. So having many children is the next best thing to a social security system. Up to a point, at least, the poor aren't poor because they have many children; they have many children because they're poor. However, it's true that too many is too many, bad for the family, bad for the country.

Prosperous Indians had learned to live with poverty around them. In the early 1970s, a doctor smoothly reassured the India Council of Medical Research that "certain hormonal changes within the bodies of the malnourished children enable them to maintain normal body functions. . . . Only the excess and non-essential parts of the body are affected by malnutrition. Such malnourished children, though small in size, are like 'paperback books' which, retaining all the material of the original, have got rid of the non-essential portions of the bound editions."

Planners had to deal with a dilemma. Should India concentrate on "durables"? That is, should it use the country's scarce resources to build factories, steel mills, power plants, and dams? That's what Russia had done two decades earlier. Nehru saw such mega-projects as the foundation of development, and he called them India's "new temples." In the long run, he was sure, they would wipe out poverty by creating wealth and jobs. (In fact, however, the state-owned companies proved to be corrupt and inefficient. India's public-sector steel mills employed ten times as many people to produce half as much steel as did the private mills of rival South Korea.)

Or should the planners concentrate on immediate human needs? Nehru and his planners knew they had to deal with India's widespread poverty not only by building mills and dams but also by improving the social infrastructure. Nehru often would remind the country that for many millions life was wretched. So India had to spend money directly on human needs: telephones and schools, roads and health.

The planners compromised. In their Five Year Plans they spent what they could both on big projects and on infrastructure. And India did make some progress in building industries and raising food production. But it scarcely touched its age-old poverty.

As Nehru reached his seventies in the early 1960s, his daughter, Indira Gandhi, started on her rise to power. (Gandhi was her married name; she was not related to Mohandas Gandhi.) As a child, Gandhi later would reveal, she dreamed of being Joan of Arc, a savior of her country. Because her father was a widower, she served as his official hostess and traveled with him everywhere. When asked, she would deny that she could ever take his place, and when he died in 1964 she took the modest post of minister of information and broadcasting. But only a year and a half after her father's death, she won the Congress Party's vote and became prime minister.

Gandhi has a special place in Indian history for this reason: she would gravely threaten Indian democracy. For the best part of a decade, she worked within that system, but she didn't find it easy. As is often said, her problem was deciding whether she was a socialist democrat, like her father, or an aristocratic dictator.

In governing, and in foreign affairs, Gandhi had her wins and losses. Among the successes were India's gains in agriculture. These were largely due to the so-called Green Revolution, the growing of high-yield wheat and rice plants, which American scientists had developed. ("Miracle rice," for example, has short and sturdy stalks that hold up many grains of rice.) The average Indian now had somewhat more to eat, although an estimated 400 million lived on the brink of starvation, spending less than twenty cents a day on food.

Gandhi got in trouble in 1975 when a court declared her guilty of campaign abuses. The court barred her from running for or holding an elective office for six years. Protestors staged a giant "sit-down" to get rid of her, and editors and politicians demanded that she resign from office.

Far from quitting, she declared a state of "national emergency." She explained, "When there is an atmosphere of violence and of indiscipline and one can visibly see the nation going down, then the time has come to stop this process." She banned political parties, "suspended" civil rights, muzzled the press, and put some armored units on alert. She jailed "subversives" by the thousands.

As a former minister of information, Gandhi knew how to justify repression. A ban on picketing and strikes was called "the fight against Fascism" (although the bans were just what Hitler and Mussolini would have done). The government's new monopoly of news and comment was "restructuring the entire newspaper industry so as to make it accountable to the people."

She chose this moment to announce a Twenty-Point Program of economic reforms, which included something to please nearly everyone. Meanwhile, good things happened: Industrial production rose. The cost of rice and barley fell. Black markets disappeared. Nervous bureaucrats showed up for work by nine.

Many Indians liked the Emergency. The parliament voted to extend it "indefinitely," and the supreme court overruled the lower court's verdict against her. This cringing to what really was a tyrant's coup was not a great surprise. Democracy was not an Indian production but a western import. In a crisis many thought it best to have a kindly tyrant solving India's problems.

What wasn't popular at all was Gandhi's shocking way of halting population growth. The population now was rising by some 13 million every year. Naturally the question rose: would that growth not cancel any gains in feeding India's millions? Believing this was true, the government began a drive to sterilize men who had two children or more. (As we saw, China soon would introduce its one-child-per-family program.) In the north of India, among the poor, officials carried out the sterilization campaign harshly, even thuggishly. They swept through neighborhoods, rounding up the men with families and requiring immediate vasectomies. The campaign frightened millions of men who feared the loss of two things that they valued highly: potency and progeny.

Gandhi once more swiftly changed her course. As abruptly as she had started the Emergency, she ended it. She freed her enemies from jail and called for elections. Why she did this isn't clear. Did some democratic-leaning generals "advise" her that she had better take her platform to the people? Was she sure that she was popular and would win another election? Or was she tired of tyrannizing?

The opposition parties joined together, promising the voters "freedom and bread." And they won, ending the Congress Party's thirty years of rule, and with it Gandhi's prime ministership. The voters clearly had rejected her quasi-despotic rule. The sterilization program, in particular, had cost her many votes.

Gandhi's day was over, so it seemed. But surprisingly she soon came back. In 1980 she ran again, using the slogan "Elect a Government That Works." And she won. She owed her victory partly to the weakness of her opposition, but also to her shrewdness and the people's fondness for the Nehru family.

In 1984, Gandhi ordered an attack on violent Sikh extremists in northwestern India. Indian soldiers violated a Sikh temple, burned a sacred tower, and killed and wounded many Sikhs. So the Sikh extremists planned revenge. One morning as Gandhi walked inside a guarded compound, two of her own Sikh bodyguards confronted her. They shot and killed her with some thirty bullets.

But the Nehru/Gandhi family hadn't lost its great mystique. Indira's son Rajiv, a former airline pilot, was now elected prime minister. He ruled five years, until the Congress Party lost in national elections. Two years later, it looked as if he might return to power, but a woman killed him with a bomb.

In the early 1990s, India, as had China, started on the path of economic change. The country dropped a good deal of its socialist system, as well as the belief that India could prosper solely by its own efforts. It encouraged foreign companies to invest in India.

Economic globalization, which we will say more about in the next chapter, helped.

At the outset of the twenty-first century, India, more slowly than the other giant, China, was just beginning to prosper. In places, Indians were making fortunes in the computer software industry. It was said that everybody but the poorest peasants now had watches, bicycles, and portable radios. But the country still was very poor. The quickening economy made the gap between the rich and poor all the harsher and more glaring. India produced abundant grain, but due to politics and corruption more than half its children under five were poorly nourished. Only one-half to two-thirds of Indians could read and write, and most of them were men. Three-quarters of the people lived in the countryside, far from where they could make more money.

Despite its many problems—corruption, the Emergency, assassinations, lingering socialism, the castes, poverty and illiteracy, and religious hatreds—India remained faithful to its democratic system. Elections took place routinely. Democracy gave a voice, at least, to Hindu, Sikh, and Muslim; men and women; north and south; rich and poor. By doing so it gave the country a resilience that may have helped it not to break apart in crises. But the system often functioned badly, like an elephant that tries to waltz. Government corruption was endemic. At one point during 1996 the presidents of all three leading political parties were under indictment for alleged wrongdoing.

DEMOCRACY MAY NOT always be the fastest road from poverty. As the twenty-first century began, semiautocratic China was doing better than democratic India. Its economy was far more vibrant. The average Chinese earned twice as much as the average Indian. Only a third as many children in China were poorly nourished. The Chinese had three times as many computers per 1,000 people as the Indians, and three times as many telephones. Literacy was far higher than in India, and not just boys but girls as well were now in school.

CHAPTER 22

Some of us do well.

HISTORIES, INCLUDING THIS one, often dwell on what is sudden, bad, and bloody—on revolutions, for example, and wars and falling empires. But these are not the matters in our lives that count the most. In the past two hundred years, the most important revolution in the life of humans was the rise in our well-being. In chapter 14 we looked at the beginnings of that rise, and here we take up the story again.

In the latter 1900s, ways of making money changed, and fast. To illustrate these changes we will focus on the story of a single firm, one that's known to nearly everyone on earth.*

During the Depression of the 1930s the foreman in a New England shoe factory lost his job. His teenaged sons, Dick and Mac, who had recently finished high school, realized they would not find work near home. They moved across the country, and in California they found sunshine, other new arrivals like themselves,

*For the McDonald's story, I have relied on John F. Love, *McDonald's: Behind the Arches* (rev. ed., 1996).

little towns, clunky cars on narrow roads, and Hollywood. For a while they made their living shifting sets for moviemakers. Then they ran a movie theater, but they barely made a living.

At this time, in spite of the Depression, a novel kind of enterprise was taking shape. To fill the bellies of a mobile people, carhop drive-in restaurants were born. These were popular with Californians, who, if they owned them, went everywhere in cars. By 1937 Dick and Mac ran a hot dog stand outside Los Angeles. Three years later they opened a bigger place, farther inland, on the edge of the Mojave Desert. It had lots of stainless steel, and roof-to-counter windows that permitted patrons to observe the kitchen. Teenaged carhop girls brought customers their food. On the front the brothers put their name: McDONALD'S.

Although their business prospered, the brothers weren't content. They were certain they could organize it better. In a small way, they could do for serving food what Henry Ford had done for making cars. In 1948 they shut their doors, and when they opened them again, three months later, they introduced a "Speedee Service System."

They had shrunk their menu down to little more than hamburgers, potato chips, milkshakes, and pie. With such a simple menu, they could use a food assembly line. This called for no employee skills and reduced their labor costs. In a break with custom, the McDonalds served their burgers—every one—with ketchup, mustard, onions, and two pickles. This uniformity made it possible to prepare the burgers in advance, and that meant Speedee service, which their patrons liked. Wrappers, bags, and paper cups and plates replaced the china, leaving nothing to be washed. And even though their carhop girls had drawn teenagers, the McDonalds fired them, saving wages. Their patrons placed their orders at a service window.

Because of all the savings, Dick and Mac could now afford to chop their hamburger price in half. Customers began to crowd the place. Other food purveyors heard of its success, and came to see it, and the brothers started, in a small way, to franchise the right to run an eatery like theirs. For the first of these, in Arizona, Dick McDonald had a bright idea. Across its shining front he set two yellow arches, visible from blocks away.

In 1954 McDonald's was still, you might say, small potatoes, but that changed after an energetic businessman named Kroc came out to see it. Ray Kroc had always been enterprising. In World War I he quit school and briefly drove an ambulance at the age of fifteen. Since then he'd been a jazz pianist, realtor, and distributor of a mixer that made five milkshakes at a time. It was when he learned that a drive-in was so busy that it had purchased eight of his mixers that he went to California to see it.

Kroc observed McDonald's at its lunch-hour rush. Much impressed, he decided to launch a chain of drive-ins, using the Speedee Service System. He bought the rights, agreeing to pay the brothers a half percent of all his earnings. Soon he was setting up his own drive-ins in some places, and selling franchises in others. He kept the name "McDonald's." It had a better ring than "Kroc's."

KROC BROUGHT THE McDonald brothers' system to all America. He introduced it to a thriving people who were ready now to buy not only goods, like clothes and cars, but also services that met their needs. They wanted speed and ease, and were happy to receive them at a bargain price. On their radios they heard McDonald's jingles: "Forty-five cents for a three-course meal / Sounds to me like that's a steal."

Kroc was fond of pithy maxims, which he posted where his workers saw them. One of them was "KISS," the acronym for "Keep it simple, stupid." Keeping food and service simple let McDonald's concentrate on (another maxim) "QSC," or "Quality, Service, Cost." To accomplish QSC, McDonald's used technology and a fussy focus on details.

Take the lowly burger. Kroc's engineers found ways of cutting every shred of beef from bones. Then, using liquid nitrogen, they froze the chopped beef patties very hard. This meant that patty packers could box them without paper, and cooks could handle them like poker chips. A way was found to grill the patties from both sides at once, saving customers some waiting time.

Or take potatoes. At one time workers in each McDonald's franchise peeled and sliced potatoes for their French fries. But this was slow and costly, and the fries were not the same in every store. McDonald's made arrangements with a giant grower in America's northwest, the perfect place to raise the right potatoes. Machines in factories scrubbed the spuds, blew their skins off, and shot them at high speed through slicing grills. Then they were frozen hard and shipped throughout the country.

Naturally, McDonald's found a thousand uses for computers, the marvels of the age. Computers gauged the moisture in each shipment of potatoes, and prescribed the frying time. Computers tested beef and set the mix of lean and fat. In the stores, employees worked with automated mixers that had stabilizers to control the quantity of ice. They turned out milkshakes faster than a barkeep draws a beer. (Later, when the company turned international, PCs and web technologies kept central management informed of every sale in every McDonald's restaurant in the world.)

Anyone who ran a franchise had to learn McDonald's methods. So Kroc set up a school called Hamburger University, with handsome buildings on a lake. After fourteen days of classes, students earned degrees in Hamburgerology.

Kroc soon quarreled with the McDonald brothers, and he purchased back their right to a half percent of sales. If they hadn't sold this right, for $2.7 million, they would have become two of the richest people in America. When the sale was done, Kroc built a new McDonald's just a block away from the brothers' original one. The new store hurt the old one's sales, and the owners closed the birthplace of McDonaldism.

Seven years from when Kroc met the McDonald brothers, 228 McDonald's stores were in business. They generated colorful, if dubious, statistics. As early as 1973, *Time* magazine reported that the flour used to bake McDonald's buns would fill the Grand Canyon, and the ketchup would fill Lake Michigan. The hamburgers McDonald's had sold could "form a pyramid 783 times the size of the one erected by [Egyptian pharaoh] Snefru."

Now McDonald's made more money than it could well invest in selling fast food in America. Kroc and colleagues thought of starting different lines, or "diversifying," as so many other firms were doing. Should they run hotels, a chain of flower stores, a football team, a huge amusement park? They decided they should stick to what they knew the best. Their wisest course was spreading overseas, selling hamburgers in huge new markets.

Many other firms, across the world, were also reaching out. The world was moving toward a single global market. It's true that since antiquity the continents had traded with each other in the goods one had and others hadn't. But that was not the same as what was now going on. Now, thanks to computers and the Internet, money, skills, and orders moved as fast as light around the world, and little firms and big ones were doing business everywhere. In many places companies would join to make a product. One firm made an integrated circuit with a label saying, "Made in one or more of the following countries: Korea, Hong Kong, Malaysia, Singapore, Taiwan, Mauritius, Thailand, Indonesia, Mexico, Philippines."

McDonald's thought Japan was promising. Its 100 million thriving people ought to be a splendid market. But would a people bred on fish and rice buy burgers, fries, and shakes? McDonald's found a partner in Japan who thought they would, provided he could get attention. He won it from the Japanese press with shocking, tongue-in-cheek remarks. "The reason Japanese are short, with yellow skins," he told reporters, "is that for 2,000 years they've eaten only fish and rice. If we eat McDonald's hamburgers and potatoes for a thousand years, we will become taller, our skin will turn white, and our hair blond."

For McDonald's, Japan turned out to be a big success. The first McDonald's in Japan opened in 1971, and after a year and a half Japan had nineteen more of them. A decade later, the biggest food purveyor in Japan was McDonald's (followed by another U.S. giant, Kentucky Fried Chicken). McDonald's stores were known to every

Japanese. It's said that when a little Japanese girl went to America and saw a pair of golden arches, she told her mother, "Look, they have Makudonardo here too!"

On a winter day in 1990, thirty thousand Muscovites lined up for the opening of the first McDonald's in Russia. China saw its first McDonald's that same year. A few years later, more than half of all McDonald's were outside the United States. And all the stores around the world were globally entwined. New Zealand cheese was flown to stores in South America. Beef went from Uruguay to Malaysia. Packaging went from Malaysia throughout Asia. Australian beef went to Japan. Russian pies went to Germany in return for packaging and soaps. American potato slices went to Hong Kong and Japan. Mexican sesame seeds went everywhere. In every store, in every place, McDonald's system was the same. As a British franchise operator said, "If you come in and challenge the system you won't last very long, because the system is the system is the system."

By the year 2000, the firm that started as a hot dog stand was big beyond belief. Around the world, its stores served fifty million people every day. They had sold 150 billion hamburgers. It's true, however, that the company's success had also bred competitors. Early in 2003, McDonald's revealed its first quarterly loss since the company became a publicly traded business.

IT WAS BIGGER, yes, but McDonald's stood for countless U.S. firms that prospered in the latter 1900s, making use of global markets and technology. America, which held one-twentieth of the people in the world, made about a fifth of the gross world product. Of the world's ten largest corporations, in 2002 American firms were numbers 1, 2, 3, 5, 6, and 9.

But many other countries also boomed. By the start of the 2000s, countries that made up only one-sixth of the world's people produced four-fifths of its goods and services. One such country was Japan, where, as we saw, McDonald's first expanded overseas. The story of Japan's successes in the postwar decades takes your breath away.

In earlier chapters we related how the Japanese built up their empire and their industries and then lost the former and much of the latter in World War II. After that, the Japanese were spent and dazed. Their cities had been gutted and their factories destroyed. To put things back together took about five years.

But then they flew. In the 1950s their economy grew about a tenth each year. They focused first on heavy goods, with the slogan "heavy, thick, long, big." By 1970, the Japanese, who had no iron ore, no oil, and little coal, were the world's third largest makers of steel and cars. Their shipyards turned out half of the world's merchant ships, and they could make a huge oil tanker in less than a year. Then they added lighter techno-industries, such as watches, cameras, and television sets. "Heavy, thick, long, big" were not forgotten, but the slogan now was "light, thin, short, small."

In 1970 Japan's gross national product reached about $200 billion. It overtook and passed the GNP of West (non-Communist) Germany, formerly the third largest in the world. Only Russia's GNP, about $350 billion, and the U.S. GNP, about a trillion, were bigger at the time, and Japan would soon surpass the Russians. The country was a marvel. People everywhere were asking how the Japanese had worked such wonders, and a Harvard scholar wrote a book he called *Japan As Number One.*

Japan's success was partly due to nimble minds. The Japanese were quick to station robots on assembly lines, they cut their use of oil (which they imported) by a quarter, and they copied U.S. quality-control techniques. Japan had once been known for shoddy goods, but now the country turned out well-made products they could sell in quantity abroad.

It also helped that Japanese were dedicated workers, and schooled enough to cope with new techniques: robotics, sophisticated electronics, and computers. They were loyal to the companies they worked for, which often promised that their jobs would last throughout their lifetimes. Devotion to the firm was intertwined with love of country. Workers at Matsushita Electrical Company sang this anthem:

For the building of a new Japan
Let's put our mind and strength together,
Doing our best to promote production,
Sending our goods to the peoples of the world,
Endlessly and continuously,
Like water gushing from a fountain.
Grow, industry, grow, grow, grow.
Harmony and sincerity.
*Matsushita Electrical.**

When McDonald's opened in Japan, the American CEO was much impressed with how his Japanese employees worked. "[U.S.] grill men don't give a damn about the system. . . . But in Japan, you tell a grill man once how to lay the patties, and he puts them there every time. I've been looking for that one hundred percent compliance for thirty years."

After 1985, Japan defied the laws of gravity and economics. The prices both of land and stocks began to soar. On paper, land around the imperial palace now was worth as much as the whole American state of California. On paper, certain companies were worth more than the GNP of many nations. On paper, the Japanese were now the richest people in the world.

But it turned out that this iridescent moment was a bubble. In early 1990 a Japanese business journal warned, "The economy is in the twilight and dusk is at hand." The price of land declined a third, and stock prices plummeted by three-fifths. Soon Japan was in a bad recession. The jobs supposed to last a lifetime, formerly the country's boast, admired throughout the world, evaporated. Many Japanese felt swindled. Was this their prize for so much work? Had their age of gold become an age of lead before they could enjoy it?

By 2003, the Japanese economy had been lying on its bottom for a dozen years. We won't examine here what caused this fall or why the country stagnated. Japan is in this chapter not to make a point about

**Time*, February 23, 1962.

a setback, but to illustrate the way, in modern times, technology and grit can make a people rich. Japan is still a wealthy country, whose gross domestic product per head is bigger than that of any other country except tiny Bermuda and tiny Luxembourg. Japan is full of able people, and in early 2004 it looked as if it might be coming back.

WHAT WE NOW would like to know is this: did the bulk of humans, in an age of economic globalism, move toward decent homes and fuller meals or homelessness and hunger?

The World Bank, a United Nations agency that sponsors economic growth, provides a rough and ready answer. Each year it publishes the worth of all the goods and services each country makes, divided by the number of its people. It thus provides rough estimates of incomes everywhere. It also tells by what percent these incomes rose or fell.

This is what the bank reported on average incomes in the final third of the 1900s. Incomes in the "low income" countries, such as India and China, rose by 3.7 percent. (But these were averages; some areas of these countries did well, while others didn't.) In middle and upper income countries incomes rose by roughly 2 percent. So it's clear that most people in the world were doing well. An income rising 2 percent a year will double in about a third of a century. An income that rises almost twice as fast, as in India and China, will double a great deal faster. In 2003 the editor of *The Economist* magazine maintained that "huge chunks of the world's population have been climbing out of poverty."*

As our incomes rose, we were eating more. "Caloric intakes," almost everywhere, rose at least a quarter. In "developing" countries they rose by almost 40 percent.

Although things got better there were some catches. One was that while the *share* of all incomes enjoyed by people in the richest countries grew a lot, this *apparently* happened at the expense of oth-

*Bill Emmot, *The Economist,* vol. 367, no. 8,330, p. 5.

ers. (We say *apparently* because there are different ways of measuring inequality.) Back in 1960, if you compared the fifth of the world's population who lived in the richest countries with the fifth who lived in the poorest countries, the incomes of the former were already a hefty *thirty* times as big as the incomes of the latter. A generation later, in 1995, the incomes of the former were *eighty-two* times as big. A gap that had been great was now immense.

At the beginning of the 2000s, the assets of the three richest people in the world totaled more than the gross domestic product of the fifty least developed nations. In areas of southern Asia, the Middle East, South America, and Africa, many people glimpsed the good life only on the foreign programs on their television sets, if they owned them. They could gaze at far-off, well-fed people driving shiny cars and housed in spacious homes. They were like the hungry child in fairy tales, with its nose pressed against the bakery window.

What should one call these poorer lands? They once were known as "backward"; then (to be gentler) as "undeveloped"; later still (and even gentler) as "less developed." Since most were near to or south of the equator, they were sometimes called the "South."

The "less developed" countries, those that failed to prosper, often were the badly governed ones. A study published in the year 2000 ranked the nations of the world according to the quality of their governance. Those who did the ranking looked at things like schools and roads, tax and labor market policies, and political environments. Not surprisingly it turned out that the ten countries that were governed best were also rich, while the ten worst ruled were all "less developed" and poor.

According to the World Bank figures, one whole region was a big exception to the general rise in incomes. This was Africa below the Sahara. By the end of the 1900s, most of sub-Saharan Africa had drifted backward. For several decades, incomes had declined each year on average by 0.3 percent.

A good example of the region's problems, and its failure to defeat them, was Nigeria, the country with by far the largest population. If the outline of Africa looks something like a snub-nosed pistol, pointing west, then Nigeria is where you would put the trigger. To explain the country's problems, we must begin 150 years ago, when "Nigeria" did not exist. Independent tribes and kingdoms filled this stretch of western central Africa. In the north was semidesert grassland; in the south were steamy mangrove swamps and forests. Before malaria pills came into use, the muggy coastland on the Bight (or bay) of Benin was notoriously deadly. A jingle warned: "Beware, beware, the Bight of Benin, Where one comes out though forty go in."

The tribes and kingdoms often fought each other, so it took a foreign conqueror to join them in a none-too-happy union. This happened in those decades in the 1800s when (as seen in chapter 15) the richer countries of the world were grabbing giant chunks of Africa and Asia.

In 1861 the British captured Lagos, an island off the coast. Later they worked east along the coast, to the region where the Niger River forms a delta just before it reaches the Atlantic. Here the farmers lived by selling palm oil, which was used for making soap and candles. A British official sailed along the delta coast aboard a ship called *Flirt.* He anchored at the steamy little ports, raised the British flag, and handed presents to the chiefs. He persuaded them to agree to treaties that joined them in a British "protectorate."

But Britain used these pacts as if they were an owner's deed, and a deed not merely to the delta but to all the land that later would become Nigeria. They wanted not merely palm oil, they declared, but the joy of bringing peace and order. Armed with cannons and machine guns, British troops and Africans (hired to conquer other Africans) battled inland, blasting holes in city walls made out of mud, knocking down the cowhide gates of villages, and spraying bullets at defenders. To conquer everything took many years. As late as 1925, a British officer wrote home: "I shall of course go on walloping them until they surrender. It's rather a piteous sight watching a village being knocked to pieces and I wish there was some other way."

Although the British governed conscientiously, they didn't try to make Nigeria prosper. After all, one doesn't build an empire to make one's colonies rich; the British wanted tin and palm oil for themselves. Why help Nigeria compete with Britain in the global market?

But Nigeria wouldn't be a colony forever. When World War II was over, as we know, European nations glumly freed their colonies. Britain freed Nigeria, which in 1960 became an independent nation. It looked as if the Nigerians would prosper, since they seemed to be prepared to rule themselves. The machinery of government was all in place. Though poor, Nigerians had higher incomes than, for instance, Indians, to whom the British had also granted independence. The farmers raised sufficient food to feed the country, and geologists had recently found gas and oil reserves.

After several years, however, self-government became a tragic failure. Soldiers killed the head of state and dumped his body in a ditch. A general seized power, and six months later other officers flogged and killed him. Another general took over, and crushed a mutiny, but that was followed by a civil war that took a million lives. The army later drove the general out of office. And so it went for decade after decade. Except for one short interlude of civil rule, eight generals in turn held power. Most of them began by promising a quick return to civil rule, but then reneged. Others threw them out of power.

Despite the turmoil, the economy at first did well. Farmers kept the country going, and manufactures prospered too; from 1965 to 1980 they rose each year. But oil (no longer palm oil, but petroleum) became the leading export.

And then, bonanza! Suddenly the oil trade boomed. This is why: Nigeria belonged to OPEC, the Organization of Petroleum Exporting Countries, and in 1973 OPEC raised the price of oil. (We'll have more to say on this below.) Then it raised the price again, again, again, again, and again. By 1980, the price of oil had risen tenfold.

Despite the political bedlam, the oil-boom times were good. The government's income multiplied by thirty-four in just a decade, and Nigeria declared that now it had so big an income that it couldn't spend it all. It spent some money wisely on mega-projects, such as major highways, universities, and a badly needed capital city. But it did too little for the country's other basic needs, such as elementary schools, country roads, and clinics.

Meanwhile, the rich and mighty stole colossal sums of public money. When one of the generals took charge, he appointed a commission to probe the doings of his predecessor. It found that $12 billion were unaccounted for. (After writing his report, the commission chairman fled the country, fearing vengeance.) But the reforming general proved equally corrupt. After this man's death (or murder), investigators searched the thirty-seven houses of a friend of his who ran the central bank. In them they discovered many millions, in various currencies, apparently withdrawn for the late general's use. Under pressure, the dead man's family disgorged astounding sums of money. Authorities asked his security chief to account for more than a billion dollars.

Those in power looted public funds, took pay for contracts that they never carried out, and put their families and friends in public jobs. (The number of public servants tripled.) When necessary they burned the buildings that housed documents that proved their corruption. It's said that the wealthy purchased golden bathtubs, and made Nigeria the world's biggest importer of champagne.

While the boom in oil made fortunes for the few, it did the millions much more harm than good. It distorted the economy, hurting industries and farming. Ordinary people's incomes rose slower and slower. Farmers flocked to cities, seeking work and finding only poverty and squalor.

But the worst was yet to come. In the early 1980s the oil-producing nations (OPEC) lost control of prices, and throughout the world oil prices dropped. Nigeria's revenues from oil declined from $25 billion in 1980 to $5 billion in 1986. This drop in revenues worsened a situation that was already bad.

Nigeria's ruling class—corrupt, inept, now short of cash—could not provide relief for all the country's hungry people. When they revised the nation's constitution in 1989, they shrank the role of government. They struck out a description of Nigeria as a "welfare state" and clauses saying that Nigerians had a right to health care and education. They cut the budgets of the universities to the point that they lacked not just computers, but chalk. The World Bank reported, "For the most part, students no longer learn, faculty no longer teach, and research . . . is largely nonexistent." Hospitals ran out of basic drugs and bandages, and declined first to the role of clinics and then morgues.

Average annual income in Nigeria shrank by more than half, from $670 to $300. Now a country rich in oil was actually short of fuel. It once had fed itself but now imported sugar, rice, and wheat. It even purchased palm oil. The costs of housing, medicines, and schools were far beyond the reach of most.

Soon the social fabric was in tatters. At the international airport, bandits raided planes. Vendors peddled phony medications, and the police connived as the drug trade boomed. Even for the wealthy, life was difficult. Those with cars found that it took them three hours to drive across chaotic Lagos, so they carried potties in their cars. It is said that murderers found it easy to kill one of Nigeria's heads of state because every morning at the same time, eight A.M., his car got stuck in traffic. Faced with all their country's problems, Nigerians wrote books with titles such as *Crippled Giant, We Are All Guilty, Another Hope Betrayed, Nigeria: A Republic in Ruins, The Trouble with Nigeria,* and *Always a Loser—A Novel about Nigeria.*

As the new millennium began, the world's sixth largest oil producer was now its thirteenth poorest nation. Nigerians were living worse than their grandparents had.

IN 1979 a new disease appeared that smashed the lives of rich and poor around the world, but especially the poor.

Doctors in the developed world first observed its symptoms in their gay male patients. Some of them were falling ill from what normally were quickly cured infections. The patients' bodies simply couldn't cope, and if a drug could wipe out one infection, another soon showed up. Patients often shrank to skin and bones and lost their minds, and in a year or two they died.

The new disease of course was AIDS—acquired immune deficiency syndrome. Searchers later found the cause of AIDS. When a deadly virus known as HIV invades the bloodstream (mainly during sex, during birth, or on a tainted needle), it enters certain cells whose job it is to guard us from the germs that cause disease. The virus turns these T-cells into factories that, far from fighting germs, turn out many copies of the lethal virus. After years the viruses cause full-blown AIDS.

Although the origin of HIV is still unknown, it almost certainly began in Africa. The first proven AIDS death occurred in the Congo in 1959. One current view is that before invading humans the virus was widespread in animals, maybe chimpanzees. From them (some think) it jumped to humans, but this may have happened long ago, and only in remote and isolated places. In recent times, however, when roads were built and commerce spread, infected villagers from such places may have come in contact with the people of the towns and cities. Perhaps in this way rare HIV infections became a full-fledged epidemic.

From Africa the virus spread around the world, with many people helping it to multiply. Here is an example from the early 1980s, when AIDS was still quite rare. AIDS investigators learned about an infected airline flight attendant, a homosexual man. They pointed out to him that he put all his sexual partners in danger. Nevertheless, he kept on having unprotected sex with perhaps 250 partners every year, in many cities, until his early death.

AIDS was above all a disease of the poor. At the start of the 2000s, about 36 million humans carried HIV or had advanced to full-blown AIDS. Of these, however, 70 percent lived in sub-Saharan Africa, the poorest region on the globe. One person out of four was infected, and life expectancy had dropped by twenty years. In Nigeria, which was suffering enough already from government incompetence, 2.5 million people were infected. Colleges in Zambia were graduating 300 new teachers each year, but meanwhile AIDS was killing twice that many teachers.

Many other victims were in Asia, which holds three-fifths of the world's poor. In the 1990s Asia had the steepest-rising curve of HIV infections. As they did in Africa, poverty and ignorance and governmental unconcern all played a role in spreading HIV and AIDS. For example, in a town in China many people earned some extra money by selling their blood. Apparently, one of them was HIV-positive. Unsupervised technicians mixed the blood of all these donors and extracted from it substances they needed. Then they reinjected all the donors with the no-longer-needed blood. In no time, everyone was positive for HIV.

GROWTH AND GLOBALIZATION helped many to prosper, but they also pushed the world toward sameness. This is how it happened. Throughout the world, the goods produced by booming nations entered local markets, and there they shaped a new consumer culture *beside* the local ones, or in their places. Before globalization, one found in Japan foods that were Japanese, such as sushi. (These are cakes of cold rice, flavored with vinegar and garnished with raw or cooked fish, egg, or vegetables.) But when McDonald's and Kentucky Fried Chicken arrived, the Japanese acquired the fast-food habit and, to some extent, abandoned sushi.

In France in 1999 police arrested the leader of an anti-globalization crusade when he trashed a McDonald's restaurant. He proclaimed before the trial that "the French people . . . are with us in this fight against junk food and globalization." In fact, McDonald's then had about 900 restaurants in France and was adding 30 to 40 new ones every year. The French, who gave the world the words *gourmet* and *haute cuisine,* were abandoning their gastronomic glory. (The judge sentenced the McDonald's trasher to ninety days in jail.)

Isolated regions of the world had once had languages and cultures known to no one else. (Keep in mind that languages and cultures overlap, since sometimes only certain words can express certain ideas.) Peoples of these isolated places had unique creation stories, foods and customs, folk arts, epic tales, and songs.

But all of these were doomed. Investors, tourists, missionaries, tax officials, radios, and asphalt reached the solitary hamlets, bringing in new tongues, new ways. The younger villagers adapted, often gladly. Soon no one but the oldsters spoke the former language, knew the former ways, or even cared about the loss. When the old folk passed away, their language and their culture also died.

The peoples in a zone of east Peru used to speak at least 100 languages, perhaps 150. But by the beginning of the 2000s, these tongues had mostly disappeared. This is how the change took place in Pampa Hermosa, a group of thatch-roofed huts beside a lake. Its people spoke a tongue called Chamicuro and knew nothing of the outside world. They had no telephones, no radios, no roads. But missionaries came and told them they must speak and write in Spanish. And radios, run on batteries, arrived, bringing Spanish talk and Spanish songs from far away. And so the younger people learned a new language and new ways. Finally, Natalia Sangama, very old, found herself the last person on earth who knew her native tongue. "I dream in Chamicuro," she said, "but I cannot tell my dreams to anyone. Some things cannot be said in Spanish. It's lonely being last."

In a Chinese mountain valley, below a snow-capped sacred peak, are the Naxi (*nah-shee*) people, a quarter million of them. For a thousand years they had their own religion and a picture-writing system. Costumed priests called *dongba* danced, led the sacrifices to the wind, and chanted from their scriptures, which taught the Naxi people everything they knew. In the 1950s China's communist regime suppressed the *dongba*. Later on, the ancient rites revived, but then they disappeared again as younger Naxi turned to modern things.

In 1981, a group of ten surviving *dongba,* all of them old men, began a labor of culture rescue. They wanted to produce a 100-volume publication of their scriptures, with pictographs, phonetic spellings, and literal translations in Chinese. They completed this great task. But it turned out that to grasp the meaning of the ancient stories it's not enough to know the literal meanings of the pictographs. They're complex, intuitive, and full of symbols. Only *dongba* understand the stories. By the year 2000 only three of them were still alive.

The English language has encouraged culture deaths. We know how English spread throughout the world. In earlier centuries, the British stamped the English language firmly on the British Isles, Australia, and North America, and they also used it in their other colonies. (India has sixteen major tongues, but even now, when India has been independent half a century, its educated people often interact in English.) In later times, the English language conquered global business, movies, television, airports, and computer-speak. Other languages gave way, and with them—here's the point—other cultures, other points of view.

The death of independent cultures has been happening almost everywhere. When villagers started hearing city voices on their radios, when missionaries spread the word of God, when cell phones reached the jungles, when television brought the World Cup matches to the Arctic Circle, when terrorists who hated western ways wore jeans and running shoes, when billions used the World Wide Web, when the French ate fries 'n' burgers, then what was only local vanished. Cultures crumbled, disappeared.

When many cultures die and only one replaces them, the world becomes a duller place.

IN THE FINAL decades of the 1900s, thoughtful people worried that earth's multiplying humans might use up resources vital to existence. How long, they asked, could so many humans shelter, feed, and clothe themselves?

In 1972, a multinational team of experts published a report they called *The Limits to Growth*. They warned that humans must stop using up their water, timber, land, and fuel. Unless they did, humanity would reach the "limits of the earth," then wither and collapse. In some parts of the world, they claimed, this already was happening. The price of food was out of reach for many, and they were starving.

About a year after *The Limits to Growth* gave this warning, the danger that the earth might reach its limits began to look more real. As we saw above (regarding Nigeria), the world went through a crisis over oil. Many countries, especially the wealthy ones, relied on large amounts of oil to fuel their generators, cars, and air conditioners, and as raw material for plastics, soap, and fertilizer. Ample oil still lay below the ground, but the Organization of Petroleum Exporting Countries knew that demand was strong and the oil wouldn't last forever. Step by step OPEC raised the worldwide price of oil tenfold.

The sharply higher prices were a shock. They hit the wealthy nations hard and the poor ones even harder. (For them, any rise in the cost of a basic need like fuel was catastrophic.) As it happened, though, the higher prices didn't last. Customers reduced their use of oil, and as demand declined, OPEC's prices fell.

Of course, the crisis caused by oil was artificial. It had not confirmed that we were near "the limits of the earth." Just the same, it did remind the world that earth's supplies of things we need do have their limits. Some day, maybe soon, these things could disappear.

And we were increasingly aware that what we needed wasn't only land and air and water. We also needed other species—animals and plants—that enhanced our lives. To some extent at least they had a right to live. Our growing numbers and (much more) our reckless passion to consume were threats to them. Just as earth might fail us humans, we were failing earth.

Amazonia was a growing worry. The Amazon jungle reaches across the shoulders of South America. It stretches from the tree line of the Andes on the west almost to the Atlantic Ocean. The forest belongs in part to other countries, but mostly to Brazil.

A thousand rivers journey through the forest. From the air the forest is a bright green carpet, crossed with lines of blue. In the rainy months the rivers swell and make the forest one enormous, island-dotted lake. These rivers, ten of which are bigger than the

Mississippi, join to form the Amazon, the biggest river in the world. Along with all its feeders, it moves a fifth of all the world's fresh water.

The three main groups involved in Amazonia's story in the 1960s scarcely knew each other. Most of those who ruled Brazil were wealthy whites whose fortunes had been made in rubber, timber, gems, and coffee. They lived in cities on the coast, along with other whites and also Indians, blacks, and the multiracial "Caboclos." The second group were the Caboclos on the edges of the endless forest, far away from cities. Nearly all of them were wretched peasants who lived (and live today) in one-room huts on bluffs above the rivers. They fished, and raised cassava in their gardens.

Deep inside the forest were the Indians, who had lived there many thousand years. Diseases of the whites were spreading through the jungle, killing many of them, and developers, who wanted gold and timber, slaughtered others. (They sometimes flew above a village, dropped some "gifts," and when the Indians had gathered, bombed them.)

To Brazil's white rulers on the coast, a thousand miles away, it was obvious that the forest had to go. Never mind that it had given greatly to Brazil and the world: rubber, timber, cassava, and drugs for cancer and malaria, or that it had much more to give. No, the time had come to master it.

To get electric power they built dams across some rivers, and to make the jungle easier to reach they pushed roads through it. One of these stretched 1,500 miles. State and federal governments encouraged global timber firms to cut the trees, and mining companies to dig up metal ores. Worst of all, perhaps, they told impoverished Caboclo peasants (in the zones around the vast forest) that Amazonia was theirs to occupy and farm.

At a thousand places, men with saws and matches poured into the virgin forest. Some had come to saw and sell the trees, and some to mine the gold, but most intended simply to clear some land and farm it. Perhaps they knew, perhaps they didn't, that the jungle soil is thin and acid, and depends on rotted leaves and branches to enrich it. Once the trees are gone, the cycle breaks, and the soil turns barren. Little grows or ever will. The trees will not return.

Typically, the settlers cleared some trees by cutting them or setting fires. They raised beans and corn and cassava until the soil was poor, then cattle till the land was absolutely bare. They transformed giant swathes of forest into wasteland, and they did this fast. As photos taken from the air reveal, in the 1970s and '80s alone burn-and-clearers ruined 10 percent of all the forest. They kept on burning in the years that followed, and in 1995 alone they razed an area the size of Belgium. Where the forest burned, nothing could be seen but smoke. The smoke blew east to Africa and south to Antarctica.

The loss of land and trees was not the only problem that the burning caused. Trees are sticks of carbon. They pull the carbon from the air around them, and if they burn, their stored-up carbon joins the air as carbon dioxide gas. Amazonia's many million burning trees added carbon dioxide to the air, and once the trees were burned, of course, nothing was left to reabsorb it. In addition to all the other carbon dioxide that we humans add to air (by simply breathing and by burning fuels), the burning forest added more. It helped to make the atmosphere of earth act like a greenhouse. The gases in the sky (mostly carbon dioxide and water vapor) trap the heat of earth as the panes of glass in a greenhouse trap the heat of the sun. (We'll have more to say below about the "greenhouse" problem.)

Far from troubled by the razing of the forest, Brazil's politicos at first were pleased. The governor of Amazonia proposed "a chain saw for each family." "I like trees and plants," he said, "but they aren't indispensable. After all, men [astronauts] have lived in space for almost a year without trees."

In time, however, Brazil thought better of it, and it changed the laws. The government decreed how many acres a person was allowed to clear, and how many fires he might set within a day. But its officials did not enforce the laws. How could they, in forests so vast and wild? So the burn-and-clearers carried on.

For the Indians, the ancient people of the woods, the burning and the highway building brought doom. We can only guess how stunned the Indians were to see intruders, men with saws and matches, raze the jungle where they had always lived. In places, they resisted. At a place in the north where contractors were building a highway, local Indians killed two hundred soldiers (guards) with poisoned arrows. Other soldiers wiped out nearly all the Indians.

Once the settlers had burned their forests and occupied the land, Indians had two choices. Some of them might stay in place, as workers on the ravaged land or diggers in the gold fields. But others fell back deeper in the forest, where other Indians lived who hadn't yet seen modern life. (Even now, perhaps three dozen tribes have never glimpsed outsiders.)

As their habitat has dwindled, and humans killed them, many plants and animals have disappeared. The jaguar, the only "big cat" in the two Americas, may go extinct. The alligator called the jacaré is slaughtered for its supple skin, used in shoes and purses. Certain monkeys are in danger: the woolly, the spider, the howler, and the bare-face tamarin. Likewise the pink porpoises that live far up the Amazon.

No one knows how many kinds of plants and beasts have perished, having lost the only habitats where they can live, and no one knows what fraction those vanished species were of all the species on the earth. Some believe the Amazon contains, or once contained, a tenth of all the species on the earth. In fact, however, no one knows how many kinds of microbes, insects, plants, invertebrates, and vertebrates exist on earth. Experts have described a million and a half, but they think the total may be anywhere from 5 to 30 million.

Many people, in Brazil and elsewhere, have campaigned to save the forest, and this struggle has had its saint and martyr. Chico Mendes was a rubber tapper in Amazonia's far west. He made his living in the forest like many others, by slashing the bark of rubber trees and gathering the latex that oozed from the cuts. When he was eighteen, he met a labor organizer hiding from the police. This man showed Mendes something he had never seen, a newspaper, and taught him how to read and write.

In the 1970s, Brazil's project to tame the Amazon drew immigrants to Mendes's region. As was happening in other places, the newly settled ranchers and loggers brought in goons to terrorize the local people. They seized the forest that local people's lives depended on (for rubber, food, and other uses) and started to destroy it. Mendes organized the local workers, and they set up human blockades around the threatened areas. The resulting standoffs rescued many thousand forest acres. Reserves were set aside, where local people still could tap the rubber and collect the fibers, fruit, and nuts.

Mendes's name soon spread throughout the world. For those concerned about the forest he became a hero, but for Amazon developers he was something else. They were used to getting what they wanted, using either bribes or bullets. Among Mendes's enemies were some local ranchers, the Alves da Silva family, who were clearing land they claimed was theirs. In 1988 Mendes sought to have this land declared a reserve. The Alves made it known they planned to kill him, and reporters came from far away to witness the event. One day Mendes blundered, setting foot outside his door without his bodyguards. From behind some bushes, gunmen shot and killed him.

Everybody knew who shot him: an Alves rancher and his son. But with their money and important friends, would they ever be tried? They were, but only after two years had passed, and because of outside pressure. The court sentenced the father to nineteen years in prison for ordering the murder, and his son to nineteen for carrying it out. The two escaped from prison, but police recaptured them.

ECONOMIC GROWTH AND nature often fought against each other. At times it seemed as if the price of economic gain was ecologic loss. If living standards rose, that happened at the cost of water, beauty, soil, and air; and animals and plants.

In the 1900s we began to warm the air around us. (Or so it seems, for not everyone blames man.) In the entire 20,000 years since the recent ice age ended, the surface temperature of earth had risen from five to nine degrees. But in the 1900s alone, it rose by one degree. What's more, the *rate* of warming quickened as the 1900s passed. The 1970s were warmer than the '60s, and the 1980s were warmer than the '70s. The 1990s were warmer still.

From the North Pole to the South Pole, glaciers melted. The Greenland ice cap shrank. Glacier Park, Montana, looked as if it might run out of glaciers. In Africa, the snows of Mt. Kilimanjaro were about to disappear. As we saw in chapter 2, it was the shrinking of a glacier in the Alps that exposed the body of the Iceman. In Peru one could sit and watch the Andes glaciers shrink sixteen inches a day. The glaciers of Antarctica have been melting, threatening to raise the level of the oceans and to flood low-lying cities everywhere.

What caused the warming? Experts disagreed, but a consensus slowly grew. Fuels that humans burned had liberated carbon dioxide and other gases. Then these gases acted like a greenhouse, or like the windshield of a car on a sunny day. They allowed the sun's rays to penetrate them so that the rays reached the earth, but then they trapped the resulting heat. A little of this kind of warming is a natural and a useful thing, and nothing new. But, the experts said, a problem rose when we began producing too much carbon dioxide. Those figurative greenhouse panes got thicker and the earth got warmer.

Some blamed the "greenhouse effect" on the burning of the Amazon forests, which do indeed make carbon dioxide. But the bigger problem was the burning everywhere of fossil fuels, mainly coal and oil. This happened most in richer countries, which generated many times more greenhouse gas than did the poorer ones.

Earth was also suffering in other ways. Everywhere—not only in Amazonia but everywhere—humans wiped out other species. This was nothing new. Several centuries ago, sailors landing on an island in the Indian Ocean came upon a kind of pigeon that was unlucky enough to be both edible and flightless. All that's left of it today is heads and bones on museum shelves, and the phrase "as dead as a dodo."

In North America in the early 1800s, many, many million bison wandered on the western plains. Indians killed them, sometimes recklessly, for food and other uses. They also sold the hides to whites for use as blankets. Later, white men joined the killing. Leather companies hired hunters to kill the bison, and "sportsmen" shot them from trains as they sped across the prairies. Soon the whites and Indians were killing two or three million bison every year. Anybody could foresee the outcome. Near the end of the century a museum expedition searched the West for bison and found a mere two hundred.

Humans often thought they had no choice but to annihilate another species. One example was the Mexican silver grizzly bear. By 1960, loss of habitat and hunters had reduced these bears to only thirty, roaming in the mountains of Chihuahua. Local ranchers, worrying about their livestock, campaigned to wipe out even this remainder, and the U.S. Department of the Interior, it is said, provided poison. By 1964 the silver grizzly bears were gone.

In part, our wiping out of other species resulted from our ignorance of nature's balance. Because mosquitoes carry malaria, officials on the Indonesian island of Borneo wanted to get rid of them. When they sprayed these pests with DDT, the poison also slaughtered wasps, which formerly had fed on caterpillars. Although saved from the mosquitoes, the local people now endured a plague of caterpillars, which ate the thatched roofs of their houses, making them cave in. Meanwhile the officials also sprayed to wipe out flies. Previously, gecko lizards had killed the flies; now they ate their corpses, which were full of DDT. As geckoes died of poison, house cats ate them, ingesting DDT, which had concentrated as it passed from flies to geckoes. The cats expired, and this prepared the way for rats. These rodents were a threat to humans' food supplies, and they also raised the threat of plague, which rats can carry. Trying to restore the balance, officials parachuted in more cats.*

*Paul Ehrlich and Anne Ehrlich, *Extinction: the Causes and Consequences of the Disappearance of Species* (1981), pp. 78–79.

While we warmed the air, polluted it, melted glaciers, and eliminated other species, humans also littered earth with trash. It sometimes seemed as if this self-inflicted damage couldn't be avoided. By the year 2000 humans numbered well over 6 billion, and in many places we could now produce so much, and so cheaply, that getting rid of everything we couldn't use had grown to be a problem. Cities had more trash than they could deal with.

In 1986 incinerator workers in Philadelphia, on America's east coast, loaded 15,000 tons of ashes on a ship. The *Khian Sea* headed east and south, searching for a landfill. No less than five small Caribbean nations and Bermuda turned the ship away, but in Haiti things went better. Having told officials they were fertilizer, the captain dumped 4,000 tons of ashes on a Haitian beach. Although these were just a quarter of his cargo, they made a mound 100 yards in length and eight feet high. When Haiti learned the truth, it ordered *Khian Sea* to leave. The captain could not or would not reload the ashes, but he left.

The *Khian Sea* was like the legendary *Flying Dutchman,* the ghostly vessel doomed to sail forever. Escorted by some Haitians, the ship returned to Philadelphia. But then, defying Coast Guard orders, it left the port. Three months later, now renamed *Felicia,* it stopped in what was then Yugoslavia. Not long after that, now re-renamed *Pelicano,* it docked in Singapore. Only now it wasn't hauling Philadelphia's ashes. The captain later testified that he had dropped them in the Atlantic and the Indian Oceans.

For several billion humans, material life got better in the latter 1900s. They had more to eat, more to wear, and better housing. But life did not improve for all, and the gap between the rich and poor got wider. Meanwhile, in return for economic gains, planet earth was made to pay an awful price.

CHAPTER 23

We walk along the brink.

AT THE END of World War II, we asked ourselves, would iron-fisted tyrants rise again, as several had between the two world wars? Would the nations that held empires turn their restless colonies free? If they did so, would the liberated peoples organize themselves as stable nations? Would the "superpowers" attack each other, use their superbombs, and end all life on earth?

In the spring of 1945, while the Russians took Berlin, delegates from fifty countries met in San Francisco and formed a global league that they named the United Nations. In its charter they declared its missions: "To save succeeding generations from the scourges of war, and to reaffirm faith in the fundamental human rights, in the dignity and worth of the human person, in the equal rights of men and women, and of nations large and small . . . to promote social progress and better standards of life in larger freedom."

In the fifty years that followed, the United Nations often proved as feeble as the League of Nations had been between the two world wars. A U.S. president (Lyndon Johnson) once declared, "It couldn't pour piss out of a boot if the instructions were printed on the heel." However, even to articulate its goals of "saving" us from war, "reaffirming" rights, and "promoting" higher living standards was a help. At the least the UN as a body offered nations higher ideal standards for behavior.

TWO "SUPERPOWERS" EMERGED from World War II. America had escaped the devastation and was rich, productive, and the sole possessor of "the bomb." Russia had been flattened in the war, but it garrisoned 3 million soldiers, the biggest army in the world. Americans believed themselves the guardians of freedom and free enterprise (or capitalism). The Russians saw themselves as leaders of a worldwide socialist (or communist) revolution. To some extent both countries' sweeping global goals were covers for their national objectives and the pleasure that they took in holding power.

Stalin, still the autocrat of Russia after twenty years, focused—no, obsessed—about the danger of encirclement by "capitalist" nations. He resolved to guard his borders, especially from Germany, which had battered Russia both in World War I and World War II. So Russia turned the smaller Eastern European nations that it held at the end of the war—Poland, Czechoslovakia, Hungary, Bulgaria, and Romania—into satellites. Each was ruled by communists, who took orders from Russia. Russia's domination of Eastern Europe shocked the western, democratic nations. Winston Churchill coined a phrase; he said an "iron curtain" had descended in the midst of Europe.

The superpowers, each fearful of the other, did the things they had to for their safety and their sense of mission. They propagandized, argued, bluffed, and blustered. Truman stated that America would "contain" communism everywhere in order "to assist free peoples who are resisting attempted subjugation." The Russians talked as tough as Truman. Nikita Khrushchev (whom we'll meet below) once told the West that "history is on our side. We will bury you." He was speaking of an economic triumph but was widely understood to mean much more.

As the superpowers quarreled, we humans walked along the brink of an abyss. But they shunned a major battle, and this caution at the highest level would persist for decades. Someone named this state of all-but-all-out war the "Cold War."

In the Cold War's early years, the biggest big-power confrontation centered on Berlin. The victorious allies had divided Germany into four zones: British, French, and American zones and a Russian one. Berlin, the former German capital, lay deep inside the eastern, Russian zone, but the city itself, like the country, was divided. Russia governed "East Berlin," and the other allies jointly governed "West Berlin." But in 1948 Russia responded angrily to a plan by the western powers to rebuild western Germany, and blocked them from Berlin. To do this, Russia cut their access to the city (through the Russian zone) by closing roads and railroads.

To the western allies, this blockade became a test of will. If they now abandoned West Berlin, they might encourage Stalin to push the Iron Curtain farther west, probably by using his huge army to seize all of Germany. So they countered the blockade by setting up an airlift into West Berlin, their section of the city. For about a year, allied airplanes flew in food and other goods to allied soldiers and 3 million West Berliners.

Truman told his top advisers that he prayed he wouldn't have to use atomic bombs to save Berlin. But if it should be necessary, Truman said, let no one doubt that he would use them. Russian planes harassed the western ones but didn't shoot them down, and the major powers stopped short of going to war. In the spring of 1949 the Russians lifted their blockade. Soon the British, French, and American zones united to form the republic of West Germany, and the Russian zone became East Germany.

In the meantime the United States performed a good—and also prudent—deed. The recent war had shattered Europe's infrastructure—railroads, factories, and ships—leaving many million people unemployed and hungry. By 1947 recovery was under way but the Europeans still badly needed aid, and America decided to support their reconstruction. Secretary of State George Marshall claimed the project was "directed not against any country or doctrine [that is, Russia and communism], but against hunger, poverty, desperation, and chaos." In Western Europe the Marshall Plan enjoyed a huge success. It speeded up recovery, and eased the hardships many people were enduring. The Eastern European countries also badly needed aid, but Russia called the plan a "venture in American imperialism." It rejected aid and forced its satellites to do the same.

Meanwhile Western European countries felt endangered by the Russian army. Russia (worried about its *own* safety) had positioned many units not in Russia but in its satellites. So Russian troops were close to Western Europe, just across the Iron Curtain. The western nations understood that safety lay in having a joint defense. So delegates from Canada, America, and Western Europe met in Washington in 1949 and agreed on a military alliance. They said that an attack on one would be considered an attack on all.

Out of this agreement rose the North Atlantic Treaty Organization, known as NATO. To serve as a nucleus for NATO's forces, the United States stationed a third of a million troops in West Germany. These soldiers also were a trip wire, a fact that reassured the Western Europeans. If the Russians tripped the wire by attacking the Americans, the United States would have to fight beside its allies.

But as it happened, war broke out in Asia, not in Europe. It happened in Korea, the offshoot of the Asian mainland that Japan had ruled until Japan was conquered at the end of World War II. The winners of that war had split the peninsula, and therefore the nation, into two "temporary" occupation zones. In the southern half, America had built a client country, South Korea. Although South Korea called itself a republic, its president tolerated no political party but his own. In "North Korea" Russia had installed a communist regime under Kim Il Sung ("Great Leader") that was more totalitarian than Hitler's Germany had been. Each of these two men burned to reunite Korea—under himself of course.

Suddenly in June of 1950 Kim Il Sung rushed his armies into South Korea. Stalin had approved the war plan, probably believing the United States would not defend the south. But the UN speedily approved the measures that America now urged, and UN forces, mostly from America, joined the South Korean side. Fifteen other nations joined them.

Armies pushed each other up and down the peninsula. First North Korea's army drove the southern forces all the way to South Korea's southeast tip, almost to the sea. But in September their commander, U.S. General Douglas MacArthur, brilliantly surprised the North Koreans. He invaded from the sea, halfway up the western coast. He thus outflanked the North Koreans. The allied forces pinched, then smashed the North Korean Army. Then they hurried north through North Korea to the Yalu River, North Korea's Chinese border.

China's CCP regarded the North Koreans as fellow communists in need and the UN forces as imperialists, insolently camping near their border. Suddenly they poured 180,000 Chinese soldiers into North Korea. Wave after wave, they stormed the UN lines, blowing bugles, many of them dying as they charged. In bitter winter fighting, China's army drove the UN forces south below the South Korean border. But then the tide reversed again. Allied bombing drove them back, and then the fighting stopped in 1951. And where did the armies face each other now? Roughly where the war had started, at the border of the two Koreas.

The war had lasted a year, but making peace took two. The upshot was that North and South Korea remained as they had been before the war. UN (mostly U.S.) help had saved the South. Together, about 1.7 million Chinese and North and South Korean soldiers and 3 million civilians had been killed or wounded. The U.S. dead numbered 36,000. As we shall see again, in modern wars the nations that have better arms and technical support suffer fewer deaths than others.

ON A WINTER NIGHT in 1953 the chief contenders for the throne of Stalin joined the aged ruler at his villa for a drinking bout. The next day someone found him on the floor unconscious, victim of a stroke or possibly of poison. (That suspicion arises because the other partiers, who were summoned to the villa, kept Stalin's condition secret for a day.) Four days later Stalin died. For the next two years the contenders struggled to succeed him until one of them, Nikita Khrushchev, won control.

In 1961 the danger of an all-out East-West war arose again. For the Russians West Berlin was still a major aggravation. Three million Germans had left the harshly governed Russian zone and moved to West Berlin. Most of them were trained and useful people. All too well the Russians knew that fleeing workers hardly proved that communism was succeeding.

At a meeting with America's new president, an exasperated Khrushchev demanded that the western powers pull out of West Berlin. When the meeting ended badly, the scent of danger filled the air. Back in the United States, John Kennedy requested that Congress increase the U.S. forces. At the same time, NATO allies gave the United States their support. But soon the crisis faded. Once again an all-out conflict hadn't taken place.

Germans still were slipping into West Berlin, and the Russians had had enough. On a summer night in 1961 they hastily built a barbed wire fence along West Berlin's boundary. Later they replaced it with a closely guarded concrete barricade that few would ever flee across. This "Berlin Wall" became a symbol of the Cold War.

Underneath the East/West conflicts lay the nuclear or atom bomb. As we saw, during World War II America had made atomic bombs and then dropped two of them on cities in Japan. For several years the fact of being sole possessor of this weapon made the United States militarily supreme, though Russia's giant armies made it too a superpower. With its wealth and weapons, the United States could influence events around the world.

The United States and the western powers were well aware that Russia too would one day have the bomb. But not, they hoped, too soon. Their advisers told them that the Russians would require from five to fifteen years to build atomic bombs. So imagine their surprise when Russia tested one in 1949. A mere four years had passed from when America had leveled Hiroshima. Battered, backward Russia (helped a little by its theft of British/U.S. weapon secrets) had worked an engineering wonder. No longer did the United States have the A-bomb to itself.

Three years later, though, the United States took the weapons lead again. On an atoll in the mid-Pacific it tried and proved a new, immensely potent "hydrogen bomb." But the U.S. lead did not last long. Within a year the Russians too had made an "H-bomb." (Great Britain, France, and China followed later.)

Hydrogen bombs explode when nuclei of atoms fuse, not when they split. The newer bombs were hundreds, even thousands of times as strong as the atom bomb that flattened Hiroshima. Their strength was measured not in thousands but in millions of tons of TNT. (In weapon-speak each of these was a "megaton.") Albert Einstein glumly told the world, "Annihilation of any life on earth has been brought within the range of technical possibilities."

To make a superbomb was one thing; to "deliver" it to a target far away at several times the speed of sound was quite another. What the superpowers needed now were intercontinental missiles. By 1958, the Russians had the lead in making them, but the United States followed soon. If it chose to, either superpower could launch its warheads from its homeland, or from planes or submarines, and slaughter far-off victims by the millions. (In weapon-speak a million deaths were a "megadeath.") Any humans whom a blast might spare would later die from radiation. As time went on, each side made bombs enough to eradicate the other several times over.

By now the statesmen of the superpowers knew better than to threaten, as they had before, "massive retaliation." They knew that if one side launched its missiles, the other side would promptly do the same. The consequence could be the end of human life on earth. In weapon-speak one called this "Mutually Assured Destruction"—for which the acronym was MAD. One could only hope the possibility of MAD would cause what military theorists called a "balance of terror" in which both sides were too scared to push their BOMB NOW! buttons.

In 1962, however, the world again came near the brink of the abyss. This time the crisis happened not in Europe but on little Cuba, an island only ninety miles from U.S. soil. In 1958 Fidel Castro won a revolution, driving out a brutal despot, a friend of the United States. Many Cubans fled to America. When Castro seized the properties of U.S. firms, the United States outlawed imports (chiefly sugar) from the island. Castro then moved closer to the Russians, now declaring that he was a communist. In April 1961, 1,500 Cuban exiles, whom Americans had equipped and poorly trained, invaded Cuba at the Bay of Pigs. They wanted to inspire their fellow Cubans to throw out Castro, but the raid was ill prepared and badly carried out. (Some of the invaders couldn't fire a rifle.) As the rebels landed they stumbled into Castro's hands.

It was those events that nearly led the world to Mutual Assured Destruction. Khrushchev recklessly declared that he would save the Cubans from a second U.S. invasion (plans for which existed), and in 1961 he secretly sent missiles to the island. As U.S. airplanes filmed them, Russians in Cuba built the sites from which to launch offensive nuclear weapons.

Since Cuba lay in what they saw as their front yard, Americans were shocked. At the least, if Russia put its missiles there, this alone would be a grave humiliation to a superpower. Much worse than that, most of the United States would be in easy range of mega-death-dealing missiles. And looking at the matter from a global point of view, if America merely became exposed to nuclear threats, its vulnerability might change the terror balance that had so far saved the world.

Since Kennedy had authorized the Cuban invasion fiasco, he faced a danger that was partly of his own making.

This was Kennedy's dilemma during thirteen scary days in October 1962. He had to keep the missiles out of Cuba, *but* if he acted rashly he could bring about an all-out nuclear war. He considered bombing the Cuban missile sites, but then he chose a safer course. To prevent more Russian arms deliveries he ordered a blockade of Cuba, and he kept his well-armed airplanes flying. Khrushchev protested the American blockade, calling it illegal. But then he ordered Russian ships that were bringing yet more missiles to turn around.

After thirteen days of confrontation, Khrushchev wavered. He wrote a note to Kennedy agreeing to take out his missiles if America would promise not to invade Cuba again. A day later Khrushchev wrote a second note, this one more demanding. He now insisted, tit for tat, that the United States also pull its short-range missiles out of Turkey. (Since the Turkish sites were as close to Russia as the Cuban ones were to America, one can see his point of view.) Cannily, the U.S. leader answered Khrushchev's *first* message, agreeing to it, but ignored the second. Privately, however, he let the premier know that the United States eventually would take its missiles out of Turkey.

Terror—fear of MAD—had saved the world. Never again would the superpowers come so close to all-out war. They learned instead to walk around each other on their toes, stiff-legged, sometimes snarling, *never* biting. They strung a "hot line" from the Kremlin to the White House to prevent a war from breaking out if some event were misconstrued and guided missiles were launched. (The leaders would have twenty minutes for discussion while the missiles flew; after that, discussion wouldn't help.) Despite such hopeful signs, both sides made more nuclear weapons, and Israel, India, and Pakistan later joined the ranks of those who had the bomb.

While the superpowers did not attack each other, they might make war on one another's friends. Such a war had happened in Korea, and another now broke out in Vietnam. This slender land of rice farms, hills, and jungles curves along the southeast coast of Asia. France had ruled it for a century, but after World War II the Vietnamese had fought the French for seven years and won their independence. Vietnam was then divided ("temporarily," like Korea) at its slender waist.

The "nationalists" who governed "South Vietnam" were people of the towns and cities and defenders of the rich. With the backing of America they declared the south an independent nation. The communists, whose longtime leader was the dogged Ho Chi Minh, took power in the north. They were nationalists as well as communists, and wanted one united nation.

The "Vietnam War" (as Americans would call it) started when communist guerrillas attacked the government of South Vietnam.

The war soon widened as North Vietnam sent armies south to fight beside the partisans. Russia gave its fellow communists in North Vietnam ample aid. China also helped the North but less, being pleased to have Vietnam, a trying neighbor, stay divided.

Americans had never heard of far-off Vietnam. (Several years later, just before he died of wounds in Vietnam, a U.S. "grunt" would say, "My mother thinks [it's] somewhere near Panama.") But the United States, under Kennedy, helped the South, convinced that this was not a little country's civil war but a major battle in the global war for freedom. If South Vietnam should fall to communism, politicians warned, other nations too would fall like rows of dominoes, each one knocking down the next. At first the United States sent military "advisers" to help the South, but these were not enough. The rulers of the South soon proved incompetent to fight a war and they were unpopular among their own people. America began to send in ground troops, joined by token forces from the Philippines, New Zealand, Thailand, and Australia.

The United States found itself in jungle muck that reached its ankles, then its waist, then its neck. Kennedy's successor, Lyndon Johnson, called the North a "little fourth-rate country," but by 1965 he knew he couldn't beat it if he didn't make a bigger effort. He told his wife, "Vietnam is getting worse every day. I have the choice to go in with great casualty lists or to get out with disgrace." In he went. He sharply raised the U.S. forces till by 1968 they totaled over half a million. Just as he had feared, many Americans (and many times as many Vietnamese) lost their lives. Americans bombed the North repeatedly in "Operation Rolling Thunder," burned villages (turning farmers into refugees), and defoliated hundreds of thousands of acres of land. Although Johnson couldn't win the war, he promised U.S. Army cadets at West Point, "Whatever happens in Vietnam, I can conceive of nothing except military victory."

Opposition to the war arose in other democratic countries, then America. Protest marchers shouted, "Hey, hey, LBJ, how many kids did you kill today?" Even several politicians dared to voice their doubts. People asked, Can we win? Can America police the world? Is the South Vietnam regime worth saving? Is there any reason for the war that justifies the loss of life? Is it true (as Johnson said) that if we lose this war the enemy will be "in Hawaii and next they will be in San Francisco"?

By 1968 Johnson had had enough. He halted bombing in the North so that peace talks could begin. And so they did, but the fighting didn't stop. Only after five more years had passed, in 1973, did the last American leave South Vietnam. And even that was not the end, since North and South Vietnam kept on fighting until 1975. Finally the war ended in a total victory for the North.

Like every other war, this one had an awful cost. At least a million, maybe three million Vietnamese—soldiers and civilians on both sides—had fallen in the war.

The richest country in the world had fought for about a decade to save the world (it thought) from communism. It had sacrificed the lives of 54,000 Americans. It used the newest weapons, and it dropped more tons of bombs, on both the North and South, than the winning side had used in World War II. But it lost the war to gritty men who lived in caves and fed on rancid rice.

IN THE 1970S Russia lost its way. Flaws in Russian socialism—perhaps in any socialism—began to hurt the country. One defect was the party's autocratic way of choosing leaders. Leonid Brezhnev, who succeeded Khrushchev, dealt with many problems by ignoring them and jailing dissidents in psychiatric hospitals. While the country slid, Brezhnev too declined in mind and body till he dribbled on himself while making speeches. To make things worse, both the men who followed him as general secretary were ill. Each tried to run the country from his sickbed and survived about a year.

Among the problems these men failed to solve, the chief one was the economy. The state-owned, state-run system simply didn't work. Yes, it could produce technology the party bosses badly wanted: space machines and missiles. Apart from that, management from Moscow stifled everyone's initiative and pride in work well done. Factories used the methods of the 1930s and were slow to automate and use computers. Steel mills built by Stalin breathed out poison smoke. Collective farms were ill equipped and badly run, so output dropped and Russia stopped releasing grain production figures. A country that had countless miles of fertile plains had to purchase grain from other countries, notably from capitalist America. Russia's economic growth rate slowed, then dropped to zero.

A paradox: Russians now lived better than they ever had. But that's not saying much. As Russia headed into bankruptcy, workers often went unpaid. Many dwelt in crowded, crumbling housing, and a Russian could purchase less than half the goods and services that an average American or European could afford. One in five lived below the poverty line, and life expectancy declined.

Russians knew their system wasn't working, and they were losing faith. "What's the difference between communism and capitalism?" went a Russian joke. "Capitalism is the exploitation of man by man, and communism is the reverse." Western TV programs gave them glimpses of prosperity in other countries, and they saw well-dressed foreign tourists on the Moscow streets. No one now believed a communist utopia was just beyond the horizon unless, another joke explained, "you understand that a horizon is an imaginary line that recedes as you approach it."

Of all the things it didn't need, Russia now began a war it couldn't win. On what was then its southern edge was bleak Afghanistan, a land of poor and quarreling peoples. Afghanistan's government was friendly to the Russians, but it faced a civil war. To support its comrades, Russia moved in troops in 1979 and began to fight the country's Muslim "holy warriors." America gave weapons to these rebels, and the Russians found they couldn't drive them from their mountain strongholds. The war went on nine years, cost a lot, and took the lives of 14,000 Russians (as well as 1.3 million Afghans).

In the meantime, the United States worsened Russia's problems when it set a challenge that the Russians couldn't meet. In the 1980s the U.S. president was Ronald Reagan, a sunny former movie star and former governor of California. Russia, Reagan told the world, was an "evil empire" whose purposes were "dark." He believed—

despite the evidence—that the Russians had begun to win the arms race, but he assumed America could win a "protracted" nuclear war with Russia. (A high U.S. military official declared that Americans could survive in such a war if they would "dig a hole, cover it with a couple of doors, and then throw three feet of dirt on top.") So Reagan doubled U.S. spending for defense, forcing Russia to use its all-but-vanished funds to try to match the U.S. buildup.

When things were at their worst, the Russians finally began to face their problems. Apparently the Politburo (the policy-making body) had decided that three decrepit leaders in a row had been enough. In 1985 they chose as general secretary Mikhail Gorbachev, who, at fifty-four, was their youngest member and the most dynamic. Did they know that Gorbachev was planning basic changes? This remains a mystery.

Under Gorbachev, from 1985 to 1991, Russians witnessed these astounding changes: The ending of the Party's monopoly of power. Freedom of the press. A president and legislature. Elections with several candidates instead of one. Reduction of central planning and collective farms. Open discussion of poverty, and of air and water pollution. People starting businesses of their own. A declaration that the union of Russian republics was voluntary. And Russian soldiers leaving Afghanistan, which was Russia's Vietnam.

Many of his friends and foes declared that Gorbachev was terminating communism, ending a vast experiment testing whether the abolition of private property would produce a better life. Gorbachev denied that he was doing this and claimed that he was saving communism by reforming it. In fact, however, he was bringing democracy and capitalism to a people who had never known them. They feared these innovations, and in the short term they were right. The first results of Gorbachev's reforms were greater poverty and widespread crime.

In 1989 and '90, as the world looked on amazed, all of Russia's European satellites freed themselves from Russia's rule. With tacit help from Gorbachev, who told his generals not to interfere, they drove their party bosses out and brought in democratic rule and open markets. In East Germany, for example, riots during 1989 forced the communists to fire their longtime leader. But merely driving out the despot didn't satisfy the huge and angry crowds. Astoundingly they smashed a hole right through the Berlin Wall, whereupon the communist regime gave in and knocked it down. The communist regime collapsed as well, and East and West Germany joined to form a single democratic nation.

But what about the fifteen republics of Russia itself, that union that now was "voluntary"? Would they want to leave it? Quite a number did. Several on both the European side, for example Ukraine, and on the Asian underside converted into independent nations. After they had left it, Russia still was huge, but much diminished, like a fat man who has dieted with some success.

In the meantime, both Gorbachev and Reagan reassessed the arms race. Both of them knew very well the most important point: their nations' rivalry could bring about a nuclear disaster. Gorbachev also knew that Russia lacked the funds to make more arms. He therefore made it known that Russia planned to drop its forceful stance throughout the world. For his part, Reagan stopped attacking Russia as an "evil empire" and claiming that America could "prevail" in a "protracted" nuclear war. He said that America now had arms enough so that it could safely cut them back if Russia did the same.

The two men held discussions, and in 1987 they signed an astounding pact. They both agreed to destroy their intermediate-range nuclear missiles within three years. And that was just the start. Four years later, Russia and America agreed to further major weapons cuts. Apparently the world no longer faced the threat of Mutual Assured Destruction, although the number of nations that had nuclear weapons was rising.

The weapons cuts were cheering news for all the world. They happened as the Wall came down, and tyrannies collapsed in many places. Readers will remember that at this time India and China, like Russia, were abandoning state socialism. That change might benefit at least a billion hungry people. China, to some degree, was moving toward democracy, and so was Latin America. Because of all these things, in 1989 and '90 euphoria was everywhere. Surely peace, democracy, and jobs for all were on the way.

PROBABLY THOSE THINGS indeed were in the future; we have good reason to believe so. But even as the Cold War ended another global problem was arising: a many-sided crisis in the Middle East. As the twentieth century ended and the twenty-first began, this crisis flared, smoldered, flared again. At this writing, we're so enveloped in its smoke that we cannot see it whole. But we shall try to sketch its outlines.

The Middle East requires defining, yet defies it, because it isn't this or that. It isn't one coherent region. It stretches from the eastern Mediterranean to the western side of India, and includes about two dozen nations in Africa and Asia. Among the more important (going from west to east) are Egypt, Turkey, Israel, Saudi Arabia, Iraq, and Iran. What do these six have in common? Three are Arabic-speaking; the others aren't. Three are rich in oil; the others aren't. Two are democracies; the others aren't. Five are Muslim; one is not.

As their crisis started to be global, nearly all the Middle Eastern states were ruled by tyrants, whether they were kings in robes or "presidents" in well-cut suits. Tyranny is the one thing that they had in common. Despotism was perhaps the major cause of turmoil. Like many tyrannies, these governments were fragile. To change regimes important people used assassinations, not elections.

Israel, in Palestine, was also a source of trouble, but for other reasons. Unlike the Muslim countries, Israel was Jewish and hectically democratic, not despotic. As Arabs saw it, Israel's Jews were intruders in an Arab, Muslim land. Yes, the Hebrews-Israelites-Jews had once ruled Palestine, but that was two thousand years ago. Arabs had lived there ever since; the land was theirs.

This was the background to this Israeli-Arab problem: After World War I, Jews from elsewhere had settled in Palestine, joining a minority of Jews already there. They viewed the region as their homeland.

More Jews came there after World War II and their sufferings in the Holocaust. The Jews declared themselves a nation, Israel. They fought a war against the Palestinian Arabs, other Arab nations, and Britain, which had governed Palestine since World War I. Despite the odds, Israel won the war. Later it had "occupied" the West Bank region between Israel and the Jordan River. Meanwhile, many of Israel's defeated Arabs fled to dismal camps for refugees. West Bank Arabs stayed where they had always lived, discontented subjects in a place they knew as home.

As the twentieth century ended, the Palestinians and other Arabs hated Israel not only because it had taken Arab land, but for another reason as well. For Muslims, as for Jews and many Christians, Palestine was sacred soil. For Jews to rule it was an abomination. And so, because of both real estate and religion, Israel and its Arab neighbors repeatedly waged brief wars against one another.

In the meantime Middle Eastern oil production had been rising, and oil became another factor in the crisis. After World War II, geologists made major finds, especially in the Muslim, mostly Arab, countries around the Persian Gulf. Among these nations were Iraq, Iran (which is Muslim but not Arab), and above all Saudi Arabia, which turned out to have stupendous oil reserves. At just this time the booming industrial countries elsewhere in the world began to use more oil than ever for their cars and air conditioners. The oil-producing Arabs found that the rising demand of others nicely matched their rising supply. So they raised the price.

Sales of oil transformed the lives of ordinary desert sheikhs who once had slept in tents and gauged their wealth by counting wives and camels. Now they dwelled in marble halls and flew in private planes. Even more luxurious were the lives of presidents and kings. Iraq's Saddam Hussein reportedly had 60 cars, 50 palaces, and somewhere between $1 billion and $40 billion. The oil-rich despots spent some of their money on things their countries needed—highways, clinics, schools—but not enough.

Certainly the rising tide of oil didn't raise all boats. In several countries—most of them, perhaps—the gulf between the hungry many and the oil-rich few became enormous. The former were like shabby sailboats floundering in the wake of new gigantic tankers. The ignorance and low horizons of the poor ripened them for harvesting by demagogues.

And then came Islamism. Not Islam but Islamism—the suffix makes a difference. Islamism was the name that *non*-Islamists gave to the rising trend among some Muslim clerics and their crowds of devotees to battle all things new. No, Islamists said, to democratic government, since only clerics (that is, Islamists), enforcing the Qur'an, may rule. No to schools that teach their students any subject but Islam. No to Muslims who refuse to do as Islamists demand. No, of course, to Israel. No to Moscow *and* to Wall Street, since both are enemies of holy truth.

Another actor figured in the Middle East. The powerful United States, though far away, tried to influence events. The major reason for its interest was its ever-growing appetite for oil. Anxious to preserve its access to the Arab (Muslim) countries' oil, the United States grew concerned about their independence. While the Cold War lasted, nearly fifty years, America sometimes smelled a Middle Eastern plot where none existed. In later years it worried over dangers that some Arabs raised for other Arabs.

America's demand for oil conflicted with another need. Despite its oil suppliers' hatred for it, America wanted Israel to survive. U.S. politicians liked the country's democracy and valued it as an ally. What mattered more, they craved the votes of U.S. Jews, often Israel supporters, who could swing elections in important U.S. states. For all these reasons America gave Israel weapons for defense and money for survival.

So much for the different factors in the crisis in the Middle East. Now: the way they intertwined.

In 1987 Palestinians in a refugee camp threw rocks at Israeli soldiers. A riot started—not the first. An Israeli soldier shot and killed an Arab boy. In no time Palestinians everywhere were burning tires, hurling rocks, and shouting insults at Israelis. Wisely, they did not use guns against a foe who could have crushed them with tanks.

The Arabs named this mutiny the *Intifada*. The word means "shake" or "shudder" but implies a shaking off or getting rid. Israelis answered them with tear gas and with plastic bullets—sometimes real ones—and they broke some arms and legs and locked up many Arabs. But these measures had no effect. To Palestinians, their prison terms and broken bones were only proof of their devotion to their cause. The Israelis who were fighting them were often torn by half-admitted guilt: should you shoot a man who wants the land that once was his?

Step by step, the Intifada grew more bitter. Palestinian terrorists (or "martyrs") set off bombs in Israel's markets and cafés. They killed Israelis by the dozens—and themselves. Israel struck back, razing homes and shooting those who gave the bombers bombs.

But the center of the Middle Eastern crisis wasn't Israel. It was farther east, in the oil-rich countries on the Persian Gulf. Before the Intifada started, two major oil-producing countries went to war. This is how it happened: In Iran in 1979 an unlikely mix of Islamists and reformers drove the shah (or king), who was both a tyrant and reformer, from his Peacock Throne. Fierce, unbending Ayatollah Ruhollah Khomeini, an Islamist, replaced the repressive shah. He and other clerics made Iran another kind of tyranny, run by men convinced they knew the will of God. The ayatollah urged the people of Iraq, next door, to overthrow Saddam Hussein and found an Islamist regime like his. For that and other reasons violent Saddam attacked Iran in 1980, certain of an easy win. Instead the fighting lasted for eight years. Khomeini and Saddam poured missiles on each other's cities, and Khomeini drove hordes of teenaged boys into suicide attacks. Both of them used poison gas. Saddam employed it even to kill his own Iraqi Kurds, who, he claimed, had helped Iran. The war changed nothing and consumed a million lives.

In 1991 Saddam, scarcely pausing after fighting with Iran, attacked a weaker neighbor. His target this time was Kuwait, a tiny Arab country floating on a sea of oil. He wanted its enormous wealth to pay his many debts. For his army, then fourth largest in the world, the conquest of Kuwait was just an easy stroll.

But this time a Middle Eastern war became a global one. Saddam had reckoned that America would not take part, but here he blundered. The U.S. president, George H. W. Bush, decided to protect the U.S. oil supply and curb the price of gasoline by fighting for Kuwait. U.S. leaders worried that Saddam might march right through Kuwait and conquer Saudi Arabia, an even greater prize. Beside protecting oil supplies, the United States had another motive. It wanted to destroy the weapons that it thought the virulent Saddam possessed: poison gas, deadly microbes of disease, and maybe nuclear weapons. With its allies, Britain, France, and others, the United States began a "Gulf War" against Iraq.

Although Saddam predicted he would win this "mother of battles," he didn't have a chance. In recent decades America had invented arms so "smart" that U.S. cannoneers (if all went well) could send a missile down a chimney miles away. The United States and its allies first destroyed Saddam's communications and his airplanes. Then, in a hundred hours, it smashed his tanks and routed the Iraqi troops. For reasons that are still unclear, the victors did not try to capture Baghdad and arrest Saddam.

Although the mother of battles had become the mother of defeats, Saddam advised his people to "Applaud your victories. . . . You have faced the whole world, great Iraqis. You have won. You are victorious. How sweet is victory!"

In the meantime, the movement we are calling Islamism had become in some respects a worldwide threat. More and more, Islamists saw the outside world, and most of all America, as "Satan."

Except in Iran (and later in Afghanistan), Islamist extremists had no armies, hence no chance of winning wars. They resolved instead to steer their enemies by planting fear. It's possible that acts of terrorism elsewhere in the world at just this time revealed to Islamists how to kill en masse and win attention. In America, for example, homegrown terrorists blew up a U.S. office building, killing many. In Japan a religious cult called "Supreme Truth" released a poison gas in a subway. In the same way Islamists now began to use dramatic acts of terror, especially against America. They downed an American airliner over Scotland, bombed two U.S. embassies in Africa, blasted a U.S. warship, and exploded a bomb beneath an office building in New York. They also slaughtered tourists in an Indonesian nightclub.

A guiding spirit of these terrorists was Osama bin Laden, a wealthy Saudi Arab in his middle forties. Bin Laden's father had made his fortune as a builder who enlarged the mosques of Mecca and Medina. Osama may have learned his Islamism first from Saudi teachers of Islamic studies. Later he was shocked and angered during the Gulf War when "infidel" U.S. troops were stationed on his country's sacred soil. He decided that the Saudi kings, who claimed to be protectors of the Muslim holy places, were really traitors to Islam. Later he would preach against American support of Israel. He left his country, built a league of terrorists, and put them to work. They performed several of the bloody deeds described above.

On September 11, 2001, Bin Laden's network carried out their most destructive attack. The murderers were mostly Saudi Arabs and Egyptians whom Bin Laden's group had planted in the United States. On September 11 they hijacked four U.S. airplanes on domestic flights and killed the pilots. Then they steered the airplanes into well-known buildings. One plane hit the west side of the Pentagon in Washington, the command center of the American armed forces. The captors of another airplane probably had planned to strike the White House, but the passengers resisted and the airplane crashed, killing all aboard.

Two other airplanes struck the World Trade Center in New York, two giant office buildings. The heat from burning airplane fuel disintegrated girders near the tops of both the towers. Nearly three thousand office workers, and police and firemen who had tried to save them, were trapped above the fires or failed to flee in time. They telephoned their families to say good-bye. Some jumped to certain death. As countless millions watched on television, first one and then the other building crumpled and collapsed.

America's response to "9-11" was to carry out a "War on Terror." (The name was misleading, suggesting that the enemy was a tactic, rather than those who used it.) The U.S. president, George W. Bush (a son of George H. W. Bush), waged a lightning war in bleak Afghanistan, whose Islamist leaders had given refuge to Bin Laden. Once again, Britain allied itself with the United States. Relying heavily on Afghan rebels, the allies crushed Bin Laden's Afghan friends, but the planner of so many murders vanished in the rugged mountains.

Then Saddam Hussein's familiar face appeared again on TV screens around the world. The United States apparently had grown to like its role of global sheriff, and it declared Saddam a black hat. It gave at least four reasons: he owned and might use weapons of mass destruction, he aided international terrorists, he was a barrier to peace in Israel, and he oppressed his own Iraqi people. America and Britain, aided by some friends, went to war against Iraq again despite UN opposition. They swiftly conquered it and then began the thorny task of helping it rebuild.

By now the Middle Eastern crisis had replaced the Cold War as a focus of concern. What A-bombs, ideologies, and superpower ambitions had been to the Cold War, despotism, zeal, and oil were to the Middle East. Like charcoal, sulfur, and saltpeter they formed a deadly mix, and could explode. In the early 2000s no one could predict what damage they might do.

IT'S OFTEN SAID that humans never were as violent as in our age. It's true that the 1900s had their share of evil: two world wars, tanks, machine guns, poison gas, labor camps, slaughter by starvation, Hitler, firestorms, A-bombs, H-bombs, the Holocaust, apartheid, the Cultural Revolution, Korea, Vietnam, the Iran-Iraq war, exploding airplanes, blown-up buildings. Other evils happened that this book hadn't space to mention: nerve gas, napalm, anthrax, massacres, killing fields, serial killings, and murderers whose very names will cause a shudder: "the Jackal," Idi Amin, Pol Pot.

But is it true that humans had never been so violent? We don't know that, and we never will. (History's sources aren't complete.) We should not forget the events of ages past that we do not know about:

Assyrians castrating and impaling captives, Mongols slaughtering whole populations, at least a quarter of the Chinese dying in civil wars in the 1600s, the horrors of Europe's Thirty Years' War, and the enslavement of Africans.

This is suggestive: when Europeans first encountered other peoples, they often found them making war. When Marco Polo's father and uncle detoured into China they found an emperor who was always making war. Cortès, in Mexico, found the Aztecs fighting with their conquered peoples. Pizarro likewise, in Peru, found the Incas waging civil war. Magellan, when he reached the Philippines, found the islanders at war. Unrecorded wars on every continent may once have been quite common. We just don't know.

Even as we humans slaughtered one another in the past one hundred years, we also made some gains in ending war and other violence. The League of Nations, and later the United Nations, dealt with crises and they sometimes helped to end them. The formation of the North Atlantic Treaty Organization (NATO) and the European Union may have ended Europe's custom of a major war or two in every hundred years. Trials of war initiators (as in Germany after World War II) gave their victims a sense of justice done, revenge not needed. If democracies are less belligerent than tyrannies, it was encouraging that in the year 2000, 140 of the world's nearly 200 nations—seven out of ten—held multiparty elections.

Atomic bombs have not been used since 1945. We hold our breath.

CHAPTER 24

We do the unbelievable.

In 1939, the year when World War II began in Europe, America's biggest city held the New York World's Fair. On a marshy piece of land (once called Corona Dumps) rose a stunning group of buildings, several in the shapes of prisms, spheres, and cones. Here the visitors, who came from all around the world, could see "The World of Tomorrow." What the planners of that fair foresaw, what they thought the future held, gives us some perspective on the triumphs of technology since then.

The planners made some big mistakes. They thought that medicine would conquer cancer soon. Machines would run on liquid air. Everyone would live in cheap and almost weightless houses that you threw away when you no longer needed them. But the planners also got a few things right. They predicted long, sleek, air-conditioned autos speeding over cities on freeways fourteen lanes across. They prophesied that television, only recently invented, would have a place in every home. (Visitors to the opening of the fair watched on tiny black-and-white TVs as U.S. President Franklin Roosevelt gave a speech.) The planners foresaw rockets, but they thought that they'd be fired from cannons and would carry travelers around the earth.

But here's the point. The planners of the fair did not foresee—in fairness who can blame them?—amazing things that World War II would bring about within the next few years. These included radar, penicillin, helicopters, and atomic bombs. Much less did they imagine later wonders: nuclear reactors, transistors, fiber optics, organ transplants . . . on and on.

Above all they did not foresee the unbelievables that are the subjects of this final chapter.

ONE OF THESE incredibles is of course the computer, whose concept is much older than you may have thought. In the 1830s and '40s an Englishman designed (but didn't make) an "Analytical Engine" that he declared would make any calculation that a person wished. In most respects Charles Babbage's engine would have had the same internal "logic" as a modern-day computer, and experts of today believe it would have worked. But Babbage didn't have the cabbage. The government supported his research a while, then stopped. Babbage blamed his problems on the English mind: "If you speak to him [an Englishman] of a machine for peeling a potato, he will pronounce it impossible; if you peel a potato with it before his eyes, he will declare it useless, because it will not slice a pineapple."

As usual, need impelled us humans to invent. In the latter 1800s businesses and governments began to grow so big they needed help in dealing with their data. Processing the 1880 U.S. census took so long—seven years—that the results were out of date when published. In 1890, therefore, the government used a punched-card tabulator to sort and count, and did the job under budget by the deadline. And they did that without electricity. In the early 1900s electricity led to speedy calculators for scientists and engineers, and business machines for billing and accounting. Babbage's vision of an "Analytical Engine" neared fulfillment.

On the eve of World War II inventors in America, Germany, and England were already working on computers. Then the needs of warfare hurried them along. Who was first to make a real computer? That's hard to say, since everything depends on how you use the word *computer.* This writer, who taught at the University of Pennsylvania for many years, gives Penn the credit.

In the early years of World War II the U.S. Army needed a device to calculate the "firing tables" that gunners used when they aimed their cannons. At Penn two engineers set out to make this calculating tool. One of them, John Mauchly, was an assistant professor who dreamed about machines that ran on nothing but electrons. Presper Eckert, a "research associate" and only twenty-two, was known (as much as anything) for having made a system to play chimes in graveyards to drown the noise of crematoriums. These two collected some researchers and began their work.

In 1946 (when the war was over) Eckert and Mauchly finished making their computer. They named it ENIAC, for Electronic Numerical Indicator and Computer. ENIAC was eight feet tall, eighty feet long, and weighed as much as eight ordinary cars. By 1940s standards it was very fast. It carried out 5,000 operations in a second, and computed an artillery shell's trajectory faster than the shell could fly. (But it could be troublesome. It's said that a moth once flew into an ENIAC and short-circuited it, giving rise to the computer expression "bug.")

As soon as brilliant people invented the computer, other brilliant people made it better. One of these was John von Neumann of the Institute for Advanced Study at Princeton. Von Neumann was a universal genius. Certain unsolved problems of the ENIAC entranced him, and he wrote a paper laying out what he believed should be the "architecture" or logic of computers. This included a control to tell the computer what to do and when. Von Neumann's paper was extremely influential, and some have (wrongly) called him "father of the computer."

The astounding new machine began to win attention. In the presidential election of 1952 a computer predicted for national television that Eisenhower would win 438 electoral votes. He won 432.

Just the same, the first computers (ENIAC and those that quickly followed) were too big and, yes, too dumb. A writer for *Popular Mechanics* magazine speculated hopefully in 1949 that the computer might shrink one day to the size of a car. The writer was too pessimistic. By the 1960s, computers were already smaller, smarter, and easier to use. IBM was making sections of computers that could fit in elevators. By 1969 NASA had a computer small enough to squeeze aboard the little craft that landed on the moon—yet smart enough to do its job. However, even in the early 1970s, most computers still were costly, big, and hard to use. Computer makers, such as DEC and IBM, had recognized these problems, and some had started making smaller, more convenient "minicomputers." But even these were difficult to use, and they cost more than ordinary people could afford.

The answer to these problems was the personal computer, or PC, which would transform many lives. The facts about its origin are much debated; we shall simplify. In 1975, a tiny firm named MITS began to sell the kits (MITS kits?) for make-it-by-yourself "microcomputers." The fact that these were hard to assemble, often wouldn't work, and did nothing useful didn't matter. What mattered was the price, which was under $400. The kits sold very well among the horde of young computer hobbyists and helped to increase them to an army, many on America's West Coast. These clever, casual people chatted at the Homebrew Computer Club, bought components at the Byte Shop and ComputerLand, and subscribed to *Byte* and *Dr. Dobb's Journal of Computer Calisthenics and Orthodontics.* People like them thought up acronyms like GUI (graphical user interface), WIMP (windows, icon, mouse, and pull-down), and POTS (plain old telephone service).

In 1975 Paul Allen, a computer whiz about to graduate from Washington State University, saw a story on the new computer kits in *Popular Electronics.* The news that these were on the market galvanized both Allen and a friend of his at Harvard, a pre-law student named Bill Gates. The two of them liked making money nearly as much as they liked designing software. They decided to write a programming language for the low-priced kits, based on an existing program known as BASIC. They finished their program in six weeks of hard work and named it GW[for Gee Whiz]-BASIC. They also formed a software firm, one of hundreds—maybe thousands—at that time, and named it Micro-Soft.

Their programming language was successful, and they sold or licensed it to hobbyists and computer firms, one of which was Apple. Apple was a tiny firm based in a garage in California owned by Steve Jobs's parents. Jobs was then a twenty-one-year-old computer hobbyist with striking confidence. (As a boy of thirteen, he had telephoned William Hewlett of Hewlett-Packard, the big electronics firm, and asked him for components. Hewlett didn't merely give this prodigy the parts he asked for; he offered him a part-time job.) Jobs's friend and partner, Stephen Wozniak, then twenty-six, was a self-taught engineer.

In 1975 Wozniak built his own computer—basically a naked circuit board. He and Jobs christened it the Apple and began to handbuild Apples in the Jobs garage. In 1976 the Byte Shop sold two hundred Apples for them.

Jobs was far from satisfied. Others too were making small computers, mainly for the hobbyists, but Jobs had vision. He foresaw a broader market for the small computers, provided they were made for ordinary people. A computer, Jobs believed, should be useful, fun, and friendly, and small enough to find a place amid the clutter on your desk. Using Jobs's suggestions Wozniak began to build a friendlier computer, Apple II. And this is where Microsoft (the former Micro-Soft) came in because, for a while, Wozniak and Jobs found GW-BASIC what they needed for their easy-to-use personal computer. By this time, 1977, Apple had outgrown the Jobs garage and employed about a dozen people.

The giant IBM Corporation, by contrast, had battalions of men in dark blue suits, and they were not asleep. The firm held back for several years to see if personal computers would replace electric typewriters, of which it sold a million every year. By early 1980 IBM could see the future, and it quickly made a prototype PC and then began production. However, rather than develop its own PC software, IBM bought from Microsoft a license to let buyers of its PCs use Microsoft's operating system. As we'll see, this action—or inaction—had significant results.

The IBM PC immediately enjoyed a huge success, and the company quickly quadrupled production. Alas for IBM, however, other firms could buy most of the same parts that IBM used, so they began to manufacture less expensive "clones" of IBM's PC. IBM did not go under, but it did retrench. The clones reduced the price, and this encouraged wider use of personal computers. Apple, meanwhile, stuck with what it had already; it wouldn't clone the IBM PC. Jobs competed with the clones by making even better software and developing an ultra user-friendly computer, the Macintosh.

By 1982 the personal computer was available to anyone who could afford its fairly modest, always falling, price. In 1983 enthusiasts could find roughly thirty magazines for users of personal computers. The triumph of the small machines was official in January 1983 when *Time* selected the PC as its "Machine of the Year."

In the meantime, though, attention shifted from computers to their software. This happened as the uses for computers rose, but could not have happened if computers hadn't grown so smart. By 1985 the "chips" inside computers that really do the work held up to 275,000 transistors that could carry out 6,000,000 instructions in a second. One little chip was at least 1,200 times as fast as ENIAC, that pea-brained dinosaur, and the processing power of chips was doubling every eighteen to twenty-four months. The growing power of their brains allowed PCs to use a lot more software.

From 1981 to 1984 the PC software market rose from $140 million to $1.6 billion, a more than tenfold surge. The clearest illustration of the crucial role of software is the tale of Microsoft. In 1980 Microsoft employees numbered only thirty-two, but luck was with Bill Gates. When IBM decided (as we saw) not to make the software for its personal computer, it first approached a firm called Digital Research, Inc. When this did not work out mighty IBM went to little Microsoft.

As we saw above, Gates and Allen agreed to license IBM to let users of IBM PCs use Microsoft's operating system. (The deal was complicated, but it turned out to be very good for Microsoft and bad for IBM.) Gates, who didn't have a system ready, bought the software from another firm for $30,000, improved it, and provided it to IBM. Eventually this system would be "bundled" into nearly every IBM PC and every clone. Microsoft would earn from $10 to $50 on each bundle, and Gates would quickly make (what else?) a bundle. At one point in the 1990s Gates's fortune equaled all the worth of the 106 million poorest Americans. By this time Microsoft was making other software that was widely used. It had 14,000 workers, and the value of its stock was nearing that of IBM's.

Americans of course were not the only gainers from computers. They stimulated economic growth on every continent. As we saw in chapter 22, the world now had a rising global economic system, and rising (though more and more unequal) incomes. Personal computers often were the fairy queens who flourished wands and changed the poorest Third World hamlets into glowing symbols of the dawning age. For example, outside Bombay, India, were villages where hunger and infanticide were common. Amid this destitution in the year 2000 was a new development known as SCEEPZ—the Santa Cruz Electronic Export Processing Zone. Here, in SCEEPZ's air-conditioned rooms, well-fed Indian computer programmers turned out software for multinationals on other continents.

The biggest change in the computer was not the way it shrank in body nor the way it grew in mind, but the ways we humans used it. In the 1940s its inventors had intended it for crunching numbers, which is why they named it a "computer." But fairly soon computer makers and business users began to change computers into data storers and manipulators.

Those who made and those who used computers now began to find a million uses for them. By the 1970s computers "processed" words, which for them was just a cinch. In the Gulf War in 1991 computers were using satellites in the sky above them to guide allied soldiers through the trackless deserts. Up to now we have not been able to create computers that can think like humans, but they have begun to master chess. In 1996 the world champion chess player Gary Kasparov defeated a powerful computer named Deep Blue, but in 1997 an improved Deep Blue beat Kasparov in the deciding game of a six-game series. In 2003 Kasparov played a six-game series with the world's best chess computer, Deep Junior, which can analyze 3 million moves per second. The match ended in a draw.

By the year 2000 ordinary humans carried digital assistants in their pockets that contained a million times the memory of the computer on Eagle, the moon lander. Chips now held a million transistors in an area slightly bigger than a postage stamp, and by the time you read this *they* will be out of date.

Now that we had made electronic brains, the next step was to link them. This was something like what happened long ago to early humans: they first developed bigger brains, and then they linked their brains to other human brains by using speech. In the same way, software engineers in the 1960s and 1970s found ways to link *networks* of computer users to other networks. The Internet (as it was later named) is thus a net of nets. In the 1990s it enormously expanded and in theory became a way for one to be in touch with all, instantly, around the world.

The linking or connecting of computers was as vital to the human story as the making of them was. And yet the linkers are as thoroughly forgotten as von Neumann, Gates, and Jobs are known and lionized. For the record, one father of the Internet was J. C. R. Licklider, an MIT professor. "Lick," as he was known, foresaw the benefits of linking nets and envisioned what he called "mechanically extended man."

As the Internet expanded, the problem of discovering the riches in it also grew. In 1989 a British physicist, Tim Berners-Lee, devised the World Wide Web, a means of sharing global information via Internet. Berners-Lee has written, "The vision I have for the Web is about anything being potentially connected with anything." With the Web one navigates among discussion groups, useful information, press releases, pornography, libraries, "club rooms," and miles and miles of junk. Computer users may post data (images, video, words, or sound), or use the data others anywhere have posted. Everywhere on earth people use the Web to buy and sell, amuse, inform and influence, and penetrate the worldwide store of knowledge.

At the latest count the Web probably contained at least five billion pages.

OUR VENTURE INTO space began in dreams. When Russians, Germans, Americans—even a Rumanian—pioneered in making rockets in the 1920s, they were not concerned about their service to their country or the money they might make. No, they dreamed of human flight in space. They had read the fantasies of earlier generations, such as Jules Verne's *From the Earth to the Moon* and H. G. Wells's *The War of the Worlds* (about invading Martians) and *The First Men in the Moon*. The rockets they were building were to be man's means to see the dark side of the moon and learn if there were men on Mars.

As late as the 1920s and 1930s many found the very thought of rockets quite absurd. In 1919 Robert Goddard, an American inventor who now is sometimes called the "father of rocketry," published his classic work, *A Method of Reaching Extreme Altitudes*. The *New York Times* responded with a jocular editorial. "That Professor Goddard . . . does not know the relation of action to reaction and of the need to have something better than a vacuum against which to react—to say that would be absurd. Of course he only seems to lack the knowledge ladled out in high schools." People nicknamed him "Moony" Goddard.

Even after World War II, the dream of human flight in space seemed much like science fiction. *Science Digest* guessed in 1948 that "Landing and moving around the moon offers so many serious problems for human beings that it may take science another 200 years to lick them."

But German engineers had run a rocket—that is, missile—program both before and during World War II. In the spring of 1944 they had fired the world's first medium-range ballistic missiles across the English Channel at anything in England they might chance to hit. At the end of the war, when the allied armies entered Germany, both Russia and America were planning missile programs of their own. The Russians reached the German rocket center first, but the finest of the German rocketeers had fled westward and surrendered to the Americans. (Stalin menacingly demanded, "How and why was this allowed to happen?")

Among the Germans whom the Americans carried off was Wernher von Braun. As a little boy in Germany, von Braun had been already space-obsessed, and one time he had fastened rockets to his wagon, causing an explosion. When he was eighteen, an article he read on travel to the moon provided his vocation. Later he was technical director at a weapons center, and he and his team of engineers developed the missiles that the Nazis made and fired across the Channel. (A satirical American, Tom Lehrer, later wrote, somewhat unfairly, "Once the rockets are up / Who cares where they come down? / That's not my department / Says Wernher von Braun." Soon the former German weapon maker was working for America.

But it was Russia that was first to enter space, and this happened in the context of the Cold War. In the decade after World War II Russian scientists and engineers had beaten the Americans in the race to build an intercontinental missile. What they made was basically a mighty rocket, impelled by twenty separate engines, carrying a two-ton atom bomb. But now the Russians used the rocket not for warfare but to enter space.

In the middle of the night on October 4, 1957, a bugle sounded, flames erupted, and, with a roar like rolling thunder, Russia's rocket lifted off. It bore aloft the earth's first artificial satellite, a shiny sphere the size of a basketball. Its name was *Sputnik,* meaning "companion" or "fellow traveler" (through space). The watchers shouted, "Off. She's off. Our baby's off!" Someone danced; others kissed and waved their arms. The rocket and the *Sputnik* disappeared and then they parted from each other. For several months *Sputnik* circled earth every hour and a half, transmitting beeps.

Now the Americans had to hurry up and match the Russians to prove the superiority of capitalism and freedom. They scheduled the takeoff of a satellite in Florida in December 1957, two months after *Sputnik*'s triumph. Reporters gathered from around the world, and everybody waited through two days of rain and "holds" until at last the moment came. The rocket rose four feet, paused, sank to earth, collapsed, and burned. The little satellite, unwisely christened *Vanguard,* rolled away, stopped, and chirped. An English newspaper headlined this fiasco as "Kaputnik," and a paper in Japan described it as a "A Pearl Harbor for American science." At the United Nations Russian envoys asked if the United States would welcome aid for less developed countries.

The Russians also had their failures, but for decades they would keep the worst one secret. In the fall of 1960 they endured the greatest horror in the history of exploration. Technicians had prepared three rockets to be sent to Mars, but two of them had fizzled after launching. When they tried the third one it would not ignite. The commander ordered his technicians to examine it up close. It suddenly exploded, killing scores and maybe hundreds of his personnel.

In the years that followed *Sputnik,* space technology advanced in leaps. The United States, having failed the first time, sent aloft a tiny satellite in 1958, and later they recovered a U.S. capsule after it returned to earth. The Russians photographed the far side of the moon, and they orbited two dogs and brought them down alive. The Americans launched a communications satellite to send TV over the Atlantic.

Satellites like this would soon permit twenty-four-hour instant global communications; they drew the continents together. They also made it possible to fully map the earth, learn much more about the earth's ecology, and discover the alarming changes in its atmosphere. (Even in the early 2000s, the most important exploration done in space so far was that regarding planet earth.)

In 1961 the Russians had their greatest triumph in the race to space: they launched a "cosmonaut" (a Russian astronaut) into orbit. Courageous Yuri Gagarin orbited the earth in an hour and a half, then parachuted to a pasture in central Russia. Three decades later an official would recall that day: "I was in school. All lessons ended. Even today I can recall that tremendous joy . . . the tears we all had in our eyes—because we were the first, because we were in space. . . . *almost everyone* believed that in twenty years we would surpass the U.S. in every way. . . . We would become Country Number One in the world—and socialism would be utterly victorious"

For America, the Russian triumph was a shock. Two days later President Kennedy held a frenzied meeting with officials. He asked them, "Is there any place where we can catch them? . . . Can we go around the moon before them? Can we put a man on the moon before them? . . . Can we leapfrog? . . . Let's find somebody, anybody! I don't care if it's the janitor over there, if he knows how." A month and a half later Kennedy addressed Congress asking Americans to bear the costs of sending humans to the moon. He proposed that America "should commit itself to the goal, before this decade is out, of landing a man on the moon and returning him safely to earth. No single space project will be more impressive to mankind, or more important for the long-range exploration of space."

As Kennedy knew very well, flying humans to the moon didn't necessarily make sense. Many experts argued then and later that it would be cheaper, safer, and scientifically more fruitful to send robots to the moon or Mars, rather than a man. But Kennedy, a skillful Cold War politician, knew that he must not permit the Russians to be first to put a man on the moon. It didn't take a rocket scientist to figure that out. With von Braun in charge, Americans prepared to make the journey. They orbited two monkeys, Able and Baker, and a chimpanzee called Ham, and then they sent up astronauts on little flights. Then the U.S. program matched the Russians' triumph with Gagarin when, in 1962, it put an astronaut, John Glenn, in space. In five hours he circled earth three times and then descended (as was planned) in the ocean east of Florida. Soon after Glenn many other astronauts orbited the earth, and in 1968 Americans made the first manned voyage around the moon. They sent pictures of its rocky surface back to earth, where they were shown on television.

But Kennedy had promised "on the moon," not around it, and "before this decade is out." In July of 1969 the United States launched a three-man crew whose destination was the surface of the moon. Four days later Neil Armstrong and Edwin Aldrin landed on it in a "lunar module" while Michael Collins orbited aloft and waited for them. Mission leader Armstrong clambered down a ladder, trod in lunar dust, and spoke his famous line, "That's one small step for [a] man, one giant leap for mankind." Aldrin joined him, and a TV camera in the module filmed the two men for the millions watching on the earth. They raised a U.S. flag, gathered soil and rocks, and photographed everything in sight. Before they left for home the earthlings set in place a plaque that reads, "Here men from the planet Earth first set foot upon the Moon July, 1969 A.D. We came in peace for all mankind."

A science writer (Wilson da Silva) tells us that after the moon landing he, then a little boy, and his grandmother stood outdoors one night gazing at the moon. She'd been born before the Wrights first flew their little plane in 1903. She said, "You know, they didn't really go to the moon." He was puzzled so he asked, "You mean, the astronauts?" She nodded. "It was done in a film studio. It had to be. How could someone get to the moon and back?"

For America and Russia both, what next? The moon was mastered, and our neighbor Mars, which might hold "life," was (then at least) too far away for human flights. So both the rivals chose the same project, and in the 1970s each built a "station" about 250 miles away in space. These stations were short-lived, but in 1986 the Russians built another, and they maintained it for a dozen years. In 1998 a group of nations led by the United States began to build an "international" station as big as the passenger cabin on a jumbo jet. To build these stations astronauts in spacecraft known as "shuttles" left initial pieces of the stations out in space. On later flights they snapped on the additions as one does with Lego parts.

Strangely, no one ever fully clarified the purpose of these stations, though the Russians did use theirs to study how to prevent the breakdown of one's weightless bones in space. The Americans also studied how to live in space, and carried out a host of small experiments involving weightlessness. Defenders of the stations claimed the stations gave humanity a foothold off the earth, a beachhead in the heavens.

Critics claimed, as they had said before about manned journeys to the moon, that the stations were not worth their cost. (One shuttle flight cost half a billion dollars.) The experiments performed in space didn't teach us much, and why prepare for flights by astronauts to other planets (if indeed that was a goal) when an interplanetary flight would last too long for humans to endure? (Even getting to our nearest neighbor, Mars, would take the best part of a year.) The money, they maintained, would be better spent, and more would be learned, by sending robots to the planets. (And this was being done, as we'll see below.) They also pointed to the danger to the astronauts of flying shuttles to the stations.

As if to prove the critics right about the danger, the shuttle *Challenger* blew up in 1986 a minute after takeoff, killing seven astronauts. In 2003 *Columbia* fell apart before a landing, also killing seven. As with other space disasters, these two troubled many humans, giving them a sense of having tempted fate (or angered God). Already humans had unleashed the atom's power, peered inside the molecule that holds our genes, invented ways to store and use a vast amount of information. God had ordered us to "Fill the Earth," not to leave it. Were we like the Babylonians who tried to build the tower of Babel "with its top in the heavens," only to have God foil their plan? Or the ancient titan who infuriated mighty Zeus by stealing fire from gods and giving it to humans? Zeus enchained Prometheus and had an eagle feed forever on his liver.

As scientists insisted, America, Russia, Europe, and Japan did send unmanned space machines to land on planets, or to fly beside and photograph them, or to otherwise explore our solar system. They visited the smaller planets, Mercury, Venus, and Mars; and those giant balls of gas, Jupiter, Saturn, Uranus, and Neptune. (Far-off Pluto still awaits us.) One craft that did this planet-touring left the solar system during 1983 (as planned), and plunged inside the cosmic ocean. According to one calculation, at its present speed it should soar near a star named Aldebaran in the year 8,001,972. It bears a plaque with drawings of a man and woman and a map of our solar system intended to help a far-off creature who might find the plaque to work out earth's location.

Astronomers throughout the world are living in their golden age. Using giant telescopes, artificial satellites; and radio, gamma, infrared, ultraviolet, and X-rays they now can peer far out in space. One of their new tools has been the Hubble Space Telescope, which was set in earthly orbit in the 1990s. Because it soars above earth's atmosphere the Hubble gave much clearer images of things in space than any telescope before it.

However, giant telescopes on earth with better eyesight (aided by computer optics) now complement the Hubble. The telescopes have found titanic happenings in the universe: (in rising order of importance) a comet hitting Jupiter, a tempest raging over Saturn, stars exploding, other stars and planets forming inside hells of gas and dust, galaxies—whole galaxies—devouring one another, and the fireball of the Big Bang when the universe began. Astronomers have arrived at bold new theories, new understandings, about the universe.

What the layman really wants to know about the universe, above all other things, is this: are there other intelligent beings out there, or are we all alone? Since the middle 1990s some astronomers have looked for planets around the nearby stars. (Their purpose was to understand the physics of the forming of planets, not to look for life.) They have found many planetlike objects. Seemingly these findings raise the statistical likelihood that intelligent life exists elsewhere in our galaxy. Other scientists have been trying to eavesdrop on possible interstellar radio communications among conjectured civilizations elsewhere in our galaxy.

If your goal is finding life outside of earth, you must discover planets that are much like ours. They must have liquid water and they can't be broiling hot, like Mercury and Venus; or mostly gas, like Jupiter, Saturn, Uranus, and Neptune; or sheathed in ice, like Pluto. The planets that astronomers have found so far don't pass these tests.

EVEN AS WE humans peered around a universe that holds a hundred billion galaxies we also probed in cells that (in the case of humans) measure one 250,000th of an inch across.

This is the story of the finding of our genes, and it begins about 150 years ago when Gregor Mendel bred his peas. Mendel lived while Darwin did and, like Darwin, in his early years he didn't show much promise. At the age of twenty-one he became a monk in what is now the Czech Republic. When he took an examination for a teaching license in his town of Brno, he failed it, and every time he took the test again he failed. Nevertheless he taught science in a local school, and went to meetings of the local science club.

In 1856 Mendel began to study the inheritance of traits, using the materials he had at hand, which were pea plants in the monastery garden. For about a decade Mendel crossbred plants some 20,000 times. He followed seven of their traits, including height, the color of their flowers, and the shape of their seeds. This is what he learned concerning height. If he crossed tall pea plants with dwarf ones, all of the resulting plants were tall. Strangely, though, if he then crossed his hybrid tall plants with each other, only three-quarters of the resulting plants were tall. The other quarter were dwarfs.

Mendel eventually concluded that inside every plant were unseen "factors" that controlled the seven traits. (In the case of height, factors existed for tallness and dwarfism.) When plants were bred, their factors didn't blend and disappear in their offspring; they kept their identity. The offspring either did or didn't manifest the factors they inherited, in accord with simple rules that Mendel figured out. These rules explained, for instance, why three of the offspring of tall hybrid plants were tall and one was dwarfed. Mendel guessed the unseen factors were located in our germ cells.

In 1866 he published a paper called "Experiments with Plant Hybrids" in the journal of the Brno science society. But publication in such a minor journal was tantamount to burial, and almost no one read it. He then was chosen abbot of his monastery and had little time for more research. When he died in 1884 he was practically unknown.

In spite of Mendel's brilliant work, therefore, our interest in the workings of our cells was probably inspired not by him but by Darwin's celebrated *Origin of Species*. Darwin didn't really know how plants and animals passed on their traits, and when he made a guess he got it wrong. But he excited everyone about evolution and inspired later scientists to learn the facts about heredity.

Unlike Mendel, most of the researchers of the following generations studied heredity on the level of the cell. Thanks to recent improvements in microscopes, they now could dimly see the nuclei of cells, and a German scientist discovered how to stain a nucleus with dyes and make what lay inside more visible. He discovered threadlike shapes and oberved that just before a cell divided these shapes split longitudinally in half.

In 1900 three botanists independently discovered the report that Mendel had published so obscurely a third of a century before. (The fact that three men found it in one year suggests how quickly cell research was moving.) Researchers quickly realized that Mendel's findings about unseen "factors" (for example, that they didn't blend) neatly matched what they were learning about those strands that split apart. Scientists soon recognized Mendel as the "father of genetics," although statisticians found that he had sinned by leaving out the data on some traits that didn't fit his model.

Between 1900 and the early 1950s scientists discovered, among many other things, that genes (Mendel's "factors") provide the recipes for making the proteins that really do the work of cells. They also learned that genes are sited in a polymer or large molecule known as DNA. But how did molecules of DNA store the genes, and how did they pass them on when the cells they were in divided? To learn the answers someone had to figure out the structure of the molecules. But this would be no easy task. A molecule of DNA is so tiny that a million of them side by side would be as wide as a normal sewing thread.

In the early 1950s British scientists used X-rays to study DNA. Their hazy photos showed that the shape of DNA crystals was a helix (or spiral). This information was important. But the X-rays offered no hint of how the compounds that form the molecule of DNA fit against each other, or how they permit the molecule to duplicate itself.

Francis Crick, a youngish Briton, and James Watson, an even younger American, got to know each other at England's Cambridge University in 1951, and started to research the structure of DNA. Instead of using test tubes in a lab, they placed and endlessly replaced bits of wire and cardboard, beads, and metal plates. They were trying to build a model of the molecule that matched what was known already about its chemistry and what the X-ray crystallographers were learning about it. Watson's splendid memoir of their research (*The Double Helix*) suggests that even though the two were quite obsessed with DNA, they tinkered with the model in the intervals between their fencing lessons, tennis, movies, skiing in the Alps, and chats with pretty women.

Bit by bit they figured out the structure. They concluded that a molecule of DNA is shaped like a double helix, or a twisted ladder. It consists of two connected strings, each of which in humans is two yards long and made up of four types of tiny compounds. Scientists would later learn that hundreds, even millions, of these compounds—just a tiny bit of the strings—form a single gene.

What a DNA molecule does when its cell divides is as clever as a zipper. Before the cell divides, the two DNA strands must duplicate themselves. So the two strings separate, and then each string collects the needed elements and makes a new partner string just like the string it lost. Then one double string goes to one of the new daughter cells, and the other string goes to the other. The two DNA molecules (in two cells) are like the original in all respects.

This solution was (said Watson) "too pretty not to be true," and on the morning of February 28, 1953, Crick and Watson decided that they had it right. (The writer is typing these words on the morning of February 28, 2003.) When they went to lunch at a favorite pub Crick announced to everyone that they "had found the secret of life."

How much we, the human race, had learned about ourselves in just a hundred years! During a little fraction of our total time on earth so far, Darwin and a host of others had discovered how we and other species had evolved. Mendel and others had learned that genes inside our cells control our traits. Now Crick and Watson (building on the work of others) had described the molecule that holds the blueprint for maintaining life. (And others soon would learn much more about our genes.)

Scientists now began to work on medical applications of this knowledge. They found that many diseases result at least in part from missing or defective genes, and they began to search for ways to treat them. Starting in 1991 doctors injected "vectors," usually prepared viruses, into a patient's cells. The vectors spliced "good" genes exactly where they should be in the patient's DNA, and the good genes would, the doctors hoped, direct his cells to make the enzyme that was needed. This was promising, but by the early 2000s gene therapy still hadn't had one clear success.

But technicians now manipulated cells in animals to make them manufacture insulin (for diabetics) and human growth hormones and a treatment for leukemia. Doctors started using genes (usually experimentally) to revive dead heart tissue and attack cancer cells. And in labs around the world scientists were attempting to grow human vessels, valves, and muscle in their labs.

And then, specific knowledge of our genes dramatically increased. In 1990 biologists in America had begun to map our genome: all the genes in human DNA. A publicly sponsored Human Genome Project competed with a private company called Celera Genomics to be the first to map the genome. In the year 2003, much faster than expected, both of them finished. The rivals agreed that their products "complemented" each other, which laymen found a little hard to understand. In any case, now that so much was known about our genes, biotech companies suddenly discovered that there were more diseases potentially treatable with genes than they could take on all at once. They had to choose which targets they should aim at first.

Meantime what once had been the dream of crazy kings and egomaniacs began to look dimly possible. Consider these events: 1953, artificial human insemination; 1978, a baby conceived in a test tube; 1984, a baby born from sperm that had been frozen; 1998, a U.S. scientist clones one of his own cells in a cow's egg; 1990s, doctors test for defective genes, so that if there is a problem the mother may abort the fetus; 2001, a human embryo is cloned (for medical research) and develops to six cells before it dies. Was other human cloning on the way? Human cloning, even in the test tube, for research, was proving hard to do.

AS THE THIRD millennium began, humankind had reached the unreachable, worked the unworkable, feased the unfeasible. Among a host of other marvels we had learned to organize and make available the information gathered by six billion human beings. We had scouted in the Milky Way, learning more about the evolution of the universe, searching for the origins of elements and life, seeking possibly for other life and for some universal scheme of things. We had found the book of life, our genome, and were on the verge of using it to treat disease and lengthen life.

And what lay in our future? Wait, you say, no one knows the shape of things to come. Yes, that's true. But the things that I will mention here are not prophecies but extrapolations. Given what has recently occurred, these things are likely to follow.

We will make computers out of molecules, yes molecules, that we turn "on" and "off." These molecules may manage tiny robots swimming in our veins and doing chores. Computers will enable engineers to "see" the stresses in a beam and surgeons to "see" inside a brain. "Books" that you hold in your hand will contain the contents of libraries. We will fire a missile at a comet. We will fly to Mars, and some of us may even choose to live there. We'll discover planets that look as if they could support life. We'll learn astounding things about the universe.

Here's more: parents won't just take what comes; they will plan their babies, adding genes for beauty, brains, and longer life. We will live two or three times as long as we do now. We will order cells to grow new kidneys, hearts, or bones as needed, just as salamanders grow new tails. We will clone humans for medical research, and some will want to do it for other purposes.

What's the point of making these extrapolations? I hope they help my readers to see the recent past (from World War II until today) in full. I would like you not only to see the changes in these decades as they happened—the usual perspective—but also to view them from the future by seeing what astounding outcomes they may lead to. This stereoscopic view should make this clear: in the decades after World War II our species crossed a line. Of course, as individual humans we didn't change; we look and feel the way we did before. But as a species we achieved a previously undreamed-of mastery of life.

One sometimes hears that "Humans weren't here at the beginning, and we won't be here at the end." But is that prediction true? One has to wonder: given our growing mastery, will our species ever let another species wipe us out? If any species does destroy us, it will surely be our own.

Epilogue: So Far So Good

From Labrador to Coral Sea
Our lives were stunted, bleak, unfree.
We shared our huts with rats and fleas
And lost our children to disease.
(Our holy men would sigh and nod
And tell us, "That's the will of God.")

But then, with steam, vaccines, and votes,
Our fortunes rose like tide-raised boats.
We'd more to eat; drew breath more years;
Dethroned (or worse) our tsars, emirs;
Sent men and mirrors as our eyes
To search the black galactic skies;
And in our cells, till then unseen,
We found our Fates, our djinns: our genes.

The world's still cruel, that's understood,
But once was worse. So far so good.

Recommended Reading

CHAPTER 1 • *We fill the earth.*

For a general view of early human life, see Roger Lewin, *In the Age of Humankind: A Smithsonian Book of Human Evolution* (1988); and recent editions of John E. Pfeiffer, *The Emergence of Humankind;* and Bernard Campbell, ed., *Humankind Emerging.* The Pfeiffer and Campbell books are textbooks but never mind, they are also good reading.

Many books have illustrations of cave paintings. A good example is Mario Ruspoli's *The Cave of Lascaux: The Final Photographs* (1987). Jean-Marie Chauvet, et al., *Dawn of Art: The Chauvet Cave: The Oldest Known Paintings in the World* (1996) is an engaging account of the finding of paintings.

CHAPTER 2 • *We gather by the rivers.*

For Sumer, Samuel N. Kramer's *The Sumerians: Their History, Culture, and Character* (1963) is a well-known scholar's enthusiastic introduction. Leonard Cottrell, *The Quest for Sumer* (1965) tells how archaeologists rediscovered Sumer under silt and sand. The Gilgamesh epic offers a glimpse at what Sumerians believed about four thousand years ago.

For Egypt, try Leonard Cottrell, *Life under the Pharaohs* (1960); Zahi Hawass, *The Mysteries of Abu Simbel: Ramesses II and the Temples of the Rising Sun* (2000); and Christiane Desroches-Noblecourt, *Tutankhamen: Life and Death of a Pharaoh* (1984). And look for books with first-rate photos of Egyptian art.

CHAPTER 3 • *The wanderers settle down.*

The first thing to read about ancient Israel is of course the Bible. I would read the Hebrew Bible, or Old Testament, through Kings 1 and 2, skipping Leviticus, Numbers, and Deuteronomy, and then the books of Job, Hosea, and Amos. *The New Oxford Annotated Bible* (1962) has helpful notes. Modern histories of ancient Israel abound. Antony Kamm, *The Israelites: An Introduction* (1999); and John Bright, *A History of Israel* (3rd ed., 1981) are good.

CHAPTER 4 • *Two ancient cities follow diverse paths.*

Before looking at any books about the ancient Greeks by others, read what they wrote about themselves. Herodotus's *History* and Thucydides's *History of the Peloponnesian War* (edited in translation by Richard Livingstone, 1943) are good reads, but they do have dry stretches. You may skip when bored. You won't be tempted to skip when reading the tragedies of Aeschylus, Euripides, and especially Sophocles. You should also read at least a dialogue or two of Plato, especially *Phaedo,* which takes place on the last day of Socrates's life, as he awaits his execution.

After sampling them, try one of these excellent short histories: H. D. F. Kitto, *The Greeks* (1951)—this is a gem; M. I. Finlay, *The Ancient Greeks* (1963); Peter D. Arnott, *An Introduction to the Greek World* (1967); or Peter Green, *Ancient Greece: An Illustrated History* (1973).

CHAPTER 5 • *China excels and endures.*

Chinese history is so vast that approaching it is like trying to find the best route up a Himalayan mountain. Two historians who show nearly the whole bulk of it are John King Fairbank, in his *China: A New History* (1992); and Charles O. Hucker, in *China's Imperial Past: An Introduction to Chinese History and Culture* (1975). Jonathan D. Spence's many books on Chinese history are interesting and readable. *When China Ruled the Seas: The Treasure Fleet of the Dragon Throne, 1405–1433* (1994), by Louise L. Levathes, delivers even more than the title promises.

See also the 1700s novel by Cao Xueqin called *The Dream of the Red Chamber.* (A translation by David Hawkes and John Minford is called *The Story of the Stone.*)

CHAPTER 6 • *Some attempt to rule us all.*

The Persians left us almost no description of their deeds, and neither did Alexander's army or the Mongols. The Romans, however, tell their story well. Try some of Plutarch's *Lives,* Cicero's letters and speeches, and Suetonius's racy *Lives* of the first emperors.

We need a good, short book about the Persian empire. Peter Green's *Alexander the Great* (1970) is short and readable, and so is W. W. Tarn's older book (1948) of the same name. For the Mongols, Leo De Hartog, *Genghis Khan: Conqueror of the World* (1989) is up to date, but Michel Hoang's first-rate *Genghis Khan* (trans. Ingrid Cranfield, 1990) is a better read. One of the most famous histories ever written is Edward Gibbon's *The Decline and Fall of the Roman Empire* (1776–88). It is very old, very fine, and six volumes long.

CHAPTER 7 • *We found the worldwide faiths.*

To get to know the "book" religions one must read the books. For Hinduism, one of these is an ancient tale of feuding gods and men, the *Mahabharata* (mah-ha-BA-ra-ta). William Buck's very free version of this epic is splendid. The *Dhammapada* is a brief anthology of sprightly Buddhist teachings. For Christianity, you should read some of the New Testament. I suggest the Gospel of Luke and Paul's letter to the Christian community at Rome. For Islam, read the earliest-written chapters of the Qur'an, which usually appear at the end.

These are good introductions: A. L. Basham, *The Wonder That Was India* (3rd. ed., rev., 1968); Michael Grant, *Jesus: An Historian's Review of the Gospels* (1977); Edith Hamilton, *Witness to the Truth: Christ and His Interpreters* (1948), which is good despite its piety; Maxime Rodinson, *Muhammad* (trans. Anne Carter, 1971); and Karen Armstrong, *Muhammad: A Biography of the Prophet* (1992).

CHAPTER 8 • *Europe prepares for its big role.*

Two famous epic poems reveal medieval customs and ideas. The *Song of Roland* (c. 1100) tells how a fictitious rearguard of French knights gave their lives to block a Muslim army, and Dante's *Divine Comedy* (c. 1310–14) is a tour of Hell, Purgatory, and Heaven. (Hell is best.) Machiavelli's *The Prince* and More's *Utopia* are short and winning. Despite its dreary title, *Pursuit of Power: Venetian Ambassadors' Reports on Spain, Turkey, and France in the Age of Philip II, 1560–1600* (1970) nicely shows how men of power saw their time. (The author of *The Human Story* chose the reports and translated them.)

Desmond Seward, *The Hundred Years War: The English in France, 1337–1453* (1978) is brief and clear. The first half of Emmanuel Le Roy Ladurie, *Montaillou: The Promised Land of Error* (1975), offers haunting glimpses of the lives of real people in a mountain village caught up in a nightmare. The mini-lives in Eileen Power's *Medieval People* (1924) are more cheerful. John Man describes the invention of the printed book in *Gutenberg: How One Man Remade the World with Words* (2002).

CHAPTER 9 • *We find each other.*

Travel and Discovery in the Renaissance (1962), by Boies Penrose, tells its story well. J. H. Parry's *The Discovery of the Sea* (1974) is dryer, yet engaging. Samuel Eliot Morison fitted and sailed a ship like one of Columbus's before writing his *Admiral of the Ocean Sea* (2 vols., 1941) about Columbus. It won a Pulitzer Prize. (A short abridgment is called *Christopher Columbus, Mariner.*)

William Manchester writes readably but not always reliably about Magellan and his times in *A World Lit Only by Fire: The Medieval Mind and the Renaissance Portrait of an Age* (1992). Alan Moorehead tells about Cook's voyages in *The Fatal Impact: An Account of the Invasion of the South Pacific (1767–1840)* (1966), and so does Alan J. Villiers in *Captain James Cook* (1967).

Daniel J. Boorstin, *The Discoverers* (1983) deals not only with Columbus et al. but also with "mankind's need to know" all kinds of things. The book is long but interesting, and its notes on other books are useful.

CHAPTER 10 • *The New World falls to the Old one.*

Frances Gillmor, *The King Dances in the Marketplace* (1964), tells about the cruel and austere grandfather of the last Aztec emperor, who had the same name. Perhaps the best account of Cortés's conquest was written by one of the Spanish soldiers who took part in it, Bernal Díaz del Castillo. His *The Conquest of New Spain* is full of blood, sweat, and wonder. For Pizarro and Peru perhaps the best thing is Birney Hoffman, *Brothers of Doom: The Story of the Pizarros of Peru* (1942), which is a much shorter retelling of William H. Prescott's classic *History of the Conquest of Peru* (1847).

Alfred W. Crosby's *The Columbian Exchange: Biological and Cultural Consequences of*

1492 (1972) manages to be short and readable and also a major contribution to our understanding of world history.

CHAPTER 11 • *We suffer famine, war, and plague.*

Most books on population history were written by experts for other experts, but E. A. Wrigley's *Population and History* (1969) is an engaging introduction. The opening chapters of Thomas Malthus's famous *Essay on the Principle of Population* (1798, but easily found) are gloomy but stimulating; they have influenced many people.

Many readable books deal with medicine, disease, and history. *Disease and History* happens to be the title of a short and anecdotal book by Frederick F. Cartright (1972). Hans Zinsser's *Rats, Lice and History* (1934) is bright and zany. W. H. Lewis, *The Splendid Century: Life in the France of Louis XIV* (1953), deals pungently with many things, including dirt and doctors.

CHAPTER 12 • *We learn who we are and where we live.*

The Sleepwalkers (1959), by Arthur Koestler, nicely covers the whole story of the Copernican revolution, though Kepler is Koestler's hero. (The abridged version *is The Watershed: A Biography of Johannes Kepler.*) *Sun, Stand Thou Still* (1947), by Angus Armitage, is a short life of Copernicus. Dava Sobel pictures Galileo, his findings, and his conflict with the Church in *Galileo's Daughter: A Historical Memoir, of Science, Faith, and Love* (1999), using letters from Galileo's daughter, a cloistered nun. Michael White writes simply (which isn't easy in this case) about *Isaac Newton: The Last Sorcerer* (1997.)

Gertrude Himmelfarb's *Darwin and the Darwinian Revolution* (1959) is among the best of many Darwin biographies. *Darwin and the Beagle* (1969), by the always winning historical writer Alan Moorehead, is short and nicely illustrated.

CHAPTER 13 • *Here and there, the people rule.*

Robert R. Palmer's *The Age of the Democratic Revolution: A Political History of Europe and America, 1760–1800,* vol. I (1959) is first-rate on the French and U.S. revolutions and the launching of democracy. Despite its silly title, Catherine Drinker Bowen's *Miracle at Philadelphia: The Story of the Constitutional Convention May to September 1787* (1966) is sensible and readable. A good book on slavery in the United States is John Blassingame's well-illustrated *The Slave Community: Plantation Life in the Antebellum South* (1979).

Georges Lefebvre's *The Coming of the French Revolution* (trans. Robert R. Palmer, 1949) is short and taut, while Simon Schama's *Citizens: A Chronicle of the French Revolution* (1989) is long and anecdotal. Irene Nicholson's *The Liberators: a Study of Independence Movements in Spanish America* (1968) concentrates on "thoughts and emotions."

CHAPTER 14 • *We make more and live better.*

For the big picture, nothing is so clear and readable as Robert L. Heilbroner's short *The Making of Economic Society* (1962). The early chapters of *Industry and Empire (From 1750 to the Present),* by Eric J. Hobsbawm (1968), are good on England's industrial revolution. Matthew Josephson's *The Robber Barons: the Great American Capitalists* (1934) recounts with verve their deeds and misdeeds. The second and third volumes of Daniel J. Boorstin's *The Americans* (1965, 1974) have readable chapters on economic life. James C. Davis's *Rise from Want: A Peasant Family in the Machine Age* (1986) tells how industrialization changed lives in what is now northeast Italy.

CHAPTER 15 • *The richer countries grab the poorer.*

Heaven's Command: An Imperial Progress (1974), by James Morris, is the first of three impressionistic volumes on the British Empire. *The Reason Why* (1953), by Cecil Woodham-Smith (a woman), is a short and perfect book about two English brothers-in-law, both of them noblemen and generals, whose stupidity led to the disastrous "charge of the Light Brigade" in a war between imperial powers.

The Scramble for Africa 1876–1912 (1991), a long book by Thomas Pakenham, tells the stories of a crowd of vivid people. In *King Leopold's Ghost* (1998), Adam Hochschild tells the story of the Belgian ruler's exploitation of the Congo. The middle third of Edwin O. Reischauer's *Japan: The Story of a Nation* (4th ed., 1990) is useful on Japanese empire building, as is Ian Buruma's brief *Inventing Japan, 1853–1964* (2003).

CHAPTER 16 • *We multiply, and shrink the earth.*

Two general histories of the twentieth century are J. M. Roberts's thoughtful *Twentieth Century: the History of the World, 1901 to 2000*; and Paul Johnson's *Modern Times: the World from the Twenties to the Nineties* (1991), which is rich in piquant details.

Cecil Woodham-Smith's *The Great Hunger: Ireland 1845–49* (1962) is splendid, nearly as good as her *The Reason Why.* The same writer's *Florence Nightingale 1820–1910* (1951) tells how one extraordinary woman changed the medical profession. *Microbe Hunters* (1926), by Paul de Kruif, tells about early discoveries of the causes and cures of diseases.

Man on the Move: The Story of Transportation (1967), by tire maker Harvey S. Firestone Jr., is a breezy overview. A book to browse in, not as heavy as its title, is James E. Vance's *Capturing the Horizon: The Historical Geography of Transportation Since the Transportation Revolution of the Sixteenth Century* (1986). *The Railway Journey: The Industrialization of Time and Space in the 19th Century* (1977), by Wolfgang Schivelbusch, is a readable academic book.

CHAPTER 17 • *We wage a war to end war.*

In *The Long Fuse: An Interpretation of the Origins of World War I* (1965), Laurence Lafore engagingly discusses the big question. Barbara W. Tuchman's *The Guns of August* (1962) describes the outbreak of the war; it won a Pulitzer Prize. *Illustrated History of the First World War* (1964), by A. J. P. Taylor, tells the story briefly, and so does Cyril Falls's opinionated *The Great War* (1959). Alan Moorehead, *Gallipoli* (1956), describes the Dardanelles campaign.

Good-bye to All That (1929) is a moving memoir by Robert Graves, who served in the war as a British army officer.

CHAPTER 18 • *A utopia becomes a nightmare.*

On Marx, Engels, and others see Robert L. Heilbroner's highly readable *The Worldly Philosophers: The Lives, Times, and Ideas of the Great Economic Thinkers* (1953); and Edmund Wilson's *To the Finland Station* (1940).

Alan Moorehead, *The Russian Revolution* (1958) is good reading. In *Nicholas and Alexandra* (1967), Robert K. Massie writes vividly about the tsar, his silly wife, and the Russian Revolution. Brian Moynahan, *The Russian Century: A History of the Last Hundred Years* (1994) is short and slashing. For an eyewitness look at Russia under Stalin see John Scott, *Behind the Urals: An American Worker in Russia's City of Steel* (1942).

In *The Anatomy of Revolutions* (1965), Crane Brinton finds similarities between the English civil wars in the 1600s, the French Revolution, and the Russian Revolution.

CHAPTER 19 • *A Leader tries to shape a master race.*

According to his biographer, Ian Kershaw, Hitler is now the subject of 120,000 articles and books. Kershaw's *Hitler* (2001) is short, dry, and up to date. I prefer Alan Bullock's older *Hitler: A Study in Tyranny* (abridged, 1971). After World War II, Milton Mayer interviewed ten German friends, and wrote a fine book with a revealing title: *They Thought They Were Free: The Germans 1933–45* (1955).

On Hitler's Italian friend, see Paolo Monelli's *Mussolini: an Intimate Life* (trans. Brigid Maxwell, 1953).

CHAPTER 20 • *We wage a wider, crueler war.*

In *The Origins of the Second World War* (1961), A. J. P. Taylor argues that the outbreak of war in Europe resulted from blunders by both Hitler and his enemies. John Keegan's military history, *The Second World War* (1989), is crisp and interesting. Cornelius Ryan, *The Longest Day: June 6, 1944* (1959) describes the Normandy invasion. Two of my former University of Pennsylvania colleagues have written readably about American airmen and airborne: Thomas Childers, *Wings of Morning: The Story of the Last Bomber Shot Down over Germany in World War II* (1995); and Martin Wolfe, *Green Light! Men of the 81st Troop Carrier Squadron Tell Their Story* (1989). H. Trevor Roper's *The Last Days of Hitler* (1962) tells what happened in the bunker as the Russians neared.

On the physics that led to the nuclear bomb, C. P. Snow, *The Physicists* (1981), is short and easy reading. Richard Rhodes won major prizes for *The Making of the Atomic Bomb* (1986), which is very long. John Hersey's *Hiroshima* (1946) tells briefly of life and death in that city just before and after the bomb was dropped.

The Holocaust: A German Historian Examines the Genocide, by Wolfgang Benz (trans. Jane Sydenham-Kwiet, 1999), is short and factual. Anne Frank, *The Diary of a Young Girl: The Definitive Edition* (trans. Susan Massotty, 1991) is unforgettable, and so are Primo Levi's memoir, *Survival in Auschwitz: The Nazi Assault on Humanity* (trans. Stuart Woolf, 1959); and Thomas Keneally's documentary novel, *Schindler's List* (1982).

CHAPTER 21 • *The Asian giants try to feed their poor.*

Jonathan D. Spence, *Mao Zedong* (1999) is a very short biography which is also a history of the Chinese Revolution. Nien Cheng, *Life and Death in Shanghai* (1986) is a Chinese woman's harrowing account of how she survived the Cultural Revolution.

Two reporters, Nicholas D. Kristof and Sheryl Wudunn, wrote *China Wakes: The Struggle for the Soul of a Rising Power* (1994), a lively, anecdotal book.

Mohandas Gandhi is the subject of well over 500 biographies. Ved Mehta, *Mahatma Gandhi and His Apostles* (1976) is more a portrait than a biography. Larry Collins and Dominique Lapierre, in *Freedom at Midnight* (1975), tell the dramatic story of the end of British rule and the birth of India and Pakistan. The Nobel Prize–winning writer V. S. Naipaul wrote *India: A Wounded Civilization* (1976). Shashi Tharoor describes his *India: From Midnight to the Millennium* (1997) as a "subjective account."

CHAPTER 22 • *Some of us do well.*

These are readable books on matters touched on in this chapter: Thomas Friedman, *The Lexus and the Olive Tree* (1999), on globalism; Karl Maier, *This House Has Fallen: Midnight in Nigeria* (2000); George Monbiot, *Amazon Watershed: The New Environmental Investigation* (1991); and Gale E. Christianson, *Greenhouse: The 200-Year Story of Global Warming* (1999).

Ray Kroc, organizer of the famous eateries, serves it up in *Grinding It Out: The Making of McDonald's* (1977).

Nigel Barley, *Innocent Anthropologist: Notes from a Mud Hut* (1983), is a captivating book on village Africa.

CHAPTER 23 • *We walk along the brink.*

Books in English on the Cold War always stress the U.S. role. For an overview of the age see Walter LaFeber, *America, Russia, and the Cold War, 1945–1990* (9th ed., 2002). George C. Herring, *America's Longest War: the United States and Vietnam, 1950–1975* (3rd ed., 1996) is critical of America's role. In *Tet!* (1971), Don Oberdorfer tells the story of the biggest battle of the Vietnam War.

Two good books on Russia in decline are Stephen Kotkin, *Armageddon Averted: The Soviet Collapse 1970–2000* (2001); and Grigori Medvedev, *The Truth about Chernobyl* (trans. Evelyn Rossiter, 1991). The latter tells of incompetence in Russia's nuclear disaster.

Avi Shlaim's pithy *War and Peace in the Middle East: a Concise History* (1994) covers 1914–1994. Efraim Karsh and Inari Rautsi, in *Saddam Hussein: A Political Biography* (1991), and Peter L. Bergen, in *Holy War, Inc.: Inside the Secret World of Osama bin Laden* (2001), introduce these two incendiaries.

CHAPTER 24 • *We do the unbelievable.*

Harry Wulforst's short *Breakthrough to the Computer Age* (1982) tells about the making of the first computers. Martin Campbell-Kelly and William Aspray, *Computer: A History of the Information Machine* (1996), is the most readable overall account. Tim Berners-Lee (the inventor of the Web) wrote *Weaving the Web: the Original Design and Ultimate Destiny of the World Wide Web* (1999), which is short and useful but somewhat technical.

James Watson has a lighter touch than Berners-Lee. His *The Double Helix: Being a Personal Account of the Discovery of the Structure of DNA* (1977) is funny and absorbing. Most other books about genetics deal with science and morality, not the way things happened.

Walter A. McDougall, . . . *The Heavens and the Earth: A Political History of the Space Age* (1997) is a splendid book which won a Pulitzer Prize. In *The Right Stuff* (1979) Tom Wolfe tells the story of the astronauts in the age of discovery. In his entrancing *Rocket Boys: A Memoir* (1998), Homer H. Hickam (a NASA engineer) relates how in his youth he and other boys made rockets.

This delightful book doesn't quite fit the subject of this chapter: Nobel Prize–winner (in physics) Richard P. Feynman with Ralph Leighton, *"Surely You're Joking, Mr. Feynman"*: *Adventures of a Curious Character* (1984).

Permissions

My thanks to the following for permitting me to use some poems, drawings, and diagrams:

University of Chicago Press: p. 19, parts of a table from Samuel Noah Kramer, *The Sumerians: Their History, Culture, and Character.* Copyright © 1963 by the University of Chicago.

Cornell University Press: p. 24, drawing by Jennifer Houiser, reprinted from Byron E. Shafer, ed., *Religion in Ancient Egypt: Gods, Myths, and Personal Practice.* Copyright © 1991 by Cornell University.

Stanford University Press, Stanford, CA, www.sup.org, for permission to use (on pages 77–78) three Chinese poems translated by Charles O. Hucker that appear in Hucker's *China's Imperial Past* (1975). Copyright © 1975 by the Board of Trustees of the Leland Stanford Junior University.

I have tried without success to find the publisher and obtain permission to use (on pages 196 and 199) two diagrams by Doris Meyer from Angus Armitage, *World of Copernicus* (1947).

Index